Public participation in the governance of international freshwater resources

Public participation in the governance of international freshwater resources

Edited by Carl Bruch, Libor Jansky, Mikiyasu Nakayama and Kazimierz A. Salewicz

United Nations
University Press

TOKYO · NEW YORK · PARIS

ENVIRONMENTAL
LAW·INSTITUTE®

The views expressed in this publication are those of the authors and do not necessarily reflect the views of the United Nations University.

United Nations University Press
United Nations University, 53-70, Jingumae 5-chome,
Shibuya-ku, Tokyo, 150-8925, Japan
Tel: +81-3-3499-2811   Fax: +81-3-3406-7345
E-mail: sales@hq.unu.edu general enquiries: press@hq.unu.edu
http://www.unu.edu

United Nations University Office at the United Nations, New York
2 United Nations Plaza, Room DC2-2062, New York, NY 10017, USA
Tel: +1-212-963-6387   Fax: +1-212-371-9454
E-mail: unuona@ony.unu.edu

United Nations University Press is the publishing division of the United Nations University.

Cover design by Rebecca S. Neimark, Twenty-Six Letters
Cover photograph by Sam Abell/National Geographic Image Collection

Printed in Hong Kong

ISBN 92-808-1106-1

Library of Congress Cataloging-in-Publication Data

Public participation in the governance of international freshwater resources /
edited by Carl Bruch ... [et al.].
    p.   cm.
  "Many of the chapters in this volume were first delivered at a symposium on
'Improving public participation and governance in international watershed
management,' held on 18–19 April 2003 in Charlottesville, Virginia."
  Includes bibliographical references and index.
  ISBN 9280811061 (pbk.)
  1. Water resources development—Management—Citizen participation.
2. Water resources development—International cooperation. 3. Water-supply
—Management—Citizen participation. 4. Watershed management—Citizen
participation 5. Water quality management—Citizen participation. 6. Fresh
water. I. Bruch, Carl (Carl E.), 1967–
  HD1691.P83   2005
  333.91—dc22   2005003471

Contents

List of tables and figures

Note on measurements

In this volume:

1 billion = one thousand million (10^9)
1 trillion = one million million (10^{12})
1 quadrillion = 10^{15}
1$ = 1 US dollar
1 mile $\cong$ 1.61 km
1 acre $\cong$ 4.05 × 10^3 m^2
1 hectare $\cong$ 1 × 10^4 m^2
1 bushel = 8 gallons $\cong$ 36.4 litres (UK) or 35.3 litres (US)

Acknowledgements

This volume has been a collaborative effort, and the editors are indebted to the numerous individuals and institutions who contributed. The publication of this volume marks the culmination of an 18-month project to foster a dialogue on specific experiences in watercourse management. This project seeks to improve public involvement in international watercourse management – and, ultimately, watercourse management itself – by collecting and exchanging experiences from specific watercourses around the world. This publication was made possible by a major grant from the Carnegie Corporation of New York.

Many of the chapters in this volume were first delivered at a symposium on "Improving Public Participation and Governance in International Watershed Management," held on 18–19 April 2003 in Charlottesville, Virginia. This workshop was co-convened by the University of Virginia School of Law, Environmental Law Institute (ELI), United Nations University (UNU), US Department of State, US Environmental Protection Agency (USEPA), United Nations Environment Programme (UNEP), and America's Clean Water Foundation (ACWF). Through their financial, technical, and in-kind contributions, these institutions conceived, shaped, and realized the symposium. Without their assistance, this volume would not have been possible.

In developing the themes of the symposium and this volume, and in identifying contributors, the editors would particularly like to recognize the intellectual contributions of Professor Jon Cannon at the University of Virginia School of Law. The editors are also grateful to Walter H. Kansteiner III, Assistant Secretary of State for African Affairs, and his office in the US Department of State Bureau of African Affairs; Judith Ayres, USEPA Assistant Administrator for International Affairs; Roberta Savage, President of ACWF; and Stephen J. Del Rosso, Jr, of Carnegie Corporation. Sarah King and Elizabeth Seeger of the ELI were indispensable in organizing the symposium and the participants.

The editors are especially grateful to Angela Cassar, who, as a Visiting Scholar at ELI, worked tirelessly with the authors and editors to finalize the manuscript.

Introduction

1

From theory to practice: An overview of approaches to involving the public in international watershed management

Carl Bruch, Libor Jansky, Mikiyasu Nakayama, Kazimierz A. Salewicz, and Angela Cassar

Clean water is essential to human survival, yet it is increasingly scarce. Despite pressures on this crucial resource, people often have little or no opportunity to participate in watershed decisions that affect them, particularly when they live along international watercourses. The United Nations has identified the rising demand for water as one of four major factors that will threaten human and ecological health for at least a generation. Over the coming decade, governments throughout the world will struggle to manage water in ways that are efficient, equitable, and environmentally sound. Whether these efforts succeed may turn, in large part, on providing the public with a voice in watershed-management decisions that directly affect them. Public involvement holds the promise of improving the management of international watercourses and reducing the potential for conflict over water issues.

Recent years, particularly the past decade, have seen a rapid growth of international law regarding the important of participatory decision-making generally and in the specific context of international watershed management (Bruch 2001, 2002). The body of emergent law ranges from provisions in international and regional declarations to binding conventions [for example on transboundary environmental impact assessment (TEIA) or international watercourses]. (The various international norms and practices are examined in more detail in chapter 2 of this volume.)

With the normative framework providing a clear set of objectives – transparency, participatory decision-making, and accountability – attention increasingly has turned to specific approaches for operationalizing these objectives. In some instances, this is done through the development of detailed conventions and protocols, especially at the regional level [for example, within the UN Economic Commission for Europe (UNECE)]. For international watercourses, operationalization has been more through policies of river basin authorities, international financial institutions, and other international organizations. In a number of instances, projects, work programmes, and other informal, less-legalistic activities provide an ad hoc approach (see chap. 2, this volume).

Through experimentation in specific instances and specific watercourses, a body of specific practices is emerging to give substance to the general objectives and requirements that have become ubiquitous. Public involvement is moving from theory to practice, from hortatory to actualized.

This volume collects many of the specific experiences and lessons learned in seeking to enhance and ensure public involvement in international watercourse management. It highlights successful mechanisms, approaches, and practices for ensuring that people have access to information about watercourses and factors that could have an effect on them; that people who may be affected have the opportunity to participate in decisions regarding the watercourse; and that people can seek redress when they are affected by activities in an international watercourse. At the same time, the volume examines conditions that facilitate or hinder public involvement, as well as contextual factors that may limit transference of experiences from one watershed to another.

The analysis in this volume draws upon experiences in various international watercourses, as well as some relevant sub-national watercourses and international institutions (see fig. 1.1). It also considers existing and emerging tools that can improve governance and public involvement.

This overview provides an introduction to the volume. It places the various chapters in the overall context and highlights some of the key lessons learned. The following section of this chapter concerns part I of the book, which examines some of the theoretical frameworks and considerations relating to public involvement in international watercourse management. The next section, corresponding to part II of the book, provides an overview of experiences in various international watersheds. The subsequent section, corresponding to part III of the book, examines the role of international institutions in promoting public involvement in international watercourse management. The fourth section, corresponding to part IV of the book, summarizes some of the innovative experiences in engaging the public in domestic watershed management, experiences that could provide conceptual or model approaches to be adapted for specific international watersheds. The fifth section, corresponding to part

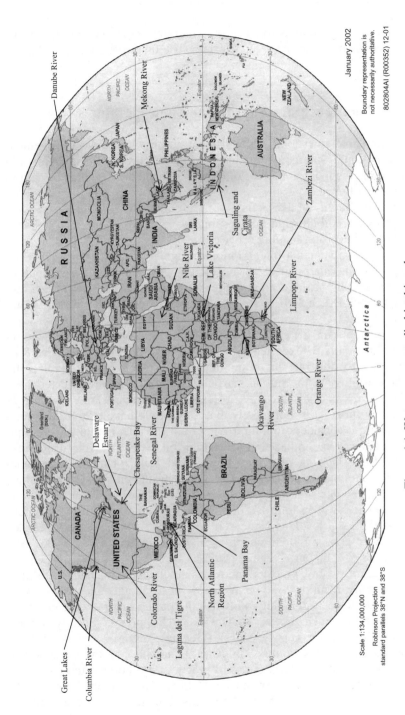

Figure 1.1 Watercourses studied in this volume

January 2002

Boundary representation is
not necessarily authoritative.

802804AI (R00352) 12-01

Scale 1:134,000,000
Robinson Projection
standard parallels 38°N and 38°S

5

V of the book, examines some of the emerging tools that could improve public involvement in the years to come. The final section of this chapter provides a few concluding remarks.

Part I: Theoretical frameworks and considerations

As a threshold question, it is worth inquiring why public involvement in international watercourse management is important. Most chapters in this volume highlight a number of reasons. Together, they may be said to entail the following:
- improved quality of decisions
- improved credibility and public support
- facilitated decision-making processes
- improved implementation and monitoring.

These reasons mirror most of the highlighted benefits of participatory decision-making in the academic literature (Benvenisti 1996; Milich and Varady 1998; Bruch 2001; Getches 2003; Avramoski 2004; see also chap. 2, this volume).

For example, in his chapter on the Mekong River Basin, Prachoom Chomchai points to experiences with the Pak Mun Dam and the Rasi Salai Dam to illustrate his point that failing to effectively involve the public can alienate the public, particularly those who are affected by a project, and can lead to costly protests. Had the decision makers consulted the public, it is more likely that they would have taken the latter's concerns into consideration, improving both the quality and the credibility of the decision. Similarly, in his chapter on the North American Great Lakes, John Jackson describes how Great Lakes United [a regional non-governmental organization (NGO) focusing on the Great Lakes] facilitated the decision-making process in a number of instances. Public involvement can also improve implementation and monitoring, particularly at the local level, as Nancy Gitonga, Roy Hoagland, and Rebecca Hanmer describe in their chapters on Kenyan and Chesapeake Bay watersheds (Cronin and Kennedy 1997).

Although the time, financial, and personnel costs associated with public involvement can deter some agencies, most scholars and practitioners assert that the costs of failing to involve the public generally are greater – and sometimes much greater. As David Getches noted, "Society can pay now or pay later for their decisions" (Getches 2003).

Chapter 2 of this volume, by Carl Bruch, traces the genesis and evolution of norms, institutions, and practices promoting public involvement in international watercourses. It highlights not only the specific approaches but also the international instruments and mechanisms advancing public involvement in environmental decision-making generally, which together have established a normative framework that seeks to ensure public access to information, participatory decision-making, and public accountability.

A variety of frameworks exist through which to construct mechanisms for engaging the public in watershed management: these are economic efficiency, participatory democracy, collective action and common property resources, integrated water resources management (IWRM), and a hydro-social contract. These different frameworks generally acknowledge the general benefits set forth above, although they rely on them to differing degrees. There are also other frameworks, such as watershed democracy, which has been advanced as a context for promoting direct democracy. The approaches examined in detail in this volume, however, tend to focus more on participatory processes, in which people have a voice in the decision-making process but the decision makers make the ultimate determination.

In his chapter on transboundary ecosystem governance, Bradley Karkkainen examines the increasing role of NGOs and members of the public in governing international resources, focusing on international watercourses. He advances the idea of a post-sovereign world in which the development, implementation, and enforcement of international law is no longer the sole province of sovereign nations. Although it is too early to pronounce the demise of sovereignty as a guiding principle of international law, Karkkainen highlights the new institutional space occupied by non-governmental actors, as well as the role of informal rules.

In the third chapter of the section on theoretical frameworks (chap. 4), Hans van Ginkel explores the meaning and limitations of public involvement in the Information Society. This chapter addresses the same general topic as Carl Bruch's later chapter (chap. 18) on Internet-based tools, but van Ginkel focuses on policy considerations of such tools, particularly in light of information overload, "data smog," and unequal access to electronic tools.

In a number of chapters the challenges of engaging lay people in making decisions for complex, non-linear, natural, social, and political systems are noted. This is particularly a theme of the chapters by John Volkman, Tomlinson Fort, and Bradley Karkkainen, and it merits mention here. Not only are there many uncertainties but also, in non-linear systems, these uncertainties mean that long-term predictions and actions are not possible (Gleick 1987). Accordingly, a flexible, responsive process is often necessary. This process is called adaptive management, and it is discussed in more detail in the section on emerging tools, below.

Part II: Experiences from international watersheds

The five chapters of part II of the book examine experiences in promoting public involvement in the management of international watersheds on four continents – Africa, Asia, Europe, and North America. Although this constitutes but a portion of the relevant experiences, it is supplemented by portions of other chapters (for example those in the parts on international institutions and emerging tools). Together, they represent many of the leading examples.

The various examples are cause for cautious optimism. Many of the case studies illustrate past failures of nations and watercourse authorities to share information with the public, to provide a venue for the public to participate in decisions regarding proposed policies or projects, or to operate in a publicly accountable manner. However, popular reactions to these failures have driven recent innovation.

There are tentative initial efforts to involve the public in a number of watercourses. Some are more successful than others. For example, Ruth Greenspan Bell and Libor Jansky examine the ongoing evolution of participatory management of the Danube River. In this basin, the construction of the Gabčíkovo–Nagymaros Dam, initially without public involvement or consultation, proved to be a key factor in mobilizing public attention and creating political space for public participation. They contrast this experience with other efforts to clean up the Danube River, and the mixed success with involving the public. There is an increased recognition by policy makers of potential difficulties in implementing a project if the public is not involved in the decision-making process; however, the specific modalities for involving the public are still evolving in a number of regards and progress is irregular. Elsewhere in Europe, public participation has been developing gradually in the management of the Dniester River (Trombitsky 2002) and other watercourses (Avramoski 2004).

In his chapter on the Mekong River Basin, Prachoom Chomchai observes that popular resistance to some large-scale hydropower dams and other projects has led the Mekong River Commission to develop policies on information exchange and public consultation. The Mekong River is particularly interesting owing to the long history of participatory governance at the local level, with a striking disconnect in the lack of public participation at the national and international levels over most of the past 150 years. Noting some of the differences in the way that historical participatory practices differ from contemporary advocacy, Chomchai highlights ways in which regional efforts to improve public involvement in managing the Mekong River could draw upon traditional local and recent national developments in transparent and participatory governance.

African watercourses are also developing participatory principles and mechanisms for governance. For example, in the Okavango River and Delta, Peter Ashton and Marian Neal highlight regional initiatives (such as "Every River Has Its People") that have enhanced transboundary governance by improving stakeholder participation in decision-making. More generally, Michael Kidd and Nevil Quinn suggest that the general lack of provisions in instruments governing specific watersheds in Southern Africa may be a contributing factor to their lack of effectiveness. At the same time, they sound a note of optimism in the recent commitment of the Southern African Development Community (SADC) to "increase[...] awareness, broad participation and gender mainstreamed in water resources development and management by 2005." Promising developments over the past few years can also be found in the Nile River Basin (Shady 2003), Lake Victoria (Ntambirweki 2003), and the Niger River and Lake Chad Basins (Namata 2003).

The North American Great Lakes have some innovative experiences with involving the public in decision-making, which John Jackson explores in his chapter. Although some of this may be attributed to provisions in the organic documents (dating to 1909) for the International Joint Commission (IJC), which governs the waters, he argues that the NGOs and community groups living around the lakes have also created the political space to be involved. Jackson explores a range of ways that a transboundary citizens' coalition can improve – and, indeed, has improved – governance of a transboundary watercourse. He also examines some of the financial and cultural challenges faced, as well as approaches taken to address these challenges.

One of the common themes running through the experiences represented in the chapters of this book, as well as elsewhere, is the growing effort to inform the public of potential project or policy developments. Some of this is influenced by the development of regional and international norms of transparency, participation, and accountability. However, public outrage over certain projects (about which they frequently were not alerted or consulted) is a substantial factor in the evolution of participatory governance in a number of watercourses. Such responses have often driven initiatives to develop more inclusive and transparent policies and to create formal mechanisms for involving the public. This dynamic holds for many of the watercourses described in this part, as well as for international institutions. For example, in their chapter on the African Development Bank (AFDB), discussed in the next section of this chapter, Aboubacar Fall and Angela Cassar examine the recent developments of transparency and participation in management of the Senegal River, arising in part in response to earlier difficulties when the public was marginalized.

Part III: International institutions

Because of their role in financing large-scale infrastructure – including dams, diversions, and irrigation systems – international financial institutions can have a significant role in the development of transboundary watercourses. In response to concerns over specific projects, multilateral development banks at the international and regional levels have developed a number of operational policies to ensure transparency, participation, and accountability in the planning and implementation of projects along such watercourses. These include policies governing dams, EIA, resettlement, indigenous peoples, and other relevant aspects. The World Bank, examined by Charles Di Leva, has been a (sometimes reluctant) leader in promoting public involvement in the realm of public financing, and it is starting to affect how private sector finance is conducted. One of the World Bank's innovations is its Inspection Panel, which allows people affected by Bank-funded projects (including dams) to seek redress if the Bank fails to follow its policies. Thus, if the proponents of a particular project and the Bank fail to consult the public, conduct an inadequate EIA, or do not provide for an adequate resettlement plan, aggrieved persons or organizations can submit a complaint to the Inspection Panel.

Regional development banks, such as the AFDB, have also developed policies and practices to improve public involvement in their projects, including those affecting international watercourses. In their chapter, Fall and Cassar examine recent experiences of the AFDB in improving such public participation. Particular emphasis is placed on the Senegal River Basin, where the AFDB has been particularly active in supporting and advancing public involvement.

In addition to financial institutions, a number of other international bodies seek to encourage public involvement in international watercourse governance. River-basin authorities, many of which are described in part II of this volume, focus on a specific watercourse. The World Commission on Dams (WCD) – which involved cooperation between the World Bank, the International Union for the Conservation of Nature, and other organizations – was a transparent, consultative process to address the controversial aspects of large-scale hydropower dams; and the Dams and Development Project of the United Nations Environment Programme (UNEP) is continuing the work of the WCD (WCD 2000; UNEP 2003; Van Dyke 2003). UNEP also is a leader in developing and making publicly available information on transboundary watercourses (Cunningham 2003). The Global Environment Facility (GEF) does much to advance public participation in the management of international waters through integrated water-resources management, trans-boundary diagnostic analyses, and strategic action programmes (Gonzalez 2003).

At the regional level, a number of bodies complement efforts by river-basin authorities to strengthen public participation in transboundary watercourse management. In his chapter, Geoffrey Garver outlines the array of mechanisms that the North American Commission for Environmental Cooperation utilizes to improve transparency, public participation, and accountability in the region, highlighting examples where these tools have been applied to watershed management. In Europe and Central Asia, the UNECE serves a similar role; in fact, it has developed a number of strong regional conventions for improving governance of transboundary watercourses (UNECE 1992), EIA (UNECE 1991), and public involvement generally (UNECE 1998). Both the SADC and the recently re-established East African Community seem poised to serve similar roles in Southern and East Africa, respectively, as outlined in the chapters by Kidd and Quinn and by George Sikoyo.

Part IV: Lessons from domestic watercourses

Domestic watersheds frequently provide a laboratory for developing management techniques that can be adapted and applied at the international level. For example, Chomchai in his earlier chapter reports that the Mekong River Commission (MRC) is drawing upon the lessons learned from the Murray–Darling Basin Commission in formulating the MRC's public participation strategy. This part of the volume highlights some novel approaches from Africa, Asia, and North America that may serve as models for improving public involvement in transboundary watercourse management.

In her chapter on management of Kenyan fisheries, Nancy Gitonga surveys a number of approaches for involving the public in decisions to manage fisheries. Experiences highlighted in her chapter show that stakeholder involvement can be instrumental to the effective implementation of control measures to rehabilitate exhausted fisheries. Perhaps the most innovative approach that she examines is the establishment of beach management units to manage Lake Victoria fisheries at a local level. This practical approach to managing shared resources in an international lake in a coordinated manner between the local, national, and regional levels has shown great promise.

In the United States, the Chesapeake Bay has been a model for participatory, interjurisdictional management for more than two decades. In their chapters, Roy Hoagland and Rebecca Hanmer examine the experiences in involving the public in management and operational decisions from the non-governmental and governmental perspectives, respectively. Notable for its size, population, and economy, the bay is also distinguished by the numerous national, regional, state, and local authorities that have responsibilities affecting it. The lessons learned over the years in involving the public in such a politically, administratively, socially, and ecologically complicated context are likely to be relevant to many transboundary watercourses. The chapter by Tomlinson Fort III, addressing standard-setting in the Delaware Estuary, explores mechanisms for involving not only the general public but particularly the regulated community in determining standards that will affect conduct (and expenses incurred) by numerous parties. His chapter draws upon experiences representing a regulated industry; it highlights some of the tensions inherent in the process and also practical ways that have helped to facilitate continued, constructive collaboration (again in a multijurisdictional context). Lessons learned in other US watersheds may also be relevant, particularly with regards to multijurisdictional experiences between federal, state, and local authorities (Griffin 1999; Hayes 2002).

The final chapter in part IV (by Mikiyasu Nakayama) addresses lessons learned regarding public involvement in developing and implementing resettlement schemes associated with dam construction in Indonesia. This chapter highlights ways in which public involvement can improve resettlement. It is also significant for its methodology: by comparing predicted impacts with actual impacts, the underlying survey presents opportunities for improving the overall assessment process. Such a comparative analysis of predicted and actual impacts also could be applied in the context of environmental and social impact assessment to improve transboundary impact-assessment processes.

One significant lesson from domestic watercourses – and one that also applies to international watercourse management – is the importance of involving the public in the correct manner. Not all approaches are equally effective: Gitonga and Fort in part IV, and Mary Orton in part V, all provide contrasting experiences of conflict and acrimony in participatory processes, and then constructive, outcome-oriented dialogue within the same watershed. The former experiences tend to be painful for the governing authorities, so that they come to dread (and avoid) public involvement. At the same time, when done constructively, public involvement can be an enriching, consensus-building process that enhances not only the substantive decisions that are made but also the working relationships among the various parties involved.

Part V: Emerging tools

In addition to experiences in domestic watercourses, a wide range of emerging tools and approaches facilitate public access to information and involvement in international watercourse management. These tools range from information development and dissemination, to public participation in decision-making, to dispute settlement.

Increasingly, the Internet presents opportunities to disseminate information on the status of transboundary watersheds and projects that could affect them, as well as providing an avenue to solicit public input regarding decisions on projects and broader policies. Technological aspects of information gathering, processing, and dissemination have become central to decision-making in water-resource systems. In fact, significant advances in natural resource management, development planning, and environmental protection could not take place without technical and methodological advances in information technology. Accordingly, information technology – and the Internet in particular – are becoming standard tools for professionals, scientists, advocates, and decision makers in their daily activities. In his chapter on Internet-based tools, Carl Bruch examines how various watercourse authorities, governmental agencies, academic and research institutions, and international organizations are utilizing the Internet to improve public participation in international watercourse management. His chapter reviews a variety of Web pages, decision support tools, chat rooms, and other innovative Internet-based approaches.

Decision support systems (DSS) provide tools for members of the public, government, and technicians alike to identify possible outcomes of a range of options facing decision makers. As such, they can help everyone to understand the trade-offs that must be made. In his chapter on DSS, Kazimierz Salewicz traces the evolution of DSS as tools for decision-making, highlighting their increasing public accessibility. Looking forward, he explores options for making DSS available over the Internet. In her chapter on alternative dispute resolution, Mary Orton also considers practical means for diverse parties to utilize DSS to understand possible outcomes and build common ground in a polarized decision-making context.

Adaptive management is another emerging tool for managing watercourses, as well as natural resources more broadly (Salafsky, Margoluis, and Redford 2001; Murray–Darling Basin Ministerial Council 2003). John Volkman considers experiences with adaptive management in the Columbia River Basin, one of the more-developed applications of adaptive management to a significant watercourse. Karkkainen and Fort also advance adaptive management as an important tool for resolving problems in a watershed. Fort highlights one of the difficulties associated with adaptive management: the iterative approach inherent in adaptive management may be resisted by parties who want more stringent (or less stringent) actions.

In light of the occasionally contentious nature of public hearings and consultations, watershed authorities are turning to alternative dispute resolution (ADR) methods to facilitate public involvement in a constructive way. Mary Orton's chapter examines the application of ADR methods to the revision of a management plan that pitted businesses against recreational users against environmental concerns. These experiences from the Colorado River are particularly striking for the contrasts between approaches to public participation that were problematic and those that were ultimately successful. In this specific example, various tools were employed to bring people together to constructively discuss and settle on a final management plan. One tool – the use of surveys – was also used successfully in the Chesapeake Bay, as highlighted in Hanmer's chapter.

Transboundary environmental impact assessment (TEIA) builds upon experiences in national-level EIA to ensure public involvement in projects with transboundary impacts (Cassar and Bruch 2004). In his chapter on the development of TEIA in East Africa, George Sikoyo focuses on the participatory process that the East African Community is undertaking to develop TEIA guidelines. In addition to addressing an emerging tool – TEIA – the process is notable for its broad, consultative nature not only in one country but across three countries.

Publicly accessible tribunals represent the final tool, and chapter, considered in this volume. Although accountability through tribunals is less developed than transparency or public participation, formal and informal mechanisms have developed rapidly over the past decade. In addition to the Inspection Panels in place at the World Bank and under development at the AFDB, and the Citizen Submission Process of the North American Commission for Environmental Cooperation, other quasi-judicial mechanisms are emerging. For example, Juan Miguel Picolotti and Kristin Crane examine experiences over a period of three years with the Central American Water Tribunal (CAWT). The CAWT is unique among these bodies in that it is a citizen-led initiative, with no formal mandate from governments. Notwithstanding this limitation, however, the CAWT has been able to provide an informal venue in which to bring public attention to violations of international law relating to water use and development in Central America.

Conclusions

Public participation makes sense. The economic, political, decision-making, and human rights bases are all well established. International agreements, declarations, and other instruments regularly attest to the critical importance of an informed and engaged public, generally as well as in the particular context of international watercourse management. Until recently, though, the practical details were lacking regarding how to involve the public in decision-making.

As the chapters in this volume illustrate, the specific standards and institutional practices are still emerging. Although implementation is still nascent in many instances, the experiences thus far are promising. Around the world – from Africa, to the Americas, to Asia, to Europe – institutions are putting in place detailed policies and institutional mechanisms to provide the public with information about the status of watercourses and factors that could affect the watercourses, to ensure that the public has a meaningful opportunity to participate in decision-making processes, and increasingly to offer a means for affected members of the public to seek redress for harm arising from mismanagement of international water resources. As the various governmental, non-governmental, intergovernmental, and research institutions develop these approaches, there is an urgent need to share these experiences, to adapt the experiences to the particular contexts of various watercourses, and to build local capacity.

This volume examines the experiences in many watercourses around the world, drawing lessons learned and highlighting areas for further development. In addition to sharing experiences, the chapters in this volume also identify some of the considerations – linguistic, political, legal, traditional and cultural, geographic, and institutional – that should be kept in mind in extending and adapting the approaches to other watersheds.

However, this is an iterative process. As practice has expanded rapidly over the past decade, there has also been an effort to update and expand the normative framework governing international watercourses. Thus, the International Law Association (ILA) found it necessary to revise its Helsinki Rules on the Use of Waters of International Rivers, which were approved in 1966 and formed the foundation for the 1997 UN Convention on the Law of the Non-Navigational Uses of International Watercourses. Its revised *Rules on the Equitable and Sustainable Use of Waters* includes an entire chapter on "Individual Rights and Public Participation," with specific provisions addressing individual rights and duties, public participation, information, education, rights of particular communities, and right to compensation (ILA 2004). Other articles of the revised Rules address impact assessment, access to courts, and remedies.

It is likely that, as the world becomes more and more interconnected, as new technologies emerge (as the Internet, computers, and wireless technologies have over the past few decades), and as economic and political integration continues, the iterations of normative and institutional development will continue. In many ways, though, the most dramatic changes are taking place now. Government and governance is increasingly open: this has long happened at the local level around the world; the quiet revolution is at the national and international levels, as governments commit to transparent and participatory processes. They are even agreeing, albeit gradually, to be accountable to members of the public for their actions. Shared rivers and lakes are likely to continue to provide a primary context in which to foster and facilitate public participation in transboundary governance.

REFERENCES

Avramoski, Oliver. 2004. "The Role of Public Participation and Citizen Involvement in Lake Basin Management." Internet: ⟨http://www.worldlakes.org/uploads/Thematic_Paper_PP_16Feb04.pdf⟩ (visited 14 April 2004).

Benvenisti, Eyal. 1996. "Collective Action in the Utilization of Shared Freshwater: The Challenges of International Water Resources Law." *American Journal of International Law* 90:384.

Bruch, Carl. 2001. "Charting New Waters: Public Involvement in the Management of International Watercourses." *Environmental Law Reporter* 31:11389–11416.

Bruch, Carl (ed.). 2002. *The New "Public": The Globalization of Public Participation*. Washington, DC: Environmental Law Institute.

Cassar, Angela Z. and Carl E. Bruch. 2004. "Transboundary Environmental Impact Assessment in International Watercourses." *New York University Environmental Law Journal* 12:169–244.

Cronin, John, and Robert F. Kennedy, Jr. 1997. *The Riverkeepers*. New York: Scribner.

Cunningham, Gerard. 2003. "UNEP Net Information Services for Watershed Management." Presentation at the Symposium on Improving Public Participation and Governance in International Watershed Management. Charlottesville, Virginia, 18–19 April.

Getches, David. 2003. "The Efficiency of Experts vs. the Chaos of Public Participation." Presentation at the Symposium on Improving Public Participation and Governance in International Watershed Management. Charlottesville, Virginia, 18–19 April.

Gleick, James. 1987. *Chaos: Making a New Science*. New York, NY: Penguin.

Gonzalez, Pablo. 2003. "Multi-stakeholder Involvement and IWRM in Transboundary River Basins: GEF/UNEP/OAS Experiences with the Strategic Action Program for the San Juan River Basin of Costa Rica and Nicaragua." Presentation at the Symposium on Improving Public Participation and Governance in International Watershed Management. Charlottesville, Virginia, 18–19 April.

Griffin, C.B. 1999. "Watershed Councils: An Emerging Form of Public Participation in Natural Resource Management." *Journal of the American Water Resources Association* 35:505–517.

Hayes, David J. 2002. "Federal–State Decisionmaking on Water: Applying Lessons Learned." *Environmental Law Reporter* 32:11253–11262.

International Law Association (ILA). 2004. "The [Revised] International Law Association Rules on the Equitable Use and Sustainable Development of Waters." Tenth Draft. February.

Milich, Lenard, and Robert G. Varady. 1998. "Managing Transboundary Resources: Lessons From River-Basin Accords." *Environment* 40:10.

Murray–Darling Basin Ministerial Council (MDBMC). 2003. Communiqué. 14 November.

Namata, Adamou. 2003. "Public Participation in Shared Waters Management in West Africa: The Niger and Lake Chad Basins." Presentation at the Symposium on Improving Public Participation and Governance in International Watershed Management. Charlottesville, Virginia, 18–19 April.

Ntambirweki, John. 2003. "People-to-People Cooperation Across Borders: An Emerging Trend in the Lake Victoria Basin, or an Illusion?" Presentation at the Symposium on Improving Public Participation and Governance in International Watershed Management. Charlottesville, Virginia, 18–19 April.

Salafsky, Nick, Richard Margoluis, and Kent Redford. 2001. "Adaptive Management: A Tool for Conservation Practitioners." Internet: ⟨http://fosonline.org/resources/Publications/AdapManHTML/Adman_1.html⟩ (visited 23 November 2003).

Shady, Aly. 2003. "Public Participation in the Nile River Basin: Past Experiences and Prospects for the Future." Presentation at the Symposium on Improving Public Participation and Governance in International Watershed Management. Charlottesville, Virginia, 18–19 April.

Trombitsky, Ilya. 2002. "The Role of Civil Society in Conservation and Sustainable Management of the NIS Transboundary Watercourse, the Dniester River." Report for the 10th OSCE Economic Forum. Prague, 28 May–1 June.

UNECE (United Nations Economic Commission for Europe). 1991. Convention on Environmental Impact Assessment in a Transboundary Context. Internet: ⟨http://www.unece.org/env/eia/welcome.html⟩ (visited 23 November 2003).

UNECE (United Nations Economic Commission for Europe). 1992. Convention on the Protection and Use of Transboundary Watercourses and International Lakes. Adopted 17 March in Helsinki. Internet: ⟨http://www.unece.org/env/water/welcome.html⟩ (visited 23 November 2003).

UNECE (United Nations Economic Commission for Europe). 1998. Convention on Access to Information, Public Participation in Decision-making and Access to Justice in Environmental Matters. Adopted 25 June in Aarhus, Denmark. Internet: 〈http://www.unece.org/env/pp/welcome.html〉 (visited 23 November 2003).

UNEP (United Nations Environment Programme). 2003. Dams and Development Project (DDP): Interim Report Covering the Period November 2001–March 2003. Internet: 〈http://www.unep-dams.org/files/DDP.Interim.report.2003.pdf〉 (visited 23 November 2003).

Van Dyke, Brennan. 2003. "Public Participation Successes and Challenges of the World Commission on Dams and Follow-up." Presentation at the Symposium on Improving Public Participation and Governance in International Watershed Management. Charlottesville, Virginia, 18–19 April.

WCD (World Commission on Dams). 2000. *Dams and Development: A New Framework for Decision-Making*. London: Earthscan Publishers. Internet: 〈http://www.dam〉.

Part I

Theoretical frameworks

2

Evolution of public involvement in international watercourse management

Carl Bruch

Introduction

Citizens, non-governmental organizations (NGOs), businesses, universities, and other members of civil society have played an essential role in developing and implementing environmental and natural-resource laws and institutions at local and national levels over the past decades. This role has extended more recently into numerous international institutions, processes, and contexts (Shelton 1994; Taylor 1994; Stec and Casey-Lefkowitz 2000; Bruch 2002; Bruch and Czebiniak 2002; Nakayama and Fujikura 2002). This chapter examines the emerging norms and practices that guarantee transparency, public participation, and accountability in the management of international watercourses. Particular attention is paid to how these norms have been, and may be, implemented to improve the management of transboundary watercourses in regions around the world.

Although there currently is no definitive statement under customary law on the topic, this chapter demonstrates that participation provisions are incorporated increasingly into waterbody-specific instruments with benefits for communities, governments, and project implementation alike. This increasingly widespread practice suggests that norms on public involvement not only are emerging but also are rapidly crystallizing. In fact, the efforts by the International Law Association (ILA) to revise its *Rules on the Equitable Use and Sustainable Development of Waters*, which reflect customary norms and practices, confirm the rapid emergence and recognition of public involvement in international watercourse management (ILA 2004). Regional context and variation can often help this process: for example, in Africa, evolution of these norms has the added benefit of a "rich tradition of participation in water management" at the local level (Sharma et al. 1996), which can form the basis for similar development at the international level.

The first main section, below, reviews the needs for, and benefits of, public involvement in managing international watercourses. This section also surveys the various international watercourses, global and regional instruments, and international institutions discussed in this chapter. The next main section (pp. 32–41) examines norms, practices, and mechanisms that enable citizens and other non-governmental actors to obtain access to information about the water quantity and quality in transboundary watercourses, as well as information about activities that could affect these waters. The subsequent main section (pp. 41–48) considers public participation in the negotiation of treaties, in the development of policies and other norms, and in the review and approval of projects. The section on access to justice (pp. 48–58) considers different venues – including domestic courts as well as international tribunals and fact-finding bodies – in which citizens may file complaints if a private or public entity is harming, or threatens to harm, international watercourses (often termed "access to justice"). In the light of the analyses of the three pillars of public involvement – access to information, participation, and justice – the penultimate section (pp. 58–63) examines factors affecting the development and implementation of participatory frameworks for managing international watercourses, strategies for advancing public participation, and some of the promising approaches and mechanisms that are emerging for promoting public involvement. The final section (pp. 63–64) provides some concluding thoughts on the evolution of public involvement in international watercourse management.

Motivation, norms, and institutions

This chapter reviews the genesis and evolution of public involvement in environmental decision-making, particularly in the context of international watercourse management. It analyses various ways in which the public can become involved in the management of international rivers and lakes. These mechanisms range from making information available to the public, to consulting the public, to empowering the public to file complaints, and they are available in both domestic and international forums.

Public involvement in managing international watercourses

Benefits of public involvement in water management

Public access to information, public participation in decision-making, and access to mechanisms for redress have synergistic benefits. Public involvement builds awareness (Shumway 1999). This insight can, in turn, build the public's capacity to participate and also their respect and support for the decision-making process. Public involvement also improves the quality of decisions because public input can supplement scarce government resources for developing norms and standards, as well as for monitoring, inspection, and enforcement (Sharma et al. 1996).

Decisions affecting international watercourses frequently are made by government officials who are located far from the waters in question. As a result, these decisions rarely reflect the interests of the border residents, who frequently are far from the sources of power. Expanding on this theme, Milich and Varady have observed that

international agreements that depend on internal political processes may fall short of achieving goals precisely because they do not sufficiently consider the local interests that ultimately determine the extent to which laws are implemented. National and international institutions rarely have incentive to heed realities of the field. Instead high-level policy makers are rewarded for setting ambitious goals without providing the appropriate understanding, tools, and capacity at the local level to implement the measures needed to achieve those goals. (Milich and Varady 1998)

They conclude that "transnational linkages that permit national agencies to speak to each other but remain deaf to local interests are destined to fail."

Similarly, decisions made in the interest of national governments do not necessarily reflect the interests of the transboundary ecosystems that are intricately connected to transboundary watercourses (Eriksen 1998). Thus, the public has a critical role to play in "represent[ing] an ecosystem over and above their national loyalties" (Sandler et al. 1994). By involving the public in the management of these waters, it is more likely that the decisions will respect the long-term ecological interest of transboundary ecosystems (Ferrier 2000).

Public involvement can identify and address potential problems at an early stage. When the public is not given an opportunity to participate, negative public reaction to unaddressed (and unresolved) issues can lead to major (and sometimes violent) protests that stall or halt projects and add significantly to their overall costs. For example, the construction of the Pak Mun Dam on a tributary to the Mekong River in Thailand did not include public participation in the assessment process. Although the dam was completed in 1994, the communities affected by the dam have objected to the compensation, which they view as inadequate, and the unexpected costs associated with the protests have increased the dam's overhead, altering the cost–benefit analysis (Kaosa-ard et al. 1998). Controversy over the Sardar Sarovar water project in India, which lacked effective public involvement, has also increased the costs of the project (Taylor 1994).

Similarly, when the public is not involved in decisions that could affect them, the simple lack of public support can impede implementation. For example, the World Bank-funded Kampong Improvement Program lacked public participation, which led to apathy on the part of the intended beneficiaries and a failure to maintain the project (Taylor 1994).

In contrast, involving the public in managing international watercourses can improve the credibility, effectiveness, and accountability of governmental decision-making processes. Initiatives by NGOs can also facilitate the decision-making process. When negotiations over international watercourses become polarized as governments become locked into their positions, NGOs with a regional focus can, "by highlighting regional and ecosystem-related perspectives, assist in breaking through barriers associated with traditional diplomacy" (Sandler et al. 1994).

Involvement also builds public ownership of the decisions and improves its implementation and enforcement, as the public is more likely to respect and abide by the final agreements (Milich and Varady 1998; UNEP 2002). Citizens and NGOs can also improve the monitoring of potential violations, particularly when they understand their rights and the standards that apply (Shumway 1999; UNEP 2002). For example, an increasing number of rivers and bays in the United States and in other countries have "riverkeepers" and "baykeepers" – individuals who investigate and report potentially illegal actions that harm the waters, such as illegal discharge of wastes (Cronin and Kennedy 1997).

A number of these different reasons for public involvement in the management of international waters were explicitly cited in the 1999 London Water and Health Protocol to the 1992 United Nations Economic Commission for Europe (UNECE) Convention on the Protection and Use of Transboundary Watercourses and International Lakes (UNECE 1999; Kravchenko 2002). Moreover, a wide range of conventions and international institutions have also sought to advance public involvement, as discussed in the next subsection.

Watercourses, conventions, and international institutions considered

In recent years, an increasing number of international conventions and institutions have strengthened the role of the public in the development, implementation, and enforcement of international commitments. Some of these have been general (relating broadly to public involvement in environmental matters), whereas others have specifically incorporated public involvement into the management of international watercourses. These initiatives are briefly introduced here and discussed in more detail later in this chapter.

Watercourse-specific instruments and institutions

The Mekong River Commission (MRC) has been a leader in developing frameworks to promote public involvement in international watercourse management. The 1995 Agreement on the Cooperation for the Sustainable Development of the Mekong River Basin established the MRC to manage river-related activities in the lower basin (Agreement 1995). Cambodia, Lao People's Democratic Republic, Thailand, and Viet Nam are parties to the MRC, which replaced earlier committees dating back to 1957, although the riparian nations of China and Myanmar have yet to join the MRC formally. The MRC is currently drafting and reviewing a public-participation strategy.

Along the United States–Mexico border, two organizations seek to manage the shared natural resources, including the Rio Grande and the Colorado River. The International Boundary and Water Commission (IBWC) was established in 1889 (initially termed the International Boundary Commission) to implement the boundary and water treaties between the United States and Mexico (US–Mexico 1889, 1944), and the Border Environment Cooperation Commission (BECC) was established by the North American Agreement on Environmental Cooperation (NAAEC 1993) in response to the North American Free Trade Agreement (NAFTA). The BECC must certify that proposed projects before the North American Development Bank that are located within 100 km of the border satisfy all the applicable environmental laws and have adequately incorporated community participation. Many of the projects that the BECC certifies can affect the transboundary river that forms much of the boundary between the United States and Mexico.

Along the Canada–United States border, the North American Great

Lakes constitute the largest inland freshwater ecosystem in the world (Famighetti et al. 1993). In 1909, the International Boundary Waters Treaty (IBWT) established the International Joint Commission (IJC) to prevent and resolve disputes over water quality and quantity in waters along the US–Canada border (US–Great Britain 1909; US–Canada 1978). With time, the IJC has also come to address transboundary air pollution as well as actually operating hydropower projects that affect transboundary water flows [Environmental Law Institute (ELI) 1995]. The original 1909 treaty also established detailed provisions for public participation and access to information that have been actively implemented.

In Europe, management of the Danube and Rhine Rivers has incorporated public involvement. The 1994 Danube River Protection Convention (Convention on Cooperation for the Protection and Sustainable Use of the Danube River 1994), signed by 11 states, has particularly strong provisions for public access to information. In 1999, the European Community and the nations of Germany, France, Luxembourg, the Netherlands, and Switzerland concluded the Convention on the Protection of the Rhine (Convention on the Protection of the Rhine 1999).

The 1990s saw the rapid rise of international commitment to involving the public in the management of Lake Victoria. The Lake Victoria Environmental Management Project, funded by the Global Environment Facility (GEF 1996), incorporates public participation in the development of projects and policies. In anticipation of the treaty establishing the East African Community, Kenya, Tanzania, and Uganda adopted a Memorandum of Understanding (MOU) on Environment Management that relies on public involvement and specifically addresses Lake Victoria (East African MOU 1998; Odote and Makoloo 2002; Tumushabe 2002). The three East African nations are currently finalizing an environmental protocol to the Treaty Establishing the East African Community, as well as guidelines on regional environmental impact assessment, both of which emphasize public participation in their development as well as in their substance (Sikoyo, chap. 22 in this volume).

Despite millennia of human use of the Nile River to meet residential, industrial, and agricultural needs, it is only recently that the international instruments governing its management have explicitly incorporated public involvement. In 1999, 10 of the 11 Nile Basin nations commenced the Nile Basin Initiative (NBI) – an informal, interim agreement to facilitate basin-wide and sustainable international management of this shared resource (NBI Secretariat 2000). The NBI's Policy Guidelines provide the framework for regional cooperation and incorporate transparency and participation to varying degrees. Although practice has yet to emerge from these recent Nile instruments, it is notable that – even in a context that is as sensitive as the discussions regarding allocation of Nile Basin waters – the riparian nations have committed to making the process more open and participatory.

Water-related instruments and institutions

The UN Convention on the Law of the Non-Navigational Uses of International Watercourses (1997) represents the culmination of decades of international dialogue on the management of international watercourses. As of 27 June 2002, 16 states had signed the Convention, and 12 had ratified, accepted, acceded to, or approved it. It sets forth basic principles for deciding how to allocate water as well as other non-navigational uses. The Convention also includes a few norms that promote public involvement.

The 1992 UNECE Convention on the Protection and Use of Transboundary Watercourses and International Lakes (known as the Helsinki Convention) and its 1999 London Protocol establish norms for public involvement in the management of international watercourses in the UNECE region – which consists of Europe, the states of the former Soviet Union, Canada, and the United States (UNECE 1992, 1999). The Convention seeks to reduce, control, and prevent transboundary water pollution and the release of hazardous substances into aquatic environments. As of April 2004, 34 states and the European Community have ratified, accepted, approved, or acceded to the Convention. The Protocol focuses on health-related issues associated with international waters. As of April 2004, there were 11 parties to the Protocol. (There is also a Protocol to the Convention that addresses civil liability, but that is not relevant to the current discussion.)

Progressively, Southern Africa has adopted a series of legal and institutional initiatives that rely on public involvement in developing and managing transboundary watercourses in the region. The 1987 Action Plan for the Common Zambezi River System recognized not only the environmental aspects of international waters but also the need for transparency and public participation in their management (ZACPLAN 1987). Difficulties in implementing the ZACPLAN (Nakayama 1997, 1999) led to a more comprehensive 1995 Protocol on Shared Watercourse Systems in the Southern African Development Community (SADC) Region, which was revised in 2000 (SADC 2000; Pamoeli 2002).

Other international instruments and institutions

In the last decade, proliferation of global and regional instruments has expanded and crystallized public involvement in environmental matters (Bruch 2002; Bruch and Czebiniak 2002). As both soft law (sometimes hortatory and sometimes reflective of general obligations under international law) and hard law (with binding obligations), these instruments apply to a wide range of international and domestic environmental contexts, including transboundary watercourses. Simultaneously, international institutions that conduct or support activities affecting these watercourses have opened up their processes to members of the public. The experiences of the international institutions are particularly illuminating, as they offer concrete examples of how public involvement can work, as well as some of the constraints that it can impose.

Perhaps the most universally agreed-upon international environmental declaration – the 1992 Rio Declaration on Environment and Development (UNCED 1992a) – crystallized the emerging public involvement norms in Principle 10:

Environmental issues are best handled with the participation of all concerned citizens, at the relevant level. At the national level, each individual shall have appropriate access to information concerning the environment that is held by public authorities, including information on hazardous materials and activities in their communities, and the opportunity to participate in decision-making processes. States shall facilitate and encourage public awareness and participation by making information widely available. Effective access to judicial and administrative proceedings, including redress and remedy, shall be provided.

To implement the principles of the Rio Declaration, states at the 1992 United Nations Conference on Environment and Development adopted Agenda 21 (the "Blueprint for Sustainable Development") (UNCED 1992b). Agenda 21 envisaged public involvement in developing, implementing, and enforcing environmental laws and policies in many areas including management of fresh waters. Specifically, chapter 18 contemplates integrated public participation in the management of domestic and transboundary water resources. Moreover, chapters 12, 19, 27, 36, 37, and 40 promote transparency, public participation, and accountability in environmental management generally.

Since Rio, regional initiatives have elaborated on these general principles, clarifying and implementing them. In the Americas, Asia, and Europe and the former Soviet Union, regional instruments have urged (and even required) nations to adopt specific measures to ensure domestic implementation.

The UNECE region – comprising the European Union, Eastern Europe, the Caucasus, Central Asia, Canada, and the United States – has developed some of the most detailed and binding provisions for public involvement. The 1998 UNECE Convention on Access to Information, Public Participation in Decision-making and Access to Justice in Environmental Matters (or the "Aarhus Convention") emphasizes three areas or "pillars" – transparency, participation, and accountability

(UNECE 1998; Kravchenko 2002). In each of these areas, the convention establishes minimum requirements for the state parties to incorporate into their laws and institutions. The Convention relies on enforceable rights of citizens, including procedural rights and the human right to a healthy environment. The convention also prohibits nations from discriminating against natural or legal persons on the basis of "citizenship, nationality or domicile," regardless of whether they are in a member state. The process leading to the Aarhus Convention was also ground breaking, as it saw an unprecedented involvement of NGOs in the conceptualization, negotiating, drafting, signing, ratification, and implementation of the convention (Wates 1999). The 1991 UNECE Convention on Environmental Impact Assessment in a Transboundary Context (the "Espoo Convention") and its 2003 Protocol on Strategic Environmental Assessment are also significant in establishing principles, approaches, and mechanisms for public access to information and participation with regard to activities with potential transboundary environmental impacts (UNECE 1991, 2003; Cassar and Bruch 2004).

In the Americas, the 2000 Inter-American Strategy for the Promotion of Public Participation in Decision Making for Sustainable Development (or "ISP") is an initiative by the Organization of American States (OAS) to implement Agenda 21 and Principle 10 of the Rio Declaration in the Western Hemisphere (OAS 2000; Caillaux, Ruiz, and Lapeña 2002). Whereas the Aarhus Convention is a binding treaty with specific obligations, the ISP is a "strategy" that encourages – but does not require – signatories to undertake legal and institutional reforms to promote transparency, participation, and accountability. Adopted in April 2000, the ISP comprises two documents – a short, general Policy Framework and detailed Recommendations for Action. These instruments urge member states to take action (and provide illustrations of possible mechanisms) to improve access to information, decision-making, and justice through legal, regulatory, policy, technical, and financial means. As with the Aarhus Convention, members of civil society helped to develop and negotiate the text of the ISP, albeit in a more modest fashion. There are also a variety of subregional instruments in the Americas promoting public involvement, which are addressed in more detail elsewhere (Bruch and Czebiniak 2002; Dowdeswell 2002).

In Asia, the Asia–Europe Meeting has been working to develop a framework for promoting good practices in public participation (Hildén and Furman 2002). Still in preparation, the draft document on "Towards Good Practices for Public Participation in the Asia–Europe Meeting Process" includes specific provisions for its members (25 states plus the European Commission) to adopt regarding access to information, public participation, and access to administrative and judicial proceedings. Many of the provisions build upon commitments that the European countries made under the Aarhus Convention, but these represent significant new commitments for the Asian member countries.

Institutional developments in different global bodies, such as the World Bank, have been essential in developing mechanisms for public involvement in the on-the-ground implementation of sustainable development, including in the management of international watercourses (Okaru-Bissant 1998). In response to significant pressure from civil society, in the early 1990s many organizations in the World Bank Group undertook efforts to improve their transparency and public consultations, as well as establishing independent mechanisms that the public may invoke to hold Bank organs more accountable to the Bank's stated policies and procedures (Bernasconi-Osterwalder and Hunter 2002). The World Bank Group includes the International Bank for Reconstruction and Development (IBRD, which lends money to governments, usually for large-scale infrastructure projects), the International Development Association (IDA, which provides long-term loans at zero interest to the poorest of the developing countries), the International Finance Corporation (IFC, which lends money to the private sector), the Multilateral Investment Guarantee Agency (MIGA, which provides investment guarantees against certain non-commercial risks to foreign investors), and the International Centre for Settlement of Investment Disputes (ICSID, which resolves disputes between member countries and eligible investors).

The World Commission on Dams (WCD) was a unique collaborative effort between international organizations (including the World Bank), governments, business, and NGOs (Dubash et al. 2001; Di Leva, chap. 10, this volume). It is particularly noteworthy for its transparent, participatory process for discussing broad policy questions, in this case relating to large-scale dams.

The Global Environment Facility (GEF) is jointly administered by the World Bank, the United Nations Environment Programme (UNEP), and the United Nations Development Programme (UNDP). It disperses funding for environmental projects in six major focal areas, including international waters. In this context, it has promoted multi-stakeholder involvement in the integrated management of water resources in various transboundary watercourses (Gonzalez 2003). Additionally, the International Waters Learning Exchange and Resource Network (IW:LEARN), a project of the GEF and other bodies, actively disseminates information and builds capacity on international water management.

Many UN bodies – including UNEP, UNDP, and UNESCO – have been instrumental in improving public involvement in international water management through research, capacity building, supporting pilot projects, and international leadership. For example, the World Water Assessment Programme (http://www.unesco.org/water/wwap/) is a UN-wide initiative that "seeks to develop the tools and skills needed to achieve a better understanding of those basic processes, management practices and policies that will help improve the supply and quality of global freshwater resources."

In addition to the World Bank Group, a number of regional and bi-national institutions have developed policies, institutional mechanisms, and practices to ensure that the public – especially potentially affected individuals – have access to information about projects that could affect them, as well as opportunities to comment on proposed decisions and mechanisms for appealing against decisions that may violate the institutions' policies (Bernasconi-Osterwalder and Hunter 2002; Fall 2002). The effectiveness of these initiatives varies from institution to institution, with bilateral export credit agencies often lagging behind regional and global institutions (Bernasconi-Osterwalder and Hunter 2002; Rich and Carbonell 2002).

The World Trade Organization (WTO) develops and administers the rules of the international trade system, and it resolves trade disputes that arise between member countries. Established in 1994, the WTO supplanted the General Agreement on Tariffs and Trade (GATT). Historically, the WTO and GATT have both been closed to civil society and held a narrow view of environmental laws, frequently striking them down as barriers to trade (Wold 1996; Wilson 2000). In recent years, however, the WTO has made efforts to include civil society (Gertler and Milhollin 2002), as evidenced by the Appellate Body decision in the Shrimp–Turtle case (discussed below on p. 53).

The International Court of Justice (ICJ) was established in 1945 to resolve disputes between nations. It also settles questions of international law that have been referred to it by UN organs, such as the UN General Assembly and the World Health Organization (which requested the court to issue an advisory opinion on the legality of the use of nuclear weapons). The ICJ has decided numerous cases that established the boundaries and use of international waters.

Various thematically or geographically specialized initiatives have also promoted public involvement in the governance of watercourses, domestic and international. The 1995 UN Special Initiative on Africa seeks to stimulate social and economic development in Africa throughout the UN system (UN 1998). The Water Component of the Special Initiative adopts a "Fair Share Strategy" with respect to fresh water, which relies on public participation in the management of domestic and international freshwater resources (UNEP/UNDP/Dutch Joint Project 1999). The New Partnership for Africa's Development (NEPAD), which was developed and launched in anticipation of the World Summit on Sustainable Development, also seeks to promote good governance and civil society engagement in environmental management and development activities (NEPAD 2001).

International professional societies have been instrumental in spurring governments and international institutions (including development banks) to take action. They have provided information, arguments, and energy that have focused attention on various causes – for example, through the International Hydrological Decade. Some of these institutions include the World Water Council (http://www.worldwatercouncil.org/), the International Association for Public Participation (http://www.iap2.org/), and the Global Water Partnership (http://www.gwpforum.org/).

Together, these various global and regional instruments, institutions, and initiatives establish a framework for public involvement in the management of international watercourses and other environmental matters. While the corpus of norms, institutions, and practices continues to evolve, there are numerous areas of common agreement – particularly with respect to the basic principles – and the practice continues to emerge, providing specific detail on how to implement the agreed-upon principles of guaranteeing public involvement. The following sections address, in turn, access to information, public participation in decision-making, and access to justice in the management of international watercourses.

Access to information

Broad access to information is the cornerstone of public involvement. It ensures that the public is able to know the nature of environmental threats and harms. This knowledge allows members of the public to decide whether a response is necessary and, if so, what would be the most appropriate and effective action. In an increasingly connected world, where actions in one nation can affect people and the environment in other nations, states have recognized the need not only to make information available to their citizens but also to share information between nations. This section discusses what type of information about international watercourses is publicly available, how the public can access it, and how water-management authorities have sought to institutionalize access to information processes.

Environmental information in general

There is a growing international consensus on the need to guarantee broad access to information at the national and international levels. Principle 10 of the 1992 Rio Declaration requires that "[a]t the national level, each individual shall have appropriate access to information concerning the environment that is held by public authorities, including information on hazardous materials and activities in their communities ... States shall facilitate and encourage public awareness and participation by making information widely available."

Various regional initiatives since then have significantly clarified the scope of access to information in environmental matters and, in doing so, they have had significant agreement on the specific requirements. "Environmental information" typically is defined broadly to include information in any form (written, electronic, visual, etc.) on the state of the environment or its components and factors that could affect the environment adversely or positively (UNECE 1998; OAS 2000). A person or organization may request information of public authorities (usually including national and sub-national, and sometimes including supranational authorities) without having to show an interest in the information. There is a presumption in favour of access: if the authority has the information requested, it must provide the information without discriminating (for example) on the basis of citizenship, nationality, gender, language, or ethnicity. The information should be provided in a timely manner and free of charge or for a reasonable fee. If the authority does not have the information, it should inform the requester where it believes the information may be found. The authority may refuse to provide the information only for specific reasons (such as national security, commercial confidentiality, and matters currently in litigation), and these exceptions are to be narrowly construed to ensure that the general principle of public access is maintained. Such refusal should be in writing, and should inform the requestor of how they can seek administrative appeal or judicial review of the refusal. If only some of the requested information is protected, the authority must separate out and make the non-exempt information available.

In addition to responding to requests for information, public authorities must affirmatively collect, assemble, and disseminate certain types of environmental information. Mandatory reporting systems may be established. Thus, nations have committed to regular state-of-the-environment reports and pollution registers (often in the form of Pollutant Release and Transfer Registers, or PRTRs). In doing so, authorities are charged with making the information available as a practical matter, considering in which language(s) and form(s) the information should be disseminated. Furthermore, harmonization of the information collected provides an opportunity to assess the state of the regional environment (Wates 1999). Authorities must inform the public of the type and scope of information that is available and how the public can access it. Additionally, authorities must inform the public of how they can access international legal instruments and national and international documents, and opportunities for the public to submit information on non-compliance to international bodies regarding environmental matters.

Information and international watercourses

Recognizing that information is essential to the sound management of international watercourses, and that states historically have been reluctant to compromise their negotiating positions by sharing information with other states or their own citizens (Okaru-Bisant 1998), international instruments and institutions increasingly facilitate (or even require) states to share information. This includes information on the status of a transboundary watercourse (such as water availability in the catchment area, rainfall data, simulated stream flows, and evaporation data, as well as water-quality data) and on factors that could affect the quality or quantity of water in the watercourse (such as ongoing or proposed projects).

The 1997 United Nations Convention on the Law of the Non-Navigational Uses of International Watercourses unambiguously mandates information-sharing among states, although public access to that information is less clear. Thus, under Article 9, states must regularly exchange hydrological, meteorological, hydrogeological, and ecological data (including information related to water quality and to forecasts). Article 11 requires states to exchange information on planned measures, and Article 12 requires prior notification (including technical data and an environmental impact assessment; EIA) to states that could be affected by proposed actions.

A number of water basins have committed to sharing information (Treaty for Amazonian Cooperation 1978). In Southern Africa, the 1995 SADC Protocol on Shared Watercourse Systems required member states to "exchange available information and data regarding the hydrological, hydrogeological, water quality, meteorological and ecological condition of such watercourse system." In order to monitor and develop shared water courses, river-basin management institutions are required by Article 5(b)(i) to "collect[ ], analys[e], stor[e], retriev[e], disseminat[e], exchang[e] and utilis[e] data relevant to the integrated development of the resources within shared watercourse systems and assist[ ] member States in the collection and analysis of data in their respective States." None of the Protocol's information-sharing provisions limit the obligations to inter-State exchanges, and one article specifically commands the river-basin management institutions to promote public awareness and participation in environmental matters.

Development of public access

The Southern African Development Community (SADC) Protocol recognizes that governments and international institutions frequently lack the financial resources, technical infrastructure, and personnel to manage shared watercourses effectively. This lack of reliable data has impeded the development, implementation, and enforcement of international agreements for transboundary watercourses (Okaru-Bisant 1998). In fact, the World Bank observed that in Southern Africa, "without hard information [on the actual annual flow of the Senque (Orange) River], Lesotho is unwilling to make a firm international agreement guaranteeing a certain quantity of flow into South Africa" (Sharma et al. 1996).

To supplement scarce resources and reduce political difficulties, international instruments and institutions frequently rely on civil society to generate, review, and utilize information necessary for the management of transboundary watercourses (ZACPLAN 1987; Sharma et al. 1996; Eriksen 1998; Okaru-Bisant 1998). Thus, the Nile Technical Advisory Committee considered projects designed to promote public participation and public information. The 1999 London Protocol specifically sought to incorporate the principles of public involvement set forth by the Aarhus Convention into the management of transboundary watercourses (UNECE 1999).

NGOs are also finding fertile ground to promote access to information on transboundary watercourses in the absence of an international mandate. For example, the International Nile Basin Association (INBA) is a voluntary, non-profit organization that disseminates knowledge, shares experiences, and provides information relating to the development of Nile water resources. The INBA constitutes an independent, alternative forum that complements the governmental forum and facilitates the generation and exchange of environmental information.

Information on status of watercourses

Knowledge about the quality and quantity of water in transboundary watercourses forms the foundation from which all decisions are made: is there enough water; is there enough water of sufficient quality; could a particular environmental or public health harm have been caused by the condition of the watercourse; is there any need to be concerned about proposed projects that might reduce the quantity of, or impair the quality of, available water?

Increasingly, international instruments establish what information needs to be made available, how frequently, and in what medium. For example, the 1992 UNECE Convention on the Protection and Use of

Transboundary Watercourses and International Lakes requires parties to sample and evaluate both ambient water quality and effluent into transboundary waters. This information must be publicly available "at all reasonable times," inspection shall be free of charge, and the public can obtain copies of the information "on payment of reasonable charges" (UNECE 1999). The 1999 London Health Protocol to this Convention expanded the information that states needed to collect and make available to the public – including, *inter alia*, drinking-water quality, discharge of untreated waste water and storm water overflow, and source-water quality (UNECE/UNEP 2000). Furthermore, "in the event of any imminent threat to public health from water-related disease, [the state must] disseminate to members of the public who may be affected all information that is held by a public authority and that could help the public to prevent or mitigate harm" (UNECE 1999).

Some of the most promising developments in promoting access to information about the status of transboundary watercourses occur through the growing practice of public and private institutions to collect information and make it publicly available. In addition to various region- and water body-specific initiatives, a number of global efforts are helping to build technical and institutional capacity to collect, store, and disseminate information on the status of freshwater resources in Africa. For example, the Southern Africa Flow Regimes from International Experimental and Network Data (FRIEND) programme is working to establish an international database on river flows, assemble data that can assist in determining flow regimes, analyse and estimate flood and drought frequency, integrate national inquiries into water resources, and model rainfall and runoff (Eriksen 1998). Similarly, the Nile FRIEND programme has strengthened flow-data collection and management along the Nile River in a non-governmental context, although not all of the riparian countries have participated in the programme. Additionally, the World Hydrological Cycle Observing System (WHYCOS) is developing a network of observatories around the world to collect high-quality hydrological data.

Along the United States–Canada border, the Great Lakes Water Quality Agreement (GLWQA) mandates the collection of information on water quality and quantity of the boundary waters and the tributaries (United States–Canada 1972). This information is to be made publicly available, unless it is proprietary under domestic law. The Geographic Information Systems Section of the Great Lakes Information Network (GLIN) provides on-line digital data and maps for the region (available at http://www.great-lakes.net/). Users can search by topic, geographic regions, organizations such as the US Army Corps of Engineers, or the GLIN Data Access (GLINDA) Clearinghouse. Information available through GLIN includes links to information on daily stream flows of

rivers feeding into the Great Lakes, annual reports by the US Army
Corps of Engineers on water quality along the Great Lakes and for
waters feeding into them (http://www.lrd.usace.army.mil/gl/wq_rpt.htm),
and the Great Lakes Environmental Research Laboratory Real-Time
Great Lakes data, which includes information on the Detroit River's
daily averaged flows (http://www.glerl.noaa.gov/data/now/).

In comparison, along the United States–Mexico border, citizens and
local officials had found it difficult to obtain information on transboundary watercourses from the IBWC. For this reason, when there were
questions regarding groundwater pollution in Nogales, Mexico, university researchers collaborating across the border conducted their own
groundwater testing rather than relying on the IBWC to do it (Ingram,
Milich, and Varady 1994). In more recent years, the IBWC has established an internet site that posts the daily and historical flow conditions
at different points along the Rio Grande (http://www.ibwc.state.gov/wad/
histflo1.htm).

The Mekong River provides another approach for ensuring public
access to information on the status of a watercourse. Since 1985, the
Mekong River Commission has undertaken baseline studies of water
quality and resources in the basin through its Water Quality Monitoring
Network. In 1999, the network consisted of 103 stations. Discharge-
measurement and sedimentation-sampling studies also have been conducted in Cambodia. Flow information is made publicly available in various media, and some commercial organizations have, in fact, established
a business of publishing water-flow data (originally appearing in news reports) for their members.

Information on factors that could affect a watercourse

The public also needs to learn about proposed and ongoing activities that
could affect transboundary watercourses. These activities could be developments such as water diversion programmes that affect the quantity of
water or industrial facilities that affect water quality. EIA is an important
mechanism for assessing the potential impacts of a project and deciding
whether and how to proceed (Cassar and Bruch 2004). EIA is discussed
in more detail in the section on public participation (pp. 41–48), but
the threshold step of informing the public of the proposed activity and
its potential ecological and social impacts merits mention here.

The SADC Protocol on Shared Watercourses specifically required
river-basin management institutions to promote EIAs for development
projects in a shared basin, and the East African MOU on Environment
Management recommends the use of EIA. Considering the shared concern expressed in these documents for the joint management of Lake

Victoria, it is foreseeable that the public will eventually have access to information about development projects that could affect Lake Victoria, whether the proposed project is in their country or another one. The twin East African Community initiatives to develop guidelines on regional EIA and on developing an environmental protocol are promising steps in this direction (Sikoyo, chap. 22, this volume).

In North America, when the IJC receives a project proposal, that Commission must provide notice to the public (a) that the application has been received; (b) the nature and locality of the proposed use, obstruction or diversion; (c) the time within which any person interested may present a statement in response to the Commission; and (d) that the Commission will hold a hearing or hearings at which all persons interested are entitled to be heard. The Border Environment Cooperation Commission (BECC) also requires public notification of projects that could affect a transboundary watercourse.

Outside the context of EIA, organizations may obtain information on activities adversely affecting a transboundary watercourse. The Rhine River Commission must "exchange information with non-governmental organizations insofar as their fields of interest or activities are relevant" (Rhine 1963, 1976). Furthermore, the Commission must inform NGOs when decisions have been made that could have an "important impact" on the organizations. Similarly, the 1909 International Boundary Water Treaty commits the IJC to making publicly available official records, including applications, response statements, records of hearings, decisions, and reports. The public may obtain copies of this information upon payment of reproduction costs.

In addition to access to those factors that negatively affect a watercourse, the public frequently has access to information on activities that seek to redress impacts on transboundary watercourses. Thus, for example, the 1992 Helsinki Convention makes publicly available information on "the effectiveness of measures taken for the prevention, control and reduction of transboundary impact." Specifically, water-quality objectives, issued permits (including the permit conditions), and the compliance assessment results must be made "available to the public at all reasonable times for inspection free of charge, and [states] shall provide members of the public with reasonable facilities for obtaining from the Riparian Parties, on payment of reasonable charges, copies of such information."

The International Boundary Water Agreement provides that the annual inventory of pollution-abatement requirements is publicly accessible. These pollution-abatement inventories include the monitoring and effluent restrictions and compliance schedules, so that the public can review who is in compliance.

Information on the development of watercourse norms, policies, and management plans

The public usually is guaranteed access to basic information on the institutional processes that relate to the development of policies and norms governing actions within the basin. These include draft policies, standards, management plans, and meetings, although internal documents reflecting the deliberative process are not always made available.

Many organizations – including the BECC and the IJC – require the public to be notified of upcoming meetings of regional bodies (Milich and Varady 1998). This notice normally states the time and place of the meeting, as well as the agenda or items to be discussed and how the public may participate.

The public frequently has the right to obtain information on proposed standards, management plans, and other means of implementing goals for the management of transboundary watercourses, so that they can review and comment on the proposals. The 1999 London Water and Health Protocol establishes a transparent framework for setting standards and levels of performance regarding protection against water-related disease. The European Water Framework Directive Proposal provides that the public must have access to river basin management plans, as well as the opportunity to submit written comments on the plans (Ferrier 2000).

In addition to notifying the public of proposed standards, institutions managing transboundary waters may establish a transparent process for making decisions regarding the policies and standards governing activities that affect the watercourse, so that the public can review the bases for the decisions made.

Institutionalizing access

Because of the importance of information and public involvement in the decision-making process, many mechanisms have evolved at the international, national, and local levels to ensure that citizens and organizations have access to information regarding transboundary watercourses (Kaosa-ard et al. 1998; Avramoski 2004). This includes information on the status of water flow and water quality; information on ongoing and proposed activities that could affect the watercourse; and information on the development of norms, policies, and management plans.

Although different watercourse institutions and instruments vary in the specifics, most incorporate both "passive" and "active" mechanisms for ensuring that the public has access to the necessary information. Passive mechanisms guarantee that the public can request information from a governmental or supra-governmental authority. Active mechanisms require authorities to collect and affirmatively disseminate information, for example on the status of the watercourse environment or on proposed projects.

Some international institutions have established units with the role of facilitating public access to information on transboundary watercourses. For example, the Public Relations and Co-ordination Unit of the Mekong River Commission Policy and Planning Division disseminates information through press releases, policy papers, annual reports, and monitoring and evaluation reports (Kaosa-ard et al. 1998).

To facilitate public dissemination of information, some transboundary water institutions have established resource centres (Nakayama and Fujikura 2002). The 1972 Great Lakes Water Quality Agreement established a Great Lakes Regional Office to assist the IJC in disseminating information on the North American Great Lakes. The Mekong River Commission also established centralized resource centres, as well as centres near an affected area, and the countries around Lake Victoria agreed to establish environmental resource centres.

Increasingly, institutions charged with the management of transboundary watercourses rely on electronic dissemination through both e-mail and websites (Bruch, chap. 18, this volume). The IJC website (http://www.ijc.org) is an example of the capacity of websites to disseminate information. The website allows one to search past and present projects, reports, and decisions of the IJC; to look up current notices of public hearings and reports; to investigate the status of projects and issues with which the Commission is dealing; and to access interim and final reports. The IJC also uses the site as a source for public comment by supporting what are known as "discussion rooms." Other transboundary watercourse websites include the Nile Basin Initiative (http://www.nilebasin.org), the Mekong River Commission (http://www.mrcmekong.org), and the BECC (http://www.cocef.org). The UNECE, which serves as the secretariat for the 1992 Helsinki Convention on Transboundary Watercourses and its 1999 London Protocol – as well as the Aarhus Convention, the Espoo Convention, and the SEA Protocol – also has an extensive website (http://www.unece.org).

Civil-society organizations frequently are integral in generating and disseminating information on transboundary watercourses. Thus, the Mekong Forum has acted as a clearing-house of information for the lower Mekong River Basin, and universities along the United States–Mexico border have monitored and sampled contaminated groundwater in the Nogales area.

At the national level, constitutions, laws, regulations, and policies can provide an enabling environment and ensure that citizens have access to information held by their government (or even by other governments or private actors) (Kaosa-ard et al. 1998; Bruch, Coker, and VanArsdale 2001). Nevertheless, in spite of the developments at the national and supra-national levels, there remain challenges in ensuring public access to information on transboundary watercourses as a practical matter. For example, in Cambodia, access to information on hydropower development projects prior to construction is not commonplace, despite policies to the contrary (NGO Forum on Cambodia 1997). One of the reasons for this in Cambodia is the lack of access to radio, televisions, and newspapers, particularly in rural areas. Another complication is language barriers, due to the fact that many of the EIA documents in Cambodia are printed in English. Many other countries face similar challenges in the practice of making EIA and other information available to the public.

It is precisely because of the challenges posed by multiple languages, illiteracy, few technical resources, and a chronic lack of financial resources that public involvement is necessary. Citizens and NGOs can complement governmental and supra-national efforts in generating, reviewing, and utilizing data relating to the management of transboundary watercourses. In doing so, public involvement can bring more resources to bear on decision-making. The next section examines how members of the public have been able to take the information available and contribute constructively to deliberations regarding the management of transboundary watercourses.

Public participation

If access to information is the predicate, participation of civil society in decision-making processes is the centrepiece of public involvement. Participation ensures that decision makers have the opportunity to consider the diversity of interests at stake, and it guarantees that citizens and organizations have an opportunity to submit information and arguments on decisions that could affect them.

Public participation in environmental matters in general

Drawing upon the experiences of many countries in promoting public participation in environmental management, Principle 10 of the 1992 Rio Declaration asserted that "Environmental issues are best handled with the participation of all concerned citizens, at the relevant level. At the national level, each individual shall have ... the opportunity to participate in decision-making processes." Considering the large number of signatories to the Rio Declaration, the commentary of eminent scholars since its adoption, and legal developments, public participation in environmental matters may be said to be a norm of customary international law. Chapter 18 of Agenda 21 – on integrated management of freshwater resources – has the following as one of its four principal objectives: "To design, implement and evaluate projects and programmes that are both economically efficient and socially appropriate within clearly defined strategies, based on an approach of full public participation, including that of women, youth, indigenous people and local communities in water management policy-making and decision-making...." Since 1992, regional instruments and national laws have helped to clarify the specific elements of public participation.

Through the development of EIA at the national, regional, and international levels, members of the public have the right to participate in decisions relating to proposed activities (Cassar and Bruch 2004). When an authority (be it national or supra-national, such as a water basin authority) is considering a proposed project that could affect the environment, EIA laws and policies usually require the authority to notify the public of the proposed activity, the nature of the decision that is to be made, and the procedure for members of the public to submit written or oral comments. The notice can also indicate some of the possible impacts of the proposed activity. Usually, the notice must be made in a manner (language, medium, location, etc.) that will ensure that people who could be affected learn about the proposed activity and the opportunity to comment on the proposal. The timing of the notice must also allow members of the public sufficient time to prepare their comments and participate in the decision-making process. Participation should be solicited at an early stage, when options are still open.

The authority must allow the public to review, free of charge, the documents and other information that the authority is considering in making its decision. Members of the public usually have the right to submit written comments, and can sometimes petition for a public hearing at which they can submit oral comments. The authority cannot make its decision until after the public has had a chance to submit its comments, and the authority must take "due account" of the public's submissions. In some cases (such as the United States), this means that before it can make a decision the authority must first prepare a "Response to Comments" document that addresses all the submissions that it received from citizens, NGOs, businesses, and other interests in the comment period. Once the decision has been made, the authority promptly must make the decision available, along with the reasons for its decision.

In addition to EIA, the public frequently has the opportunity to participate in administrative hearings on proposed activities, such as the granting of permits. Public participation in these activities can be important, as individual permits might not have a sufficiently significant impact to warrant an EIA whereas, in the aggregate, these permits can greatly affect the environment (e.g. siltation and eutrophication of waterways arising from urban sprawl that could be addressed at the stage of approving building permits).

International instruments increasingly recognize the rights of the public to participate in the development of plans and policies, and even more binding norms contained in regulations, laws, and international instruments (Bruch 2002; Bruch and Czebiniak 2002). These rights are still evolving, so that many of the provisions are clear but lack the specific requirements found for public participation relating to specific activities. The Aarhus Convention obliges member states to allow the public to participate in the development of plans and programmes at an early stage and to take due account of the public participation. However, the Convention requires parties only to "endeavour to provide opportunities" to "the extent appropriate" for the public to participate in preparing policies bearing on the environment. When it comes to the more binding normative instruments of binding rules and regulations, the Aarhus Convention is even more circumspect, urging states to "strive to promote" public participation and suggesting fixed time-frames, publication of draft rules, an opportunity for the public to comment, and taking public participation "into account as far as possible." Nevertheless, these provisions constitute a significant step forward in empowering the public to contribute directly to the overall environmental management framework. Indeed, they formed the impetus for the Protocol on Strategic Environmental Assessment to the Espoo Convention (Kravchenko 2002; UNECE 2003). Furthermore, the last decade has seen a marked increase in the participation of civil-society organizations in the negotiating, ratification, and implementation of international environmental agreements (Giorgetti 1998; Bruch and Czebiniak 2002).

Public participation in decisions relating to activities affecting transboundary watercourses

With the development of EIA as a standard tool of environmental management, international agreements on transboundary watercourses increasingly incorporate EIA procedures (Danube River Protection Convention 1994; East African MOU 1998; Cassar and Bruch 2004). National EIA laws also frequently provide a framework for guaranteeing that the public has access to information about proposed projects that could affect the quality or quantity of water in transboundary watercourses (Kaosaard et al. 1998). As mentioned earlier, in Africa, EIA is evolving as a tool in environmental management, and institutions charged with managing transboundary watercourses are incorporating and promoting EIA.

For Lake Victoria, EIA is emerging as a key tool in protecting the shared water and ensuring that the public has an opportunity to participate in its management (Sikoyo, chap. 22, this volume).

Since 1909, the IJC has guaranteed public participation in decisions on specific activities that could affect the North American Great Lakes. When a party or a person seeks to use, obstruct, or divert waters falling within the IJC's jurisdiction, they must submit an application to the IJC to do so. The IJC notifies the public of the application by publishing a notice in the *Canada Gazette* and the [US] *Federal Register* and once a week for three weeks in newspapers that are circulated "in or near the localities which ... are the most likely to be affected" by the proposed activity. The notice must include information on the application, the "nature and locality of the proposed" activity, and the opportunity for the public to submit written or oral comments. Within 30 days of the filing of the application, any "interested person" other than the project applicant may submit a statement supporting or opposing the proposed activity. Furthermore, "persons interested in the subject matter of an application, whether in favour of or opposed to it" are entitled to speak or have an attorney speak on their behalf at an open hearing before the ICJ. The verbatim transcripts of the hearings, exhibits filed, briefs and formal statements, and the IJC decisions and orders are all available to the public.

The public also has the opportunity to participate in discussions by the ICJ on matters that have been referred to them by either member state. Although these hearings may be sensitive – and thus more likely to be subject to constraints imposed by the parties – the process is similar to the hearings for applications.

In practice, the IJC has utilized a variety of types of public hearings. For example, the IJC has conducted "mini meetings," large public forums, virtual conferences via the Internet, conference calls, and video conferences. IJC public hearings have addressed issues ranging from management of water levels, the effects of large-scale aquaculture on the water quality of the Great Lakes, and bulk removals of Great Lakes water.

NGOs have had an important role in convening dialogues on the management of transboundary watercourses. For example, twice in the past decade, Great Lakes United have invited government and industry representatives to a series of approximately 20 public hearings around the North American Great Lakes region for citizens to voice concerns about the management of the Lakes (Jackson, chap. 6, this volume). Government and industry representatives also participated. This was the first time that citizens were able to voice their thoughts publicly on the topic, and it built bridges among NGOs who had been working toward the same goals but had not been actively collaborating. The discussions culminated in a report entitled "Unfulfilled Promises." Again, in 1998, citizens and NGOs convened another 10 hearings around the region and published another report. Finally, when the US government developed a draft management for Lake Superior and proposed to publish its plan on the *Federal Register* and give citizens 60 days to respond, the Lake Superior Alliance decided that this period was insufficient to obtain meaningful public input from those who would be most affected. As a result, the NGO convened a series of hearings on the topic. Again, government officials attended, ultimately deciding to hold similar public hearings of their own. These experiences highlight the unique role that NGOs can have in bridging political boundaries to focus on the watershed and the shared interests of those who depend on transboundary watercourses.

For the past decade, the BECC has required the inclusion of affected communities in the process for certifying proposed environmental infrastructure projects along the United States–Mexico border (Milich and Varady 1998). Thus, the affected public is able to participate in decisions affecting the Rio Grande, which runs along approximately half of the border. In addition to the standard notice to the public about an application for project certification and the opportunity for the public to submit comments, the BECC requires that projects have public support. In fact, applicants must submit a Community Participation Plan – which includes meetings with local organizations, two public meetings, public access to information about the project, and a steering committee that includes local representatives. Once the Community Participation Plan has been carried out, applicants are required to submit a report that shows public support for the project. In fact, the BECC has respected public comments on proposed projects so much that "[o]n several occasions, projects thought to be all but approved were sent back for redesign following the public-comment period" (Milich and Varady 1998).

In the Mekong River Basin, Vietnamese university academics and newspaper reporters held a series of public seminars on a government proposal to dyke the major river banks in the Mekong Delta to control flooding. As a result of the consultations, the "government accept[ed] an alternative proposal which suggested flood evacuation to the Western Sea, as opposed to the original plan of absolute flood control." (Kaosaard et al. 1998). NGOs also have been important in fostering participation by citizens in decision-making processes regarding specific projects. MekongForum, an NGO with an academic and student membership base, has been providing the public with information on development proposals along the Mekong River and its tributaries in order to raise public participation and awareness of the impacts of development (Kvær-nevik 1994; Kaosa-ard et al. 1998). Other Thai NGOs also have worked to raise awareness of the environmental impacts of large-scale hydro-power development on the Mekong.

Public participation in setting norms, policies, and plans

In participating in the establishment of norms, policies, and plans, the public can efficiently avoid an interminable series of piecemeal battles and go straight to the root of the issue. The watercourse institutions, in turn, are able to benefit from the on-the-ground experience and expertise of civil-society members, as well as avoiding repeated conflicts over projects that drain resources and delay projects. Thus, for example, the Rhine Convention empowers the Rhine Commission to recognize NGOs as observers, to exchange information with NGOs, to invite NGOs to participate in Commission meetings, and to consult specialists.

Citizens have served on commissions, boards, and task forces for transboundary watercourses. For example, people from NGOs, business, and state and local governments have served on the BECC (Milich and Varady 1998), and citizens who were both specialists and non-specialists have served on IJC boards and task forces.

For most citizens and NGOs, the priority simply is to submit information and arguments, rather than actually serving on the decision-making body. The Mekong River Commission has affirmed that all Mekong riparian states, project supporters, project opponents, national Mekong committees, and representatives of indigenous populations should take part in developing sustainable policies for the basin. In addition to resource users and occupational groups in the basin, people living outside the Mekong River Basin who may be affected by the impacts of a project may participate (Mekong River Commission Secretariat 1999). This stakeholder participation is to occur in all aspects of MRC activities, including project and programme planning, implementation, monitoring, and evaluation. Nations have committed to similar participation provisions in other regional initiatives in East Africa, for the Nile Basin, and in the UNECE (Bruch 2001; UNECE 2003). Such declarations and normative development constitute important steps toward incorporating public participation in establishing policies and plans; however, the practice to a large degree has yet to be realized.

Public participation can compel transboundary institutions to comply with their own stated policies and procedures. For example, the Internet discussion group BECCnet has influenced decisionmaking about a half-dozen times [by early 1998]. When the [BECC] commission failed to adhere to self-imposed guidelines for a forthcoming meeting, for instance, e-mail protests were so numerous that the directors rescheduled the meeting. Similarly, at another meeting attended by about 200 people, the chairman gaveled the proceedings closed before allowing public comment; the cascade of protests on BECCnet led to a public apology and a binding modification of procedures for such comment. (Milich and Varady 1998)

Similarly, the public can help to review compliance by parties to an agreement on a transboundary watercourse (UNECE 1999). The transparency of this review can encourage compliance and strengthens the credibility of the institution.

Public participation in the development of transboundary watercourse agreements

As mentioned above, civil-society organizations have participated in the development of a number of international environmental agreements. For example, NGOs played a key role in negotiating the 1987 amendments to the GLWQA (Sandler et al. 1994). NGO representatives served on the national delegations, reviewed draft position statements, and participated in decision-making at the national and bilateral levels. Through the process, they helped to establish trust between the governments and civil society. The NGO representatives complemented the government representatives, as the NGO representatives had technical knowledge that often exceeded that of their counterparts, particularly the official delegations representing the foreign ministries of Canada and the United States.

Following the adoption of the Agreement, the NGO representatives have worked to implement the agreement. The IJC subsequently noted that:

these [non-governmental] organizations are important in focusing political attention on the integration of Agreement objectives into domestic priorities and programs. They are instrumental in encouraging governments to provide the resources necessary to implement the agreement and actively promoting environmentally conscious behavior among their own membership and the public at large.... (Sandler et al. 1994)

Similarly, NGOs were actively involved in the development of the 1999 UNECE Water and Health Protocol. As illustrated in these examples, vigorous NGO support can greatly enhance the effectiveness of an agreement, and such support often can be gained by involving NGOs in the process of developing the international agreements.

Implementing public participation

Full public participation involves all sectors of society. It may vary in particular instances, depending on the interests at stake (Avramoski 2004). In order to realize broad and effective participation, it is necessary to address challenges posed by historical, geographical, and financial constraints.

One way to improve public access to decision-making processes is to take the process to the people to make it easier for them to participate. For example, along the Mexico–United States border, the BECC holds quarterly meetings in different cities. Nevertheless, although these meetings are open to the public, the great distances associated with the border region hinder public attendance (Milich and Varady 1998).

Cultural and historical contexts can also make public participation difficult. For example, in Cambodia, there is a distrust of public participation since "'public participation' was used during the Khmer Rouge regime to gather villagers in coercive activities" (Kaosa-ard et al. 1998). Attempts to adopt a "participatory approach" have been more successful, but this may require a different approach to that often used, particularly since "during the Khmer Rouge era, people attending public meetings could be killed or forced into hard labour" (Kaosa-ard et al. 1998).

Education and training of the public and of public officials are essential to establishing trust in the value of public participation and in understanding how to participate in the management of water resources (UNCED 1992b). Reliable enforcement mechanisms are also important in providing avenues to ensure that public participation is given its full due (UNECE 1999).

Access to justice

Citizen access to administrative and judicial review mechanisms – commonly termed "access to justice" – provides a third pillar in the governance of international watercourses. Access to information and public participation depend on enforcement and review mechanisms for their guarantee. Additionally, these review mechanisms can help to ensure that substantive norms are complied with – for instance, that there is not undue degradation of water quality or illegal extraction of water (Ferrier 2000).

Although work remains to be done to improve the transparency and participatory nature of governments and international institutions, discussions surrounding environmental law and international water management increasingly turn to implementation and enforcement. It is not enough to provide information, to allow the public to participate, or to have strong norms in theory; these legal rights and obligations must be backed by enforcement mechanisms that provide recourse for violations.

Over the last decade, governments have made significant strides toward involving citizens in the enforcement procedures relating to international watercourses. This section discusses proceedings initiated by citizens in many forums, including domestic courts and international fact-finding and investigative bodies such as the North American Commission for Environmental Cooperation (CEC) and the World Bank. In addition, although sometimes not able to bring cases on their own behalf in certain venues, citizens may be able to participate in proceedings between countries before such international bodies as the WTO and the ICJ through the submission of *amicus curiae* ("friend of the court") briefs.

Access to national courts and agencies

Citizens may be able use their domestic laws, courts, and administrative bodies to challenge activities that are resulting in international watercourse degradation. This can provide a familiar venue for aggrieved parties, although there might be difficulties associated with the extraterritorial application of domestic law.

In addition to utilizing their own domestic venues, citizens may also be able to participate in the judicial or administrative proceedings of another country as intervenors or affected parties (plaintiffs). However, this approach can be complex: cases involving transboundary harm often require complicated procedural and political issues to be addressed – such as sovereignty, the presumption against the extraterritorial application of national laws, jurisdiction, and *forum non conveniens*.

Cases in Europe and North America have established precedents for affected people to invoke the jurisdiction of another country. Building on these cases and the growing recognition of the role that private parties can have in the management of international waters, recent conventions have incorporated access-to-justice principles.

Cases in Europe and North America have established precedents for affected people to seek redress in domestic courts of another country. Two good examples of national cases are the Rhinewater Case and the High Ross Dam Controversy (Bruch 2001). The Rhinewater Case involved a suit in 1976 by an NGO from the Netherlands against a French mining company on behalf of individuals suffering from chloride pollution. The European Economic Community Court of Justice held that the plaintiffs could sue either where the damage occurred (the Netherlands) or where the damaging act took place (France); the plaintiffs chose the Netherlands. The case caused both countries to conduct new chloride studies and ratify a modified Chloride Convention by 1985 (Kiss 1985; Darrell 1989; de Villeneuve 1996).

The High Ross Dam Controversy arose from a proposal by the City of Seattle to increase the height of a dam on the Skagit River by over 120 feet in 1970. Under the US National Environmental Policy Act, a Draft Environmental Impact Statement (EIS) was prepared, and environmental groups in the United States and Canada were involved in the hearings. Ultimately, the courts upheld the EIS, but the IJC encouraged British Columbia and Seattle to reach a negotiated settlement, which they did in 1984. This established a precedent for allowing citizens and organizations from another country to intervene in a US case dealing with transboundary water management (Parker 1983).

Several cases in the United States have also dealt with foreign aid to projects that could affect an international watercourse, although the courts have been more reluctant to recognize the legal standing of US citizens to bring these cases (Bruch 2001). Indeed, when contrasting these cases over the past three decades, US courts appear to be more willing to entertain claims by alien or international parties who have been (or will be) directly and clearly affected by a proposed project in the United States.

Building on these cases and the growing recognition of the role that private parties can have in the management of international waters, many recent conventions have incorporated access-to-justice principles.

Treaty provisions

International treaties, conventions, and protocols increasingly seek to ensure fair, equitable, and effective access to courts and administrative agencies. These provisions build upon the experiences of citizens and NGOs in using national judicial and administrative forums to protect international watercourses and recognize the important role that these institutions can play in enforcing environmental norms.

Some of the conventions simply call for non-discrimination in providing access to justice, particularly at the national level. Thus, Article 32 of the 1997 UN Convention on the Law of the Non-Navigational Uses of International Watercourses provides:

Unless the watercourse States concerned have agreed otherwise for the protection of the interests of persons, natural or juridical, who have suffered or are under serious threat of suffering significant transboundary harm as a result of activities related to an international watercourse, a watercourse State shall not discriminate on the basis of nationality or residence or place where the injury occurred, in granting to such persons, in accordance with its legal system, access to judicial or other procedures, or a right to claim compensation or other relief in respect of significant harm caused by such activities carried on in its territory.

Although states can agree otherwise (in specific instances), the general rule is that citizens and organizations must have non-discriminatory access to legal recourse. This principle represents the culmination of three decades of negotiation and agreement among legal experts and decision makers around the globe, and, as such, it may represent an emerging norm of customary international law that is being codified in the convention. As Professor John Knox has observed, the principle of non-discrimination is a significant factor in promoting public participation in a variety of contexts, including public involvement through EIA (Knox 2002).

Under the East African MOU, Kenya, Tanzania, and Uganda are obliged to ensure public access to their administrative and judicial proceedings. Through the MOU, the partner states committed to building capacity for access to justice by developing broad policies and laws for people affected by environmentally harmful activities within the subregion. Article 16(3) further expands the non-discrimination principle by providing that "[t]he Partner States agree to grant rights of access to the nationals and residents of the other Partner States to their judicial and administrative machineries to seek remedies for transboundary environmental damage." Considering these two provisions in the context of managing and protecting Lake Victoria (Article 8), the MOU lays out a normative framework for ensuring open, non-discriminatory access to justice in the management of this shared waterbody. In considering ways to develop and harmonize the environmental laws and institutions governing Lake Victoria, a UNEP/UNDP/Dutch joint project recommended that

"[b]road principles of *locus standi* should be adopted to allow private suits as a tool for the enforcement of environmental obligations." (UNEP/UNDP/Dutch Joint Project 1999)

In the UNECE region, a number of conventions provide for access to justice in environmental matters generally as well as specifically for international watercourses (UNECE 1999, 1998, 1991; Kravchenko 2002). The Aarhus Convention contains explicit and exhaustive provisions for access to courts and administrative remedies to ensure access to information, public participation, and compliance with national environmental laws (UNECE 1998). The Convention even calls on the public to supplement the enforcement role that is traditionally the realm of governments. Access to justice is required to be fair, effective, and open, and decisions must be in writing and made available to the public. To facilitate the use of administrative and judicial review, the Convention requires each state party to endeavour to "remove or reduce financial and other barriers" and to provide information on the procedures. The Aarhus Convention represents the most detailed elaboration of access to justice by an international treaty; however, other global and regional instruments have also recognized the importance of broad access to environmental justice (Bruch 2002).

The 1999 London Protocol to the 1992 UNECE Convention on the Protection and Use of Transboundary Watercourses and International Lakes specifically extends many of the provisions of the Aarhus Convention to international watercourse management. Despite the extensive access to information and public-participation provisions discussed above, the Protocol has only a few general provisions urging parties to ensure access to justice. For example, Article 5(i) provides that "access and participation should be supplemented by appropriate access to judicial and administrative review of relevant decisions," but it does not clarify the nature of the review.

Access to international courts and tribunals

Aggrieved citizens and organizations increasingly find that, in addition to the national bodies, international courts are willing to entertain their briefs on the matter before the court. By the terms of their organic statutes, the ICJ, the WTO, and other international tribunals usually are empowered to entertain cases brought by nations – and, occasionally, by international organs such as the United Nations and its subsidiary bodies. Increasingly, these bodies allow civil-society organizations to provide separate briefs that lay out additional facts and legal arguments.

In a dispute between Hungary and Slovakia over the proposed Gabčíkovo–Nagymaros dam, the ICJ for the first time accepted a position paper or "Memorial" from a coalition of NGOs, including the Natural Heritage Institute, Greenpeace, International Rivers Network, Sierra Club, and the World Fund for Wildlife (WWF) (Okaru-Bisant 1998). The ICJ recognized the NGO coalition as *amicus curiae*, or friends of the court. The coalition's memorial argued for restoration of the Danube ecosystem and sought international protection for the planet's natural treasures (Liptak 1997). More than two years later, the ICJ ruled that the proposed dam was illegal (A-Khavari and Rothwell 1998; Sands 1999).

In another high-profile case, the WTO allowed NGOs to submit *amicus curiae* briefs for the first time (Pyatt 1999; Guruswamy 2000; Gertler and Milhollin 2002). In the Shrimp–Turtle Dispute, the Center for International Environmental Law and the Center for Marine Conservation submitted an *amicus* brief to the WTO panel in support of the US conservation measures. The panel rejected the NGOs' brief, but the Appellate Body accepted the NGO brief and reversed the panel's ruling that submissions by civil society could not be considered. The Appellate Body held that the panel may, but need not, consider non-party submissions such as *amicus* briefs.

Regional human rights commissions and courts provide civil society with a venue for vindicating fundamental human rights. In the Americas, Europe, and Africa, these commissions and courts can accept and investigate petitions filed by citizens and organizations alleging abuses of human rights (Weston, Falk, and D'Amato 1990; Scott 2000; Jean-Pierre 2002). Although these bodies have yet to be utilized in the context of international watercourses, significant alteration of the quantity or quality of water by an upstream actor could impinge on the rights to life, health, and environments to such a degree as to establish a basis for jurisdiction by one of these bodies. The moral suasion of a public decision by a human rights commission or court could embarrass a state into complying with its international legal obligations.

Fact-finding and investigative bodies

In addition to international courts and tribunals, the public increasingly can gain access to international bodies with the authority to investigate alleged violations. In fact, a number of these bodies were established precisely to ensure that citizens and NGOs have the ability to review actions of nations and international bodies (such as the World Bank) and file complaints when actions violate procedural or substantive norms. Although these bodies generally lack the authority of a legal body, they have been moderately effective in promoting compliance through their public findings.

World Bank (IBRD/IDA) Inspection Panel

The World Bank Group consists of five separate institutions that seek to promote development (Bernasconi-Osterwalder and Hunter 2002; Di Leva, chap. 10, this volume). The International Bank for Reconstruction and Development (IBRD) lends money to governments to develop typically large-scale infrastructure projects, such as hydroelectric power projects (Okaru-Bisant 1998). In 1993, the IBRD and the IDA created the Inspection Panel to increase transparency and accountability, as well as to respond to complaints regarding the environmental and social impacts of its projects. The panel is not a judicial or enforcement body, but it can influence and improve compliance with Bank policy (Udall 1999; Di Leva, chap. 10, this volume).

The Inspection Panel Operating Procedures authorize the Panel to entertain "Requests for Inspection," in which a claimant has been (or will be) affected by the failure of "the Bank to follow its own operational procedures during the design, appraisal and/or implementation of a Bank financed project." These requests may be made by a group of two or more people from the country of the Bank-financed project, or by other specified individuals. The two examples outlined below highlight the opportunities (and limitations) for citizens to invoke the Inspection Panel to protect their interests in an international watercourse.

In October 1999, RECONCILE (Resources Conflict Institute), a Kenyan NGO, submitted a Request for Inspection to the Panel concerning the Lake Victoria Environmental Management Project. The requesters claimed that the individuals whom they represented were likely to suffer harm as a result of the failures and omissions of IDA and the IBRD (the implementing agency of the GEF) in the design and implementation of the water hyacinth management component of the project. The requesters alleged that this method – which entailed mechanical shredding of water hyacinths and allowing the shredded material to sink to the lake bottom to decay – was chosen without conducting an EIA or adequate community consultation, would cause environmental degradation, and endangered the lake's communities. The request cited violations of several World Bank Policies and Procedures, particularly those dealing with environmental assessment and economic evaluation of investment projects. The World Bank management responded to the request by stating that, while it disagreed with the claims in the request, it believed that it should more thoroughly inform the public about its chosen management plan. In reviewing the request, the Panel visited the site and met with representatives from RECONCILE, other NGOs, community-based organizations, and fishermen. As a result of its review, the Panel recommended that an investigation be approved and, on 10 April 2000, the Bank's Board of Executive Directors approved the recommendation. On 14 May 2001, the World Bank Management accepted the Panel's findings and recommended six actions, including continued monitoring, heightened community participation, and cross-country participation in supervision missions. Notwithstanding the Lake Victoria case, the Board of Directors has allowed very few investigations to proceed.

In another instance, the World Bank did authorize an Inspection Panel investigation regarding a transnational watercourse (Fragano and Jorge 1999; Udall 1999). In 1996, the NGO Sobrevivencia (Friends of the Earth-Paraguay) filed a Request for Inspection for the Yacyretá Dam – a dam on the Paraná River between Paraguay and Argentina that was financed primarily by the World Bank and the Inter-American Development Bank (IDB). The request was based largely on the fact that, 25 years after the dam had been constructed, the required environmental mitigation and resettlement plans had still not been fully executed. Sobrevivencia's request noted significant environmental, social, and cultural impacts of the dam. It also cited ways that the project had violated many of the Bank's policies, including those on hydroelectric projects, environmental assessment, and project monitoring and evaluation. An Inspection Panel visited the dam site, met with local citizens and organizations, and ultimately recommended to the Bank's Board of Directors that an investigation of the allegations be conducted. The board approved a limited review of Sobrevivencia's claims and an assessment of the project's management. This review and assessment attracted media attention to the project and increased the level of involvement in the project by the Yacyretá Bi-national Entity, as well as World Bank and IDB supervision.

IFC/MIGA Office of the Compliance Advisor Ombudsman

In contrast to the IBRD, which lends money to governments, the International Finance Corporation (IFC) is the arm of the World Bank that is responsible for making loans to the private sector, and the Multilateral Investment Guarantee Agency (MIGA) provides investment guarantees against certain non-commercial risks to foreign investors in member countries. In 1999, the IFC and MIGA established the position of Environmental and Social Compliance Advisor/Ombudsman (CAO) to "respond[ ] to complaints by persons who are affected by projects and attempt[ ] to resolve the issues raised using a flexible, problem solving approach" (IFC/MIGA 2000; Bernasconi-Osterwalder and Hunter 2002).

The Operational Guidelines allow "any individual, group, community, entity or other party affected or likely to be affected by the social and/or environmental impacts of an IFC or MIGA project" to file a complaint with the CAO (IFC/MIGA 2000). This may be done directly or through a representative. The complaints must be in writing but can be in any language. The guidelines allow complaints that address the "planning, implementation or impact of projects," including the adequacy and implementation of social and environmental mitigation measures and the "involvement of communities, minorities and vulnerable groups in the project." To resolve the complaints, the CAO can investigate, convene a dialogue, or pursue more formal arrangements such as conciliation, mediation, and negotiated settlements.

North American Commission for Environmental Cooperation

The North American CEC promotes access to justice at the national and regional levels. Through the CEC's organic statute, the North American

Agreement on Environmental Cooperation (NAAEC), member states have committed to ensuring that interested persons may petition national authorities to investigate alleged violations of environmental legislation, providing persons who have legally cognizable interests with access to judicial, quasi-judicial, or administrative bodies in order to enforce the environmental legislation, and ensuring that proceedings are "fair, open and equitable" (NAAEC 1993; Dowdeswell 2002).

At the regional level, citizens and organizations can file complaints alleging that a member state is not enforcing its environmental laws (Tuchton 1996; Knox 2001; NACEC 2001; Markell and Knox 2003; Garver, chap. 12, this volume). The decisions adopted by the CEC are not binding, but the independent third-party review provided by the CEC can help to compel governments to comply with and enforce their environmental laws (Markell 2001; Markell and Knox 2003).

A number of citizen submissions have related to water resources (Garver, chap. 12, this volume). One citizen submission has involved the shared watercourse of the North American Great Lakes. CEC submission SEM-98-003 was filed by Canadian and US environmental and public health groups and a Canadian individual who were all concerned about the effects of the fallout of persistent toxic-substance emissions from incinerators on Great Lakes water quality. The complainants alleged that the United States violated both US domestic laws and United States–Canada treaties, including the Great Lakes Water Quality Agreement of 1972. Specifically, they alleged that the United States was violating its laws governing airborne emissions of dioxin/furan, mercury, and other persistent toxic substances falling into the Great Lakes from solid waste and medical waste incinerators.

The initial complaint was submitted on 27 May 1998. The Secretariat reviewed the complaint and determined that SEM-98-003 did not meet the standards set forth, stating that, "Article 14(1) reserves the Article 14 process for claims that a Party is 'failing to effectively enforce its environmental law ...,'" and concluded that the underlying issue did not qualify as "enforcement" because it related to standard-setting and not to a failure to enforce an environmental law. The CEC determined that standard-setting is outside the range of Article 14, and the US inaction was not subject to review.

In response, a revised submission was filed in January 1999. In this second attempt, the submitters were more successful in obtaining Article 14 review. The Secretariat's review found that the revised submission required Article 14 review on two issues – the asserted inspection-related failures, and failure to effectively enforce the Clean Air Act (CAA). The Secretariat noted that the CAA requires the Environmental Protection Agency (EPA) Administrator to notify the Governor of the State in which such emission originates, whenever the Administrator receives reports from any duly constituted international agency such as the IJC or CEC, that air pollution or pollutants emitted in the United States can "be reasonably anticipated to endanger public health or welfare in a foreign country."

The Secretariat's response on this issue is notable because it involves the adverse impacts of an action by one party (the United States) on another country (Canada).

The Secretariat received the US Government's response to the submission. On 24 March 2000, the Secretariat requested additional information from the United States to complete its determination regarding whether preparation of the factual record was warranted, generating a two-part response from United States under Article 21(1)(b). The Secretariat determined not to recommend the preparation of a factual record and the case was terminated on 5 October 2001. The decisions to date suggest that it would be difficult, although not impossible, for a future submission to successfully raise issues in respect of a party's international obligations that would meet the criteria of Article 14(1).

Border Environment Cooperation Commission (BECC)

The BECC certifies projects along the United States–Mexico border to ensure that the projects comply with all applicable environmental laws and involve the public. As part of the process, the BECC Board of Directors holds quarterly public meetings, where it is authorized to receive complaints from groups affected by BECC-assisted or certified projects. In order to file a complaint, two or more of the complainants "must reside in the area where the project(s) causing the effects is(are) located or in an area where the project(s)' effects are manifested or likely to be manifested based on the evidence" (BECC 2000a). Substantively, the complaint "must be based on the health or environmental effect(s) of a project(s)" or on a threat of such effects that is supported by evidence.

If the BECC Board of Directors accepts the complaint, the Board may request additional information from "the complainant, the [BECC] Advisory Council, [or] any other public or private institution it deems appropriate." The advisory council then prepares a report, "provid[ing] its recommendations regarding the complaint and the basis for such recommendations." The Board makes the final determination in writing, providing "a clear statement of the conclusion," "a full statement of the reasons supporting the conclusion," and "steps, if any, the Board intends to take as a result of the complaint, including a timetable for undertaking such steps." The determination must be made publicly available.

The BECC also has developed procedures by which certain interested parties may obtain an Independent Assessment to determine whether the provisions of chapter I of the agreement (dealing with BECC operations and project certification) or the procedures adopted by the board of directors pursuant to that chapter have been observed (NAAEC 1993; BECC 2000b). A request can be made by any NGO, group, or border community "through a duly appointed representative," or a state or local authority along the border. The complaint must be in writing, promptly follow the non-compliance, and contain sufficient information and arguments to evaluate the claim. The process for assessing the merits of the complaint and whether to proceed are open to public scrutiny, and the eventual report is also publicly available. Again, this mechanism remains largely dormant.

Advancing public involvement in the management of transboundary watercourses

Access to information, public participation, and access to justice are now included in several international and regional agreements concerning transboundary watercourses. Experiences with these norms, institutions, and practices are likely to affect how other watercourses involve the public in decision-making. However, the success and full implementation of such provisions depend on several factors.

Factors affecting the development and implementation of participatory frameworks for managing international watercourses

In developing and implementing norms and mechanisms for public involvement in the management of transboundary watercourses, it is important to look at the context of each particular watercourse. This can include an analysis of the bordering countries, local legal systems, and existing national or regional initiatives on public participation.

Experiences in transboundary watercourses vary greatly, depending on a range of geopolitical, historical, and social factors. When there are only a few riparian nations, agreements on transboundary watercourses are more likely to include the public. For example, the 1909 agreement between Canada and the United States on the management of their boundary waters and the North American Great Lakes included public-participation provisions that remain unmatched in many contemporary agreements. Conversely, rivers with numerous riparian nations (such as the Nile) are likely to raise more conflicts, and public participation often lags. Similarly, where communities straddle a watercourse, there may be more incentive to develop a management system that accounts for the interests of the counterparts on the other side of the watercourse (Milich and Varady 1998).

A related factor is the degree to which nations share a cultural, historical, and social background. With this common basis there is greater trust, not only at the government level but also at the popular level. As a result, one notices that the United States–Canada and Kenya–Tanzania–Uganda agreements evolve more rapidly and include stronger provisions for public participation than those for many other watercourses.

A highly sensitive international context can make international agreements more difficult to reach, and governmental officials more reluctant to open the door to third parties whom they perceive as posing a danger of either compromising their own position or of confusing the relationship. A context can become sensitive through economic or political instability, including warfare (Eriksen 1998). The international context could also become sensitive owing to actual, imminent, or prospective overburden of the available water, particularly where there is a historically dominant water user. In contrast, areas such as Southern Africa generally present a relatively stable economic and political environment in which the demand for available water is not yet as severe as, for example, with the Nile River. As a result, there can be more room to negotiate and to involve the public.

Existing regional initiatives on public involvement can also be of assistance in furthering participation in transboundary watercourse management. Although some of these initiatives are non-binding, they may provide nations with a framework for addressing the governance of watercourses. These initiatives promote several specific tools that advance public participation, many of which are discussed below. The initiatives recommend practices such as EIA (including transboundary EIA), public meetings early on in a project, free access to public records, regular reports by the government on the status of projects that may affect the public, and access to environmental information by citizens of neighbouring countries that may be affected by local decisions. These tools have been accepted widely for public participation domestically and may significantly increase public involvement and, ultimately, the success of projects, in international watercourses.

Strategies for advancing public involvement

Eriksen suggests a general approach when starting cooperative management of transboundary watercourses that also may apply to the context of public involvement:

... focus[ing] on water quality issues avoids contention around water allocation. Water quality is also usually a concern shared by all riparians in some way. Co-operation on scientific assessments on a drainage basin and processes within it has been a starting point for basinwide co-operation. (Eriksen 1998)

It might also be prudent to start with transboundary watercourses that flow between two (or perhaps three) nations only and are not politically sensitive.

In many contexts, public involvement may be viewed by both governments and civil society as a means by which opponents of particular projects or activities may seek to stall or halt the proposed action. This view has some basis in experience: where the public does not have formal channels for providing input, or for having decision makers incorporate or respond to their input, protest and confrontation often are the primary avenues remaining for people to express themselves. In developing and implementing approaches to facilitate civil-society engagement, consideration should be paid to ways to facilitate more constructive forms of public involvement. This may take the form of a participatory priority-setting exercise or co-management. Such constructive participatory processes can foster a more congenial and collaborative relationship between governments and civil society.

This is not to say that confrontational approaches need to be eschewed; rather, there is a spectrum of participatory processes from collaborative to confrontational. To the extent that there are clear benefits of public involvement, as illustrated through collaborative processes, governments may be more willing to provide information and opportunities for public participation – even if confrontation sometimes results.

Seeking constructive and collaborative approaches for public involvement has implications for both governments and international institutions on the one hand and for civil society on the other. For collaborative participation to work effectively, decision makers need to seek the input of civil society early in the process, when the decision can be changed or modified to reflect the various perspectives of civil society. It may be obvious, but in order for civil society to believe that their participation will make a difference (and therefore to become engaged in the process) the decision makers need to listen to civil society and they need to be willing to modify the proposed action to reflect the priorities of civil society. At the same time, civil-society institutions must show that they are willing to work constructively with the institutions, not just as critics but as collaborative stakeholders. This may mean, for example, a focus on finding alternatives and solutions rather than criticism.

Access to information can be promoted through a number of discrete mechanisms, many of which are relatively low cost. Making information available upon request obviates the need for a sizeable staff and infrastructure, and the imposition of a reasonable fee (to cover copying, for example) can further reduce the burden on the authority. Establishing a resource centre is a more expensive endeavour, but it might form a project that foreign donors would support and, in the long run, could reduce the overall burden on staff who may otherwise have to respond seriatim to requests that could otherwise be addressed through a resource centre. Another, less expensive, option is developing a website. Producing a periodic "state of the river" report poses certain difficulties; however, these can be overcome: for example, as it is expensive, the report could be kept brief. There is also the possibility of publishing the report every two years rather than annually, again reducing the production and printing costs. Such a report could focus on water-quality issues, draw upon a modest number of sampling points, and grow from there.

As a first step to developing public participation in the management of international watercourses, EIA can be developed at the national level and harmonized through the region or along watercourses (Cassar and Bruch 2004; Sikoyo, chap. 22, this volume). As it is unlikely that the river-management bodies will have the funds necessary to conduct detailed EIAs or lengthy public hearings on them, the riparian nations through the watercourse authority could require project proponents to conduct an EIA for projects likely to have a significant environmental impact and then open the discussion to the public. This is the case for projects financed by most international financial institutions (Bernasconi-Osterwalder and Hunter 2002). One easy step is to open meetings of river-management authorities to the public; this costs relatively little, and the public could participate as either silent observers or as participating, but non-voting, observers.

Access-to-justice measures can be difficult because they often require national judicial systems to be altered. Initially, however, nations in a region can establish broad interpretations of standing to facilitate access to their courts, both by their nationals and by others who may be affected, particularly those living in other riparian nations.

In developing these norms – which give a voice to citizens, NGOs, and local governments – it will be necessary to balance the roles of international, national, and local actors in the management of transboundary watercourses (Milich and Varady 1998; Avramoski 2004). Moreover, it is important to develop culturally appropriate approaches (Kaosa-ard et al. 1998; Faruqui, Biswas, and Bino 2001; Avramoski 2004). The national and international actors are essential to ensuring that local control does not lead to parochial dominance and unsustainable abuse of natural resources; and the participation of local actors is necessary for the norms and institutions to be relevant (and thus implemented) on the ground.

Promising approaches and mechanisms

This chapter has highlighted many ways in which nations and international institutions have developed and implemented mechanisms for promoting and ensuring public involvement in the management of international watercourses. In addition to the some of the more established mechanisms, a variety of approaches are emerging that are likely to improve public involvement in the years to come. These may be refinements or extensions of established mechanisms, while in other cases they are new mechanisms (Bruch 2004).

As mentioned above and elsewhere in the volume (Bruch, chap. 18, this volume), Internet-based tools have become important for disseminating information relating to international watercourses. Additionally, tools such as e-mail, listservs, and chat rooms increasingly provide avenues to solicit public comment and otherwise engage the public in the decision-making processes. As Internet connectivity continues to grow, particularly in developing nations, Internet-based tools are likely to gain more relevance and prominence.

Decision support systems (DSSs) provide another tool for improving public access to information about proposed effects of decisions on international watercourses and for engaging the public in the decision-making process. A particularly innovative approach to making DSSs publicly available is to develop Internet-based DSSs, which has been facilitated by the development of faster computers, servers, and broadband Internet access (Salewicz, chap. 19, this volume).

While EIA is well established in national laws and international declarations, and the institutions to conduct EIAs continue to develop, there are a few particular ways in which EIA is likely to improve, particularly with regard to international watercourses. First, the expansion of EIA norms and methodologies to address transboundary impacts explicitly is an important step towards improving basin-wide management of international watercourses (Knox 2002; Cassar and Bruch 2004). In many instances, international instruments and institutions have called for the development of TEIA, and TEIA has been applied in a variety of circumstances. Considering the diverse experiences thus far, a comprehensive review of TEIA experiences could improve the ongoing development and operationalization of TEIA norms, institutions, and methodologies.

Another way in which EIA is being extended is to provide a participatory framework for analysing possible impacts of proposed plans, policies, programmes, and regulations. Many regions and countries are in the comparatively early stages of developing and implementing strategic environmental assessment (SEA) generally, which could also provide a framework for improving public involvement in the development of norms governing international watercourses (Kravchenko 2002; Sikoyo, chap. 22, this volume).

A third way in which EIA can be improved is by examining the effectiveness of EIA methodologies. There is a growing body of literature examining the accuracy and effectiveness of EIA, particularly in domestic contexts (Nakayama et al. 1999; Nakayama, Yoshida, and Gunawan 1999, 2000; Bruch 2004). By comparing predicted impacts with actual impacts, EIA methodologies can be improved and made more effective. Applying the lessons learned from such comparative analysis could improve EIA at both the national and transboundary levels, and this constitutes a continuing research need.

Developments in access to justice are likely to be more incremental. Initiatives such as the Aarhus Convention, which liberalize standing requirements and impose the obligation of non-discrimination in granting standing to citizens of other countries, provide a framework for opening-up domestic courts. However, in many instances, such opportunities are only starting to be utilized. Granting public access to international tribunals is another development on the horizon. While many significant developments have been made in the past decade (Bernasconi-Osterwalder and Hunter 2002; Gertler and Milhollin 2002; Jean-Pierre 2002; Di Leva, chap. 10, this volume; Garver, chap. 12, this volume; Picolotti and Crane, chap. 23, this volume), the initial progress has slowed or even stalled. In some instances, it is simply a matter of the mechanisms maturing; in other instances, countries have been cautious about opening up dispute-resolution processes too far to the public (Gertler and Milhollin 2002). Nevertheless, considering the substantial momentum and continuing pressure for transparent, participatory, and accountable governance, it is likely that international institutions will continue to develop approaches to ensure public access to tribunals and fact-finding bodies.

There are a number of other experiences, particularly at the national and sub-national levels, in promoting public involvement in water management that could be adapted and applied to different international watercourses (Bruch 2001; Avramoski 2004). Indeed, part IV of this volume (Lessons from domestic watercourses) examines a variety of such experiences.

Conclusions

While public involvement in the management of transboundary watercourses goes back decades, if not millennia (Kaosa-ard et al. 1998), the last decade has seen a remarkable proliferation of international agreements and institutional practice. At the same time, national laws and institutions charged with the management of freshwater resources, frequently crossing national borders, have incorporated transparency, public participation, and access to justice.

In the context of transboundary watercourses, access to information ensures that citizens and other members of civil society have the ability to request from governmental and intergovernmental authorities information on the status of the watercourse and its tributaries (including water flow and water quality); factors that could affect the watercourse or its tributaries; and norms, policies, and management plans that shape activities relevant to the watercourse. Public participation should include the opportunity for members of the public to submit comments (and have the authority take due account of the information) regarding specific activities that could affect the watercourse; the development of norms, policies, and plans that govern the watercourse; and even in the development of the transboundary watercourse agreements themselves. Access to justice entails resort to national courts and agencies, international courts, and fact-finding and legislative bodies.

Regional initiatives on public involvement are developing in Europe, the Americas, Asia, and East Africa. Already the commissions charged with managing watercourses in many of these regions have begun to include more public participation, to the benefit of both the local communities and the projects. The steps to further public involvement can, in some cases, be simple and inexpensive, and at other times may require the restructuring of national environmental governance. In determining the best path toward greater public access to information, participation, and justice, nations will need to consider the specific context of the particular international watercourse. However, much can be learned – and adapted – from experiences in other watersheds. Nevertheless, in developing and refining the participatory frameworks, nations may wish to consider involving the various sectors and major groups of the public in the process to ensure that the ultimate norms, institutions, and mechanisms for public involvement are effective, relevant, and accessible.

Acknowledgements

The author gratefully acknowledges research assistance from Mark Beaudoin, Angela Cassar, Dorigen Fried, Samantha Klein, Molly McKenna, Turner Odell, Seth Schofield, Julie Teel, Elizabeth Walsh, and Jessica Warren. Support for this research was provided by the United States Agency for International Development, the John D. and Catherine T. MacArthur Foundation, and the Richard and Rhoda Goldman Fund. This chapter builds upon ideas first developed in Carl Bruch (2001, 2004).

REFERENCES

Agreement on the Cooperation for the Sustainable Development of the Mekong River Basin. 1995. Entered into force on 5 April. Internet: ⟨http://www.mrcmekong.org/about_us/agreement_1995.htm⟩ (visited 13 April 2004).

A-Khavari, Afshin, and Donald R. Rothwell. 1998. "The ICJ and the Danube Dam Case: A Missed Opportunity for International Environmental Law?" *Melbourne University Law Review* 22:507.

Avramoski, Oliver. 2004. "The Role of Public Participation and Citizen Involvement in Lake Basin Management." Internet: ⟨http://www.worldlakes.org/uploads/Thematic_Paper_PP_16Feb04.pdf⟩ (visited 14 April 2004).

BECC (Border Environment Cooperation Commission). 2000a. "Procedures Regarding Complaints from Groups Affected by Projects." Internet: ⟨http://www.cocef.org/brules.htm⟩ (visited 14 April 2004).

BECC (Border Environment Cooperation Commission). 2000b. "Procedures for Independent Assessments." Internet: ⟨http://www.cocef.org/brules.htm⟩ (visited 14 April 2004).

Bernasconi-Osterwalder, Nathalie, and David Hunter. 2002. "Democratizing Multilateral Development Banks." In Carl Bruch (ed.). *The New "Public": The Globalization of Public Participation.* Washington, DC: Environmental Law Institute.

Bruch, Carl. 2001. "Charting New Waters: Public Involvement in the Management of International Watercourses." *Environmental Law Reporter* 31:11389–11416.

Bruch, Carl (ed.). 2002. *The New "Public": The Globalization of Public Participation.* Washington, DC: Environmental Law Institute.

Bruch, Carl. 2004. "New Tools for Governing International Watercourses." *Journal of Global Environmental Change – Human and Policy Dimensions* 14(1):15.

Bruch, Carl E., and Roman Czebiniak. 2002. "Globalizing Environmental Governance: Making the Leap from Regional Initiatives on Transparency, Participation, and Accountability in Environmental Matters." *Environmental Law Reporter* 32:10428.

Bruch, Carl, Wole Coker, and Chris VanArsdale. 2001. "Constitutional Environmental Law: Giving Force to Fundamental Principles in Africa." *Columbia Journal of Environmental Law* 26:131–211.

Caillaux, Jorge, Manuel Ruiz, and Isabel Lapeña. 2002. "Environmental Public Participation in the Americas." In Carl Bruch (ed.). *The New "Public": The Globalization of Public Participation.* Washington, DC: Environmental Law Institute.

Cassar, Angela Z., and Carl E. Bruch. 2004. "Transboundary Environmental Impact Assessment in International Watercourses." *New York University Environmental Law Journal* 12:169–244.

Convention on Cooperation for the Protection and Sustainable Use of the Danube River (Danube River Protection Convention). 1994. Done at Sofia, entered into force 22 October 1998. Internet: ⟨http://ksh.fgg.uni-lj.si/danube//envconv/index.htm⟩ (visited 13 April 2004).

Convention on the Law of the Non-Navigational Uses of International Watercourses. 1997. Done 21 May. UNGA Res. 51/229, UN Doc. A/RES/51/229, *International Legal Materials* 31:700. Internet: ⟨http://www.un.org/law/ilc/texts/nonnav.htm#⟩ (visited 13 April 2004).

Convention on the Protection of the Rhine. 1999. Done at Berne on 12 April. Internet: ⟨http://www.internationalwaterlaw.org⟩ (visited 13 April 2004).

Cronin, John, and Robert F. Kennedy, Jr. 1997. *The Riverkeepers*. New York: Scribner.

Darrell, Andrew H. 1989. "Killing the Rhine: Immoral, But Is It Illegal?" *Virginia Journal of International Law* 29:421.

de Villeneuve, Carel H.V. 1996. "Western Europe's Artery: The Rhine." *Natural Resources Journal* 36:441.

Dowdeswell, Elizabeth. 2002. "The North American Commission for Environmental Cooperation: A Case Study in Innovative Environmental Governance." In Carl Bruch (ed.). *The New "Public": The Globalization of Public Participation.* Washington, DC: Environmental Law Institute.

Dubash, Nawroz K., Mairi Dupar, Smitu Kothari, and Tundu Lissu. 2001. *A Watershed in Global Governance? An Independent Assessment of the World Commission on Dams.* Washington, DC: World Resources Institute.

East African MOU (Memorandum of Understanding Between the Republic of Kenya and the United Republic of Tanzania and the Republic of Uganda for Cooperation on Environmental Management). 1998. Done at Nairobi on 22 October.

ELI (Environmental Law Institute). 1995. "An Evaluation of the Effectiveness of the International Joint Commission." Washington, DC: Environmental Law Institute.

Eriksen, Siri. 1998. *Shared River and Lake Basins in Africa: Challenges for Cooperation.* Nairobi, Kenya: African Centre for Technology Studies (ACTS).

Fall, Aboubacar. 2002. "Implementing Public Participation in African Development Bank Operations." In Carl Bruch (ed.). *The New "Public": The Globalization of Public Participation.* Washington, DC: Environmental Law Institute.

Famighetti, Robert, June Foley, Thomas McGuire, Christina Cheddar, Michael Northrop, and Donald Young (eds). 1993. *The World Almanac and Book of Facts 1994.* Mahwah, New Jersey: Funk and Wagnals.

Faruqui, Naser, Asit K. Biswas, and Murad Bino (eds). 2001. *Water Management in Islam.* Tokyo: United Nations University Press.

Ferrier, Catherine. 2000. "Towards Sustainable Management of International Water Basins: The Case of Lake Geneva." *Review of European Community and International Environmental Law* 9:52.

Fragano, Francis, and Christie Jorge. 1999. *Access to Process in International Organizations: The Yacyretá Hydroelectric Dam Project and the World Bank Inspection Panel.* Washington, DC: Organization of American States. Internet: ⟨http://www.ispnet.org/Documents/paraguay.htm⟩ (visited 13 April 2004).

GEF (Global Environment Facility). 1996. *Kenya, Tanzania, Uganda: Lake Victoria Environmental Management Project.* Project Document, Report No. 15541-AFR. Washington, DC: GEF. Internet: ⟨http://www-wds.worldbank.org/servlet/WDS_IBank_Servlet?pcont=details&eid=000009265_3961219144602⟩.

Gertler, Nicholas, and Elliott Milhollin. 2002. "Public Participation and Access to Justice in the World Trade Organization." In Carl Bruch (ed.). *The New "Public": The Globalization of Public Participation.* Washington, DC: Environmental Law Institute.

Giorgetti, Chiara. 1998. "The Role of Nongovernmental Organizations in the Climate Change Negotiations." *Colorado Journal of International Environmental Law and Policy* 9:115.

Gonzalez, Pablo. 2003. "Multi-stakeholder Involvement and IWRM in Transboundary River Basins: GEF/UNEP/OAS Experiences with the Strategic Action Program for the San Juan River Basin, Costa Rica and Nicaragua." Paper presented at the Symposium on Improving Public Participation and Governance in International Watershed Management (University of Virginia 18–19 April).

Guruswamy, Lakshman. 2000. "The Annihilation of Sea Turtles: World Trade Organization Intransigence and U.S. Equivocation." *Environmental Law Reporter* 30:10261.

Hildén, Mikael, and Eeva Furman. 2002. "Towards Good Practices for Public Participation in the Asia–Europe Meeting Process." In Carl Bruch (ed.). *The New "Public": The Globalization of Public Participation.* Washington, DC: Environmental Law Institute.

IFC/MIGA (International Finance Corporation/Multilateral Investment Guarantee Agency). 2000. *Operational Guidelines for the Office of the Compliance Advisor/Ombudsman.* Washington, DC: IFC/MIGA.

ILA (International Law Association). 2004. The [Revised] International Law Association *Rules on the Equitable Use and Sustainable Development of Waters.* Tenth Draft. February.

Ingram, Helen, Lenard Milich, and Robert G. Varady. 1994. "Managing Transboundary Resources: Lessons from Ambos Nogales." *Environment* 6:33.

Jean-Pierre, Danièle M. 2002. "Access to Information, Participation and Justice: Keys to the Continuous Evolution of the Inter-American System for the Protection and Promotion of Human Rights." In Carl Bruch (ed.). *The New "Public": The Globalization of Public Participation.* Washington, DC: Environmental Law Institute.

Kaosa-ard, Mingsarn, K. Rayanakorn, G. Cheong, S. White, C.A. Johnson, and P. Kongsiri. 1998. *Towards Public Participation in Mekong River Basin Development.* Bangkok: Thailand Development Research Institute Foundation.

Kiss, Alexandre. 1985. "The Protection of the Rhine Against Pollution." *Natural Resources Journal* 25:629–635.

Knox, John H. 2001. "A New Approach to Compliance with International Environmental Law: The Submissions Procedure of the NAFTA Environmental Commission." *Ecology Law Quarterly* 28:1.

Knox, John H. 2002. "The Myth and Reality of Transboundary Environmental Impact Assessment." *American Journal of International Law* 96:291.

Kravchenko, Svitlana. 2002. "Promoting Public Participation in Europe and Central Asia." In Carl Bruch (ed.). *The New "Public": The Globalization of Public Participation.* Washington, DC: Environmental Law Institute.

Kværnevik, Trond Inge. 1994. "The Mekong Region." In *Association for International Water and Forest Studies. 1994. Power Conflict.* Internet: ⟨http://www.solidaritetshuset.org/fivas/pub/power_c/k7.htm⟩ (visited 14 April 2004).

Liptak, Bela. 1997. *Precedent for the 21st Century: The Danube Lawsuit.* Internet: ⟨http://www.hartford-hwp.com/archives/63/005.html⟩ (visited 14 April 2004).

Markell, David L. 2001. "The Citizen Spotlight Process." *The Environmental Forum* 18:32. March/April.

Markell, David L., and John H. Knox (eds). 2003. *Greening NAFTA: The North American Commission for Environmental Cooperation.* Washington, DC: Stanford University Press.

Milich, Lenard, and Robert G. Varady. 1998. "Managing Transboundary Resources: Lessons From River-Basin Accords." *Environment* 40:10.

MRC (Mekong River Commission) Secretariat. 1999. *Public Participation in the Context of the MRC.* Internet: ⟨http://www.mekonginfo.org/mrc_en/doclib.nsf/⟩ (visited 14 April 2004).

NAAEC (North American Agreement on Environmental Cooperation). 1993. Done 8–14 September. *International Legal Materials* 32:1482.

NACEC (North American Commission for Environmental Cooperation). 2001. *Guidelines for Submissions on Enforcement Matters Under Articles 14 and 15 of the North American Agreement on Environmental Protection.* Internet: ⟨http://www.cec.org/citizen/guide_submit/index.cfm⟩ (visited 14 April 2004).

Nakayama, Mikiyasu. 1997. "Success and Failures of International Organizations in Dealing with International Waters." *International Journal of Water Resources Development* 13:367.

Nakayama, Mikiyasu. 1999. "Politics Behind Zambezi Action Plan." *Water Policy* 1:397.

Nakayama, Mikiyasu, and Ryo Fujikura. 2002. "Information Sharing for Public Participation in Water Resources Management." In I.H.O. Unver and R.K. Gupta (eds). *Water Resources Management – Crosscutting Issues.* Ankara: METU Press.

Nakayama, M., T. Yoshida, and B. Gunawan. 1999. "Compensation Schemes for Resettlers in Indonesian Dam Construction Projects – Application of Japanese 'Soft Technology' for Asian Countries." *Water International* 24(4):348–355.

Nakayama, M., T. Yoshida, and B. Gunawan. 2000. "Improvement of Compensation System for Involuntary Resettlers of Dam Construction Projects." *Water Resources Journal* 80–93 (September).

Nakayama, M., B. Gunawan, T. Yoshida, and T. Asaeda. 1999. "Involuntary Resettlement Issues of Cirata Dam Project." *International Journal of Water Resource Development* 15(4):443–458.

NBI (Nile Basin Initiative) Secretariat. 2000. *The Nile Basin Initiative Background.* Jinja, Uganda: NBI.

NEPAD (New Partnership for Africa's Development). 2001. Internet: ⟨http://www.nepad.org/⟩ (visited 14 April 2004).

NGO Forum on Cambodia. 1997. "Mekong People: The Role of Local Communities in Hydro-Power Planning; Towards Public Participation in S/EIA." *Conference Proceedings. Cambodia, UN/ESCAP/E7 Regional Workshop on EIA for Hydropower Development Emphasizing Public Participation*, held 20–24 November in Bangkok, Thailand.

OAS (Organization of American States). 2000. Inter-American Strategy for the Promotion of Public Participation in Decision Making for Sustainable Development. Adopted 20 April. Internet: ⟨http://www.ispnet.org⟩ (visited 14 April 2004).

Odote, Collins, and Maurice O. Makoloo. 2002. "African Initiatives for Public Participation in Environmental Governance." In Carl Bruch (ed.). *The New "Public": The Globalization of Public Participation.* Washington, DC: Environmental Law Institute.

Okaru-Bisant, Valentina. 1998. "Institutional and Legal Frameworks for Preventing and Resolving Disputes Concerning the Development and Management of Africa's Shared River Basins." *Colorado Journal of International Environmental Law and Policy* 9:331.

Pamoeli, Phera. 2002. "The SADC Protocol on Shared Watercourses: History and Current Status." In Anthony Turton and Roland Henwood (eds). *Hydropolitics in the Developing World: A Southern African Perspective.* Pretoria, South Africa: African Water Issues Research Unit.

Parker, Paul M. 1983. "High Ross Dam: The IJC Takes a Hard Look at the Environmental Consequences of Hydroelectric Power Generation." *Washington Law Review* 58:445.

Pyatt, Suzanne. 1999. "The WTO Sea Turtle Decision." *Ecology Law Quarterly* 26:815.

Rhine. 1976. Additional Agreement to the Agreement Signed in Berne, 29 April 1963, *Concerning the International Commission for the Protection of the Rhine Against Pollution.* Done at Berne on 3 December.

Rhine. 1963. Agreement of 29 April 1963, *Concerning the International Commission for the Protection of the Rhine Against Pollution*, done at Berne.

Rich, Bruce, and Tomas Carbonell. 2002. "Public Participation and Transparency at Official Export Credit Agencies." In Carl Bruch (ed.). *The New "Public": The Globalization of Public Participation.* Washington, DC: Environmental Law Institute.

SADC (Southern African Development Community). 2000. "Revised Protocol on Shared Watercourses in the Southern African Development Community (SADC)." Done 7 August. *International Legal Materials* 40:321 (2001).

Sandler, Deborah, Emad Adly, Mahmoud A. Al-Khoshman, Philip Warburg, and Tobie Bernstein (eds). 1994. *Protecting the Gulf of Aqaba: A Regional Environmental Challenge.* Washington, DC: Environmental Law Institute.

Sands, Phillipe. 1999. "International Environmental Litigation and Its Future." *University of Richmond Law Review* 32:1619.

Scott, Inara. 2000. "The Inter-American System of Human Rights: An Effective Means of Environmental Protection." *Virginia Environmental Law Journal* 19:197.

Sharma, Narendra, Torbjorn Damhaug, David Grey, Valentina Okaru, Daniel Rothberg, and Edeltraut Gilgan-Hunt. 1996. *African Water Resources: Chal-*

lenges and Opportunities for Sustainable Development. World Bank Technical Paper No. 331. Washington, DC: World Bank.

Shelton, Dinah. 1994. "The Participation of Nongovernmental Organizations in International Judicial Proceedings." *American Journal of International Law* 88:611.

Shumway, Caroly A. 1999. *Forgotten Waters: Freshwater and Marine Ecosystems in Africa*. Washington, DC: Biodiversity Support Program.

Stec, Stephen, and Susan Casey-Lefkowitz. 2000. *The Aarhus Convention: An Implementation Guide*. Budapest: Regional Environment Center. Internet: ⟨http://www.unece.org/env/pp/acig.htm⟩ (visited 13 April 2004).

Taylor, Celia R. 1994. "The Right of Participation in Development Projects." *Dickinson Journal of International Law* 13:69.

Treaty for Amazonian Cooperation. 1978. Done at Brasilia on 3 July. *International Legal Materials* 17:1045.

Tumushabe, Godber. 2002. "Public Involvement in The East African Community." In Carl Bruch (ed.). *The New "Public": The Globalization of Public Participation*. Washington, DC: Environmental Law Institute.

Tutchton, Jay. 1996. "The Citizen Petition Process Under NAFTA's Environmental Side Agreement: It's Easy to Use, but Does it Work?" *Environmental Law Reporter* 26:10018.

Udall, Lori. 1999. "Review of World Bank Inspection Panel." Submitted to the World Commission on Dams for Thematic Review, Institutional and Governance Issues, Regulation, Compliance, and Implementation.

UN (United Nations). 1998. "UN System-wide Special Initiative on Africa." *Africa Recovery On-Line* 11(4). March. Internet: ⟨http://www.un.org/ecosocdev/geninfo/afrec/vol11no4/march98.htm⟩ (visited 14 April 2004).

UNCED (United Nations Conference on Environment and Development). 1992a. Rio Declaration on Environment and Development. Done at Rio de Janeiro on 14 June, reprinted in *International Legal Materials* 31:874.

UNCED (United Nations Conference on Environment and Development). 1992b. *Agenda 21*. A/CONF.151/26 (Vol. I–III). Done at Rio de Janeiro on 14 June.

UNECE (United Nations Economic Commission for Europe). 1991. "Convention on Environmental Impact Assessment in a Transboundary Context." Adopted at Espoo, Finland on 25 February, entered into force on 10 September 1997. Internet: ⟨http://www.unece.org/env/eia/eia.htm⟩ (visited 14 April 2004).

UNECE (United Nations Economic Commission for Europe). 1992. "Convention on the Protection and Use of Transboundary Watercourses and International Lakes." Entered into force 6 October 1996. U.N.T.S. 33207, *International Legal Materials* 31:1312. Internet: ⟨http://www.unece.org/env/water⟩ (visited 13 April 2004).

UNECE (United Nations Economic Commission for Europe). 1998. "Convention on Access to Information, Public Participation in Decision-making and Access to Justice in Environmental Matters." Adopted at Aarhus, Denmark on 25 June 1998, entered into force on 30 October 2001. U.N. Doc. ECE/CEP/43. Internet: ⟨http://www.unece.org/env/pp/welcome.html⟩ (visited 13 April 2004).

UNECE (United Nations Economic Commission for Europe). 1999. "Water and Health Protocol to the Convention on the Protection and Use of Transboundary Watercourses and International Lakes." Done at London on 18 June. U.N. Doc. E/ECE/MP.WAT/AC.1/1998/10. Internet: ⟨http://www.unece.org/env/water/text/text_protocol.htm⟩ (visited 13 April 2004).

UNECE (United Nations Economic Commission for Europe). 2003. "Protocol on Strategic Environmental Assessment." Done at Kyiv, Ukraine on 21 May. Internet: ⟨http://www.unece.org/env/eia/sea_protocol.htmg⟩ (visited 14 April 2004).

UNECE/UNEP (United Nations Economic Commission for Europe/United Nations Environment Programme) Network of Experts on Public Participation and Compliance. 2000. *Guidance on Public Participation in Water Management and Framework for Compliance With Agreement on Transboundary Waters.* Geneva: UNECE.

UNEP (United Nations Environment Programme). 2002. "Guidelines on Compliance with and Enforcement of Multilateral Environmental Agreements." Adopted February 2002.

UNEP/UNDP (United Nations Environment Programme/United Nations Development Programme)/Dutch Joint Project on Environmental Law and Institutions in Africa. 1999. *Development and Harmonization of Environmental Laws: Report on the Legal and Institutional Issues in the Lake Victoria Basin.* Nairobi: UNEP/UNDP/Dutch Joint Project.

United States–Canada. 1978. "Agreement Between Canada and the United States of America on Great Lakes Quality." Done 22 November.

United States–Canada. 1972. "Great Lakes Water Quality Agreement of 1972." Amended in 1978 and 1983.

United States–Great Britain. 1909. "Treaty Between the United States and Great Britain Relating to Boundary Waters, and Questions Arising Between the United States and Canada." Done 11 January. *Statutes* 36:2448.

United States–Mexico. 1889. "Convention between the United States of America and the United States of Mexico to Facilitate the Carrying out of the Principles Contained in the Treaty of November 12, 1884 and to Avoid the Difficulties Occasioned by Reason of the Changes which take Place in the Beds of the Rio Grande and the Colorado Rivers," done 1 March. *Statutes* 26:1512. Internet: ⟨http://www.internationalwaterlaw.org/RegionalDocs/Co_Tj_RioG.htm⟩ (visited 13 April 2004).

United States–Mexico. 1944. "Treaty between the United States of America and Mexico, Utilization of Waters of the Colorado and Tijuana Rivers and of the Rio Grande." Done 3 February 1944, entered into force 8 November 1945. *Statutes* 59:1219. Internet: ⟨http://www.usbr.gov/lc/region/pao/pdfiles/mextrety.pdf⟩ (visited 13 April 2004).

Wates, Jeremy. 1999. "Introducing the Aarhus Convention: A New International Law on Citizens' Environmental Rights." Background Paper Distributed at the Pan-European ECO Forum Conference on Public Participation, Chisinau, Moldova. 16–18 April.

Weston, Burns H., Richard A. Falk, and Anthony D'Amato. 1990. *International Law and World Order* (2nd edn). St Paul, Minn.: West Publishing Co.

Wilson, Jessica. 2000. "Why Does the WTO Need Civil Society?" In Peider Konz (ed.). *Trade, Environment and Sustainable Development: Views from Sub-Saharan Africa and Latin America, A Reader*. Geneva: United Nations University/International Centre for Trade and Sustainable Development.

Wold, Chris. 1996. "Multinational Environmental Agreements and the GATT: Conflict and Resolution?" *Environmental Law* 26:841.

ZACPLAN. 1987. "Agreement on the Action Plan for the Environmentally Sound Management of the Common Zambezi River System." Done at Harare, Zimbabwe on May 28. *International Legal Materials* (1988) 27:1109.

3

Transboundary ecosystem governance: Beyond sovereignty?

Bradley C. Karkkainen

Introduction

This chapter begins with a bald and intentionally provocative claim: as we look ahead to the challenging and complex environmental problems that remain, conventional state-centric regulatory rules will turn out to be a less important part of the environmental management toolkit than is commonly supposed by legal scholars, environmental non-governmental organizations (NGOs), and many others. This applies both in the domestic environmental policy arena as well as in the realm of complex transboundary environmental problems, including that of transboundary watershed management, the central topic of this volume.

What is meant by that?

It is conventionally assumed that management of domestic natural resources and environmental problems is the prerogative of the sovereign state, which holds exclusive competence to impose binding rules on its subjects (Birnie and Boyle 2002). It is also assumed that transboundary environmental management is conducted principally, and of necessity, through international legal instruments consisting of mutually binding rules of obligation owed by sovereign states to other sovereign states (Weiss 2000; Birnie and Boyle 2002).

Thus, in both the domestic and international arenas, sovereign states are presumed to be the central, or even the exclusive, authors of environmental policy. Moreover, in both arenas it is assumed that policy will be expressed and carried out mainly through fixed, legally binding rules that regulate the behaviour of both states and non-state parties. Yet there is growing recognition in scientific and policy circles that, for purposes of addressing complex environmental problems generally, and for purposes of managing complex and dynamic ecosystems in particular, this conventional state-centric, fixed-rule regulatory approach is a poor fit (Holling and Meffe 1995; Tarlock 2003).

Concerted, conscious efforts to manage large-scale environmental problems through regulatory law are a relatively recent phenomenon, for the most part going back only 30–40 years or so. To be sure, some important precursors to contemporary environmental law can be found in domestic and international conservation efforts focusing on specific natural resources, and in tort (or tort-like) doctrines in the common law and, to a limited extent, customary international law (Hunter, Salzman, and Zaelke 2002; Percival et al. 2003). In the United States, the big push for comprehensive environmental protection came in the "environmental decade" of the 1970s, when the most important federal environmental statutes (such as the Clean Air Act and Clean Water Act) were adopted (Coglianese 2001). At the international level, the 1972 Stockholm Conference is generally regarded as the watershed event that ushered in the modern era of international environmental law, developed progressively through a series of subject-specific, legally binding inter-sovereign agreements (Kelly 1997; Birnie and Boyle 2002; Hunter, Salzman, and Zaelke 2002).

These legal efforts, both domestic and international, have tended to proceed piecemeal, through promulgation of regulatory-type rules aimed at particular, narrowly defined environmental problems (Fiorino 1996; Pezzoli 2000; Stewart 2001). The central notion was that, with the aid of sound science and technocratic expertise, government experts could identify and isolate the most important environmental threats and effectively ameliorate them by crafting and enforcing binding rules aimed at curbing the behaviours giving rise to the problems (Karkkainen, Fung, and Sabel 2000).

This approach reflects a familiar idea of what law is: an authoritative law-giver makes a binding rule to which all subject to its sovereign authority are bound. This view of law is grounded in nineteenth-century legal positivism and, more specifically, John Austin's "command theory" of law (Bix 2001). However, more recent jurisprudential theories, such as H.L.A. Hart's theory of law as a system of "primary" and "secondary" rules and Ronald Dworkin's theory that law includes not only rules but also principles, substantially embrace this view of the centrality of sovereign authority (Patterson 2003).

In domestic environmental policy, direct application of this state-centric, command-oriented approach typically leads to "command-and-control" regulation. Such a system entails commands by the state to its subjects to undertake certain measures, or (more typically) to refrain from taking certain kinds of actions that are judged to be environmentally detrimental (Fiorino 1996).

In the international arena, this approach leads to treaty arrangements in which sovereign states undertake mutual, legally binding contractual obligations to exercise their sovereign authority to control specified kinds of environmentally harmful action – for example, to ban or restrict production and consumption of specified ozone-depleting substances within areas subject to their territorial jurisdiction, or to regulate international trafficking in listed endangered species by following specified procedures to monitor and control exports and imports at their borders (Wirth 1993).

As appealingly familiar as this conventional, rule-based regulatory approach may be, it has several important limitations. These include, most prominently, scale mismatches and capacity mismatches.

Scale mismatches

The first limitation is the familiar problem of scale mismatches. As Eyal Benvenisti argues in his recent book *Sharing Transboundary Resources*, political boundaries are typically badly mismatched with the scale of the resource to be managed (Benvenisti 2002). In the case of transboundary watersheds, for example, sovereign states are too small to fit the full hydro-geographical scale of what is, ecologically and hydrologically, a single, indivisible resource; indeed, this dimension of the scale mismatch is what gives the problem its transboundary character. As a consequence, both the problem and the solution lie, in important part, beyond the territorial reach of any sovereign state (Benvenisti 2002). This, of course, gives rise to the need for transboundary cooperation.

Less often noticed, however, is that sovereign states may be too large to fit the geographical scale of the ecological problem. For example, management of even a very large watershed such as the North American Great Lakes, which have more than 17,000 km of shoreline, drain a basin of 766,000 km^2, and are home to some 33 million people (USEPA and Environment Canada 1995), tends to be seen as a problem of regional (sub-national) rather than truly national concern by the federal governments of the United States and Canada, and is consequently afforded a relatively low status in each nation's list of diplomatic priorities. In addition, each nation's Great Lakes policy is subject to influence or capture by an array of domestic political constituencies both within and outside the Great Lakes Basin. The US and Canadian federal governments consequently have suboptimal incentives to be fully attentive to the problems of the Great Lakes ecosystem (Benvenisti 2002).

This sort of mismatch problem extends to "domestic" ecosystems as well. Most ecological problems are poorly matched to conventional territorially defined political boundaries. Even the greatest of US watersheds – the Chesapeake Bay and the Columbia, Colorado, and Mississippi rivers – are regional rather than truly national in scale, yet they transcend the boundaries of sub-national political subdivision such as states, counties, and municipalities (Adler 1995).

The problem of scale mismatches, although perhaps obvious, is hardly trivial (Esty 1999). It points to a crucial question: how can we best match the capacities and incentives of governance institutions to the scale of the resource to be managed? To ask that question is implicitly to acknowledge that conventional political jurisdictions – sovereign states and their standard political subdivisions – may have both inadequate incentives and inadequate capacities to attend to important categories of environmental problems.

Capacity mismatches

More fundamentally, ecosystem management is bedevilled by a broader set of capacity mismatches. Quite simply, we face a crisis in the capacity of sovereign states to address complex environmental problems through the familiar tools of fixed-rule regulatory approaches, or in transboundary contexts through fixed-rule agreements among sovereign states.

Broadly, we can say that environmental regulation for most of the last 30 years has been rule-based and, more often than not, prohibitory in character (Stewart 1996). That should come as no surprise, for prohibitory regulation is what states know how to do best as a matter of domestic policy (Stewart 2001). And it also turns out to be what states know best how to agree to do at an international level. Recently, however, both scientists and policy makers have begun to appreciate some of the limitations of that rule-based approach for environmental problem-solving.

First, conventional regulatory rules tend to be negative rather than affirmative in character. It is generally easier to specify, monitor, and enforce rules prohibiting harmful actions than to mandate affirmative duties to undertake environmentally beneficial actions. But often the most environmentally beneficial actions require creativity and initiative on the part of the actor. Pollution prevention and habitat restoration, for example, work best when the polluter or land manager is positively motivated to discover and implement innovative techniques (Noss, O'Connell, and Murphy 1997; Rondinelli 2001). It is virtually impossible to mandate that kind of creativity, invention, and goodwill through regulatory rules.

Second, regulatory rules tend to be piecemeal and fragmentary rather than broadly integrative (Fiorino 1996; Stewart 2001). The standard approach to regulatory standard-setting proceeds reductively, breaking complex environmental problems into smaller, putatively manageable components and attacking them seriatim. In this scheme, relatively little attention is paid to synergies among environmental stressors or to the complex ecological interrelationships between and among various components (Holling and Meffe 1995).

Third, the rule-based approach tends to be rigid and inflexible, slow to incorporate and adjust to new learning. As a result, it tends to freeze in place old learning, tried-and-true technologies, and familiar regulatory techniques (Ackerman and Stewart 1985).

Fourth, regulation generally proceeds by lumping problems into broad categories, and then seeking categorical solutions or "one-size-fits-all" rules (Stewart 2001). As a result, regulation tends toward rough approximation across a range of facially similar situations, none of which it fits perfectly. Rules, in short, are not context sensitive.

Finally, standard regulatory approaches do not account well for ecological complexity (Holling and Meffe 1995; Levin 1999). Ecosystems, the ecologists tell us, consist of complex webs of mutual causal interdependence among physical and biological components, processes, and stressors (Holling, Berkes, and Folke 1998; Levin 1999). They are dynamic, not static, and do not necessarily tend toward stable equilibria, exhibiting important non-linear threshold effects (Holling, Berkes, and Folke 1998; Tarlock 2003). They are also characterized by inherent stochasticity and high levels of scientific uncertainty (Noss, O'Connell, and Murphy 1997). The presumption of the conventional rule-based approach was that if we study the problem long enough and hard enough, we will come to understand it well enough to craft an optimal rule, or at least a satisfactory one. But, as leading ecologists now assert, ecosystems turn out to be "not only ... more complex than we think, but more complex than we CAN think," with the result that it is impossible to have enough information to be certain that we are doing the right thing (Noss, O'Connell, and Murphy 1997).

In short, the conventional piecemeal, top-down, prescriptive, regulatory approach – the favoured tool of sovereign states – appears badly mismatched to the complex demands of contemporary ecological understanding. In response, a growing interest in integrated and adaptive ecosystem management is evident in both scientific and policy circles (Walters and Hilborn 1978; Holling and Meffe 1995; Brunner and Clark

1997; Wilkinson 2000). This approach seeks to manage particular ecosystems in an integrated way, using a place-based strategy that tailors management measures to context-specific needs and conditions, while seeking to coordinate management of entire suites of interrelated resources and environmental stressors (Clark 2001). For example, advocates of ecosystem management argue that fisheries cannot be effectively managed solely by regulating fishing effort or catch levels (Craig 2002). It is also important to consider fishing's effects on higher and lower trophic levels (predator and prey species), as well as its impact on the physical environment and habitats of non-target species. In addition, fisheries management must take into account a range of environmental stressors including pollution, habitat alteration, and fluctuating populations of non-target species that play critical roles in determining populations of the target species itself. As a corollary of this place-based approach, management measures and regulatory requirements will necessarily vary from place to place and will require high degrees of inter-agency, intergovernmental, and public–private coordination and collaboration.

Scientists and policy makers have come to realize, however, that integrated management of complex and dynamic ecosystems is an undertaking fraught with multiple layers of uncertainty (Holling and Meffe 1995). Even the most technically, legally, and administratively capable states find that they lack the competence to specify rules to carry out the task. The problems are simply too complex and too dynamic, and scientific understanding is too limited.

The upshot is that conventional fixed-rule approaches – commands by sovereign to subject, or rules of mutual legal obligation owed by sovereign states to other states – turn out to be extremely blunt, limited, and inflexible tools that are poorly matched to the subtle, complex, and ever-changing demands of ecological management. States can no longer rely on management by fixed rules – not merely because they can never know what are the right rules, but because, given the complex and dynamic character of the problem, there are no timeless rules to be found (Holling and Meffe 1995; Frampton 1996).

Collaborative ecosystem governance

In response to this crisis of state competence, a new style of governance is emerging in both domestic and transboundary contexts (Karkkainen 2002a). In this approach, state, sub-national, and non-state actors actively collaborate to fashion provisional solutions, and jointly devise adaptive learning and management strategies that allow them to adjust management measures and regulatory requirements in light of new learning and changing environmental conditions (Karkkainen 2002b).

Although sovereign states remain important actors in these hybrid governance arrangements, their role is radically redefined. Adaptive eco-system management tends to require high degrees of interagency, inter-governmental, and public–private collaboration, pooling the information, expertise, and institutional capabilities of a variety of state and non-state actors (Imperial 1999; Lee 1999). A central component of these institu-tional arrangements is a joint commitment to ongoing programmes of re-search and monitoring to better understand the ecosystem and to im-prove the scientific models upon which decision-making is based (Lee 1999; Karkkainen 2002b).

As a corollary, these hybrid governance arrangements tend to blur the usual distinctions between state and non-state, sovereign and subject. Non-state parties – including environmental NGOs, independent scien-tists, industry groups, sub-national governments, and ordinary citizens – assume prominent roles as co-authors and co-implementers with state agencies of a set of policies that jointly comprise the management effort (Karkkainen, Fung, and Sabel 2000).

Ecosystem-governance efforts also tend to transcend familiar jurisdic-tional barriers. Regional co-management of the Baltic Sea, for example, has become a joint exercise in identification and remediation of the most important environmental stressors wherever they occur throughout the Baltic basin, without regard to territorial jurisdiction (Darst 2001). Eco-system governance, in short, is effectively carried out at an appropriate regional scale tailored to that of the ecological resource to be managed, rather than being shoehorned into arbitrarily configured sovereign terri-torial boundaries.

However, the challenges of ecosystem management extend beyond attention to ecological scale and the need for inclusive collaboration – or "stakeholder governance," as it is sometimes called (Tarlock 2003). As already seen, precisely because "we never have enough information" (Noss, O'Connell, and Murphy 1997), ecosystem management demands an "adaptive management" or "adaptive learning" approach that treats policy interventions as provisional, generates and expects new learning through ongoing science and continuous monitoring, and adjusts policies periodically in response to that new learning (Ruhl 2003).

Notice that this is a distinctly non-rule-based approach – or, at any rate, it does not rely on fixed rules. Instead, it employs a strategy of roll-ing policy interventions and adjustments (Holling and Meffe 1995). Al-though these may sometimes take the form of mandatory requirements, at other times they are undertaken through voluntary measures, memoranda of understanding, or other legally unenforceable good-faith commitments (Sabel, Fung, and Karkkainen 2000). The general approach seems to be pragmatic commitment to do "whatever works" to achieve a generally stated goal or ecosystem restoration, while continuously refining and redefining the contents of that commitment as greater understanding and experience are gradually attained (Sabel, Fung, and Karkkainen 2000). Here again, however, we see a blurring of the familiar sharp distinctions between "law" and "not-law," and between "plan" and "implementation" in a characteristically pragmatic, mutual, periodic readjustment of ends and means (Sabel, Fung, and Karkkainen 2000).

Leading models

This distinctive style of hybrid governance arrangements is clearly discernible in ecosystem management efforts in the Chesapeake Bay region in the United States, and in the Baltic Sea region where a major transnational effort is under way to manage multiple environmental stressors on the marine ecosystem and its freshwater tributaries.

Each of these efforts has been held out as a model for others to emulate. The Chesapeake Bay Program is widely regarded as the premier model of aquatic ecosystem management, and especially of large-scale estuarine and watershed management (Costanza and Greer 1995; Chesapeake Bay Program 2002). The joint US–Canadian Great Lakes ecosystem management effort is viewed as a leading model of successful transboundary watercourse management (Birnie and Boyle 2002; Hunter, Salzman, and Zaelke 2002). The Baltic Sea regime has won widespread acclaim as an exemplary model of management of enclosed and semi-enclosed regional seas (UNEP 1997). Increasingly, there are also institutional linkages among these three efforts, as each has come to recognize the others as important parallel efforts from which much can be learned through scientific and technical exchange and assistance programmes (USEPA Great Lakes Program Office 1999; Boesch 2000b).

Although the particular institutional arrangements vary depending upon local ecological and institutional background conditions and possibilities, we can say broadly that these efforts exhibit the following set of core characteristics.

First, each has adopted an ecosystem focus. For example, the Chesapeake 2000 Agreement reaffirms the commitment of Bay Program partners to "protect and restore the Chesapeake Bay ecosystem," recognizing that "each action we take, like the elements of the Bay itself, is connected to all the others" (Chesapeake Bay Program 2000). The 1987 Protocol Amending the Great Lakes Water Quality Agreement of 1978 states that "the purpose of the Parties is to restore and maintain the chemical, physical, and biological integrity of the waters of the Great Lakes Basin Ecosystem." Similarly the 1992 Convention on the Protection of the Marine Environment of the Baltic Sea states in Article 3 its purpose to "promote the ecological restoration of the Baltic Sea Area and the preservation of its ecological balance." In each case, the goal is to manage the ecosystem, not merely as a collection of independent parts but as an integrated whole.

This in turn implies integrated and coordinated management of a suite of ecological stressors and resources – fisheries, non-target fish and wildlife, pollution, aquatic and riparian habitats, land use, non-point source pollution, and so on (USEPA 2000; Chesapeake Bay Program 2002).

Although the language used to describe the process varies, each proceeds through an iterative and adaptive management approach, relying on continuous feedback from joint monitoring and ongoing programmes of scientific investigation to continuously refine both the ecological models upon which policy is based and, ultimately, the policies themselves, which are seen as inescapably provisional and experimental (Boesch 2000a).

Each has also adopted a broadly collaborative and participatory management style, involving hybrid governance arrangements involving inter-agency, intergovernmental, and public–private collaboration. For example, in the Great Lakes region, a binational collaborative process involving government, industry, and NGO participants led to the adoption of a 1997 Great Lakes Binational Toxics Strategy, embracing both regulatory measures and voluntary initiatives to be periodically reassessed and revised in light of their effectiveness in achieving the goal of "virtual elimination of persistent toxic pollutants" (USEPA and Environment Canada 2002).

Finally, each is constructed less from well-defined rules of inter-sovereign obligation, than from broad and open-ended commitments to do "whatever it takes" to improve ecosystem health (Sabel, Fung, and Karkkainen 2000).

"Post-sovereign" governance

These arrangements have been termed "post-sovereign" environmental governance (Karkkainen 2002a). The term is intentionally provocative, intended to stimulate thought and discussion. Others have offered equally provocative characterizations, variously labelling this family of developments "governance without government" (Rosenau and Czempiel 1992) or "post-internationalism" (Rosenau, Czempiel, and Durfee 2000). However, these alternative formulations may overstate the case, in so far as they appear to suggest that state sovereignty is displaced in its entirety.

In contrast, the expression "post-sovereign" better captures the flavour of the new institutional arrangements described here. It recognizes that, although state sovereignty remains a bedrock feature of the international landscape, new forms of multi-party transboundary collaboration are emerging alongside it. These new hybrid institutional forms supplement, and only partially substitute for, more familiar exercises of state sovereignty. Such governance arrangements do not fit comfortably within the classical Westphalian model of an international system populated exclusively by sovereign states – a model that is increasingly inaccurate as a descriptive account (Rosenau, Czempiel, and Durfee 2000; Weiss 2000) and, in the eyes of some leading academic observers and many participants in the new governance arrangements, is increasingly self-limiting as a normative aspiration (Rosenau, Czempiel, and Durfee 2000). These new hybrid institutional forms may be considered "post-sovereign" in three distinct senses.

First, in place of exclusive sovereign authority, governance rests in the hands of multi-party collaborative governance institutions. Participation in the governance process extends well beyond sovereign states to include sub-national levels of government, local communities, NGOs, the independent scientific community, and key economic actors (Karkkainen, Fung, and Sabel 2000). Sovereign states are by no means excluded from the governance process; indeed, in crucial respects they remain the institutional backbone of the new arrangements, because their participation, financial and technical support, and legal sufferance remain essential to the success of the new hybrid institutions now being spawned. Nevertheless, as non-state actors take on new roles as co-authors and co-implementers of environmental policy, sovereign states' long-standing claims to exclusive competence to negotiate transboundary agreements and to determine domestic environmental and natural-resource policies are quietly undermined. While continuing to cloak their actions in the familiar language of state sovereignty, states have, in practice, abandoned the fiction that they are the only parties at the table and the only parties that matter. Crucial public-management decisions are being made, reviewed, and revised in hybrid multi-party settings. The role of the state, in short, is in the process of being redefined and downsized into something strikingly different from the familiar model of exclusively sovereign lawmaking to which we have long been accustomed.

Second, these arrangements are "post-sovereign," in so far as transnational cooperation has come to extend well beyond the familiar sorts of mutually agreed inter-sovereign rules of obligation that lie at the heart of public international law as we conventionally know it. Rather than one-time-only negotiations to a set of fixed rules, the new transboundary ecosystem governance arrangements are built around ongoing, continuous, open-ended commitments to "do what it takes" to restore particular ecosystems. The goal is not merely to establish fixed rules of obligation – binding at the topmost, state-to-state level – but, rather, to integrate and coordinate policy responses at multiple levels through complex networks that extend to sub-state and non-state actors, and deep into the civil society of an emerging transboundary polity that comes to define itself by its relation to the regional ecosystem. The goal, then, is that a broad range of discretionary authorities held by a variety of state and non-state parties will be exercised in accordance with an agreed (although always provisional) plan, of which a similarly broad range of state and non-state actors are co-authors and co-implementers. Conventional inter-sovereign legal agreements do play some fairly modest role in establishing overarching institutional frameworks and solemnizing major substantive commitments, but they are typically neither the driving force nor the defining feature of these complex transboundary collaborative arrangements.

Third, these arrangements are "post-sovereign" in the sense that, at the level of implementation, the measures undertaken often go well beyond traditional exercises of state sovereignty through hierarchical imposition of rules binding on those subject to the state's jurisdiction. Although some rules may be of this hierarchical and binding character, the decision-making and implementation processes are generally collaborative and polyarchic. Transboundary ecosystem management thus typically embraces a rich mix of non-hierarchical tools – such as voluntary, cooperative, and quasi-contractual commitments – that may have little or no formal legal consequence but, none the less, may have significant practical effects in directing and constraining the behaviour of both states and non-state parties.

Conclusions

Rather than reflexively falling back on familiar concepts, assumptions, and regulatory approaches, this chapter urges that we take a closer look at the world in which complex ecosystems are now being managed. New institutional arrangements are rapidly unfolding – arrangements that look startlingly different from the familiar paradigms of state sovereignty, in which individual subjects are bound by rules issued by their sovereign and sovereigns are bound to each other by mutually agreed rules of obligation. This emerging post-sovereign world is populated and governed not only by states but also by hybrid institutional entities that now perform (sometimes on a transboundary basis) many of the governance functions traditionally claimed to be the exclusive prerogative of states. They do so using the unconventional tools and techniques of adaptive, integrated, ecosystem management.

These sorts of institutional arrangements, admittedly, are not yet widespread and, even where they exist, they remain fragile. It is premature (and, perhaps, simply inaccurate) to claim for them unqualified success, even by their own self-defined yardsticks. Each of the regional arrangements discussed here – the Chesapeake Bay, the North American Great Lakes, and the Baltic Sea – has thus far fallen short of the ambitious ecosystem-restoration targets it has set for itself.

Yet it would be equally foolish to dismiss these efforts. They represent the leading edge in a wave of institutional innovation that, at a minimum, must be carefully monitored, evaluated, and analysed for what it can teach us about how better to manage a complex and dynamic world: the limitations of conventional, sovereignty-based approaches to environmental regulation have been exposed, and it is difficult to imagine ever turning back the clock.

REFERENCES

Ackerman, Bruce A., and Richard B. Stewart. 1985. "Reforming Environmental Law." *Stanford Law Review* 37:1333.

Adler, Robert W. 1995. "Addressing Barriers to Watershed Protection." *Environmental Law* 25:973.

Benvenisti, Eyal. 2002. *Sharing Transboundary Resources: International Law and Optimal Resource Use*. Cambridge: Cambridge University Press.

Birnie, Patricia, and Alan Boyle. 2002. *International Law and the Environment*, 2nd edn. Oxford: Clarendon Press.

Bix, Brian. 2001. "John Austin" in *Stanford Encyclopedia of Philosophy*. Internet: ⟨http://plato.stanford.edu/entries/austin-john/⟩ (visited 26 June 2003).

Boesch, Donald F. 2000a. "Accomplishments and Challenges in the Chesapeake Bay Program." Proceedings of the Second Joint Meeting, Coastal Environmental Sciences and Technology Panel of the United States–Japan Cooperative Program in Natural Resources. Silver Spring, MD. October 25–29.

Boesch, Donald F. 2000b. "Bay Has a Lot to Learn from European Efforts to Reduce Nutrients." *Bay Journal* 10(1).

Brunner, Ronald D., and Tim W. Clark. 1997. "A Practice-Based Approach to Ecosystem Management." *Conservation Biology* 11:48.

Chesapeake Bay Program. 2000. *Chesapeake 2000 Agreement*. Internet: ⟨http://www.chesapeakebay.net/agreement.htm⟩ (visited 7 February 2005).

Chesapeake Bay Program. 2002. *Chesapeake Bay Program, The State of the Chesapeake Bay: A Report to the Citizens of the Bay Region*. CBP/TRS 260/02, EPA-903-R-02-002. Annapolis, Md.: USEPA.

Clark, William C. 2001. "A Transition Toward Sustainability." *Ecology Law Quarterly* 27:1021.

Coglianese, Cary. 2001. "Social Movements, Law, and Society: The Institutionalization of the Environmental Movement." *University of Pennsylvania Law Review* 150:85.

Costanza, Robert, and J. Greer. 1995. "The Chesapeake Bay and its Watershed: A Model for Sustainable Ecosystem Management?" In Lance H. Gunderson, C.S. Holling, and Stephen S. Light (eds). *Barriers and Bridges to the Renewal of Ecosystems and Institutions*. New York: Columbia University Press.

Craig, Robin Kundis. 2002. "Taking the Long View of Ocean Ecosystems: Historical Science, Marine Restoration, and the Oceans Act of 2000." *Ecology Law Quarterly* 29:649.

Darst, Robert G. 2001. *Smokestack Diplomacy: Cooperation and Conflict in East–West Environmental Politics*. Cambridge, Mass.: MIT Press.

Esty, Daniel C. 1999. "Toward Optimal Environmental Governance." *New York University Law Review* 74:1495.

Fiorino, Daniel J. 1996. "Toward a New System of Environmental Regulation: The Case for an Industry Sector Approach." *Environmental Law* 26:457.

Frampton, George. 1996. "Ecosystem Management in the Clinton Administration." *Duke Environmental Law and Policy Forum* 7:39.

Holling, C.S., and Gary K. Meffe. 1995. "Command and Control, and the Pathology of Natural Resource Management." *Conservation Biology* 10:328.

Holling, C.S., Fikret Berkes, and Carl Folke. 1998. "Science, Sustainability and Resource Management." In Fikret Berkes, Carl Folke, and Johan Colding (eds). *Linking Social and Ecological Systems: Management Practices and Social Mechanisms for Building Resilience*. Cambridge, UK: Cambridge University Press, 342.

Hunter, David, James Salzman, and Durwood Zaelke. 2002. *International Environmental Law and Policy*, 2nd edn. New York: Foundation Press.

Imperial, Mark T. 1999. "Institutional Analysis and Ecosystem-Based Management: The Institutional Analysis and Development Framework." *Environmental Management* 24:449.

Karkkainen, Bradley C. 2002a. "Post-Sovereign Environmental Governance: The Collaborative Problem-Solving Model." In Frank Biermann, Rainer Brohm, and Klaus Dingwerth, eds. *Global Environmental Change and the Nation-State:* Proceedings of the 2001 Berlin Conference on the Human Dimensions of Global Environmental Change. Potsdam: Potsdam Institute for Climate Impact Research.

Karkkainen, Bradley C. 2002b. "Collaborative Ecosystem Governance: Scale, Complexity and Dynamism." *Virginia Environmental Law Journal* 21:189.

Karkkainen, Bradley C., Archon Fung, and Charles F. Sabel. 2000. "After Backyard Environmentalism: Toward a Performance-Based Regime of Environmental Regulation." *American Behavioral Science* 44:692.

Kelly, Michael J. 1997. "Overcoming Obstacles to the Effective Implementation of International Environmental Agreements." *Georgetown International Environmental Law Review* 9:447.

Lee, Kai N. 1999. "Appraising Adaptive Management." *Conservation Ecology* 3(2):3.

Levin, Simon A. 1999. "Towards a Science of Ecological Management." *Conservation Ecology* 3(2):6.

Noss, Reed F., Michael A. O'Connell, and Dennis D. Murphy. 1997. *The Science of Conservation Planning: Habitat Conservation Under the Endangered Species Act*. Washington, DC: Island Press.

Patterson, Dennis. 2003. "Fashionable Nonsense." *Texas Law Review* 81:841.

Percival, Robert B., Christopher H. Shroeder, Alan S. Miller, and James P. Leape. 2003. *Environmental Regulation: Law, Science and Policy*, 4th edn. New York: Aspen Publishers.

Pezzoli, Keith. 2000. "Environmental Management Systems (EMSS) and Regulatory Innovation." *California Western Law Review* 36:336.

Rondinelli, Dennis A. 2001. "A New Generation of Environmental Policy: Government–Business Collaboration in Environmental Management." *Environmental Law Reporter* 31:10891.

Rosenau, James N., and Ernst-Otto Czempiel (eds). 1992. *Governance without Government: Order and Change in World Politics*. New York: Cambridge University Press.

Rosenau, James N., Ernst-Otto Czempiel, and Mary Durfee. 2000. *Thinking Theory Thoroughly: Coherent Approaches to an Incoherent World*. Boulder, Colo.: Westview Press.

Ruhl, J.B. 2003. "Is the Endangered Species Act Eco-Pragmatic?" *Minnesota Law Review* 87:885.

Sabel, Charles, Archon Fung, and Bradley Karkkainen. 2000. "Beyond Backyard Environmentalism." In Joshua Cohen and Joel Rogers (eds). *Beyond Backyard Environmentalism* 28. Boston, Mass.: Beacon Press.

Stewart, Richard B. 1996. "United States Environmental Regulation: a Failing Paradigm?" *Journal of Law and Commerce* 15:585.

Stewart, Richard B. 2001. "A New Generation of Environmental Regulation?" *Capital University Law Review* 21:21.

Tarlock, A. Dan. 2003. "Slouching Toward Eden: The Eco-Pragmatic Challenges of Ecosystem Revival." *Minnesota Law Review* 87:1173.

UNEP. 1997. "Needs and Approaches to Improve Access to Environmental Information for Transboundary Decision-Making in the Baltic Sea Region." UNEP/DEIA/MR.

USEPA and Environment Canada. 1995. "Great Lakes Factsheet No. 1: Physical Features and Population." *Great Lakes Atlas*, 3rd edn.

USEPA and Environment Canada. 2002. "Great Lakes Binational Toxics Strategy 2002 Progress Report 1."

USEPA Great Lakes National Program Office. 1999. "The Great Lakes/Baltic Sea Partnership Program Record of Decision," Nov. 6.

USEPA Great Lakes National Program Office. 2000. "Great Lakes Ecosystem Report 2000."

Walters, Carl J., and Ray Hilborn. 1978. "Ecological Optimization and Adaptive Management." *Annual Review of Ecology and Systematics* 9:157.

Weiss, Edith Brown. 2000. "The Rise or the Fall of International Law?" *Fordham Law Review* 69:345.

Wilkinson, Charles F. 2000. "A Case Study in the Intersection of Law and Science: The 1999 Report of the Committee of Scientists." *Arizona Law Review* 42:307.

Wirth, David A. 1993. "The Uneasy Interface Between Domestic and International Environmental Law." *American University Journal of International Law and Policy* 9:171.

4

Implications of the Information Society on participatory governance

Hans van Ginkel

Introduction

Thomas Jefferson once observed that "when ever the people are well-informed, they can be trusted with their own government ..." (Fischer 2000). Deconstructing this statement, to ask what exactly is meant by the term "well-informed," the analysis turns to the old debate about information and participation that goes back to the turbulent times of civic protest in the 1960s and 1970s. In the midst of this period, Sherry Arnstein introduced us to the ten-step ladder of participation, in which she identified the provision of information as the most important first step to legitimate participation (Arnstein 1969). The key phrase here is "first step." The other, higher, steps (in ascending order of importance) include consultation, partnership, delegated power, and citizen control.

In more recent years, another perspective has come to the fore: this relates to concerns about information overload in the post-modern world. Social commentators argue that the acceleration of societal change has been accompanied by an increase in the information needed to keep up with all these developments. The flow of information within society has been greatly assisted by a variety of media: first, there were newspapers and the telegraph, then radio and television, and now the Internet. We live in the Information Society, and many people find themselves over-informed rather than well informed. For instance, a 1996 global survey by Reuters found that two-thirds of managers suffer from increased tension and one-third from ill health because of information overload (Reuters 1996). The situation may be getting worse, not better, and there are new expressions that capture the essence of the socio-information problems, such as Information Fatigue Syndrome. The symptoms of this syndrome include increased anxiety, poor decision-making, difficulties in memorizing and remembering, and reduced attention span. Another often-heard term is "data smog," which relates to an overabundance of low-quality information (Shenk 1998).

It is interesting that, in the context of information overload, an environmental term – smog – is used to describe the problem. This leads to the second cause for concern. Environmental problems on both the global scale (for example, climate change) and local scale (e.g. endocrine disruptors) are increasingly complex to identify, analyse, and address. Faced with complex environmental issues and ever-growing information flows, the central question becomes (and the main concern of this chapter is) can democracy and public participation flourish in today's complex technological information society? How will the hundreds of millions of people currently online benefit from the use of the Internet to improve their understanding of complex issues at the global, regional, and local levels and their relations? And how can this tool be used to enhance public participation?

This situation should not be seen as a problem but as an opportunity – or, perhaps, rather as a "virtual opportunity." This term embodies all of the benefits of being online and, at the same time, a sense of being increasingly distanced from reality.

This chapter examines some of the challenges and opportunities for participatory governance presented by the rise of the Information Society. The first section (pp. 89–91) examines the Information Society as a technocratic ideology and tool that could promote public participation in decision-making. The next section (pp. 91–94) focuses on the potential role that the Internet may have in inclusive governance. The subsequent section (pp. 94–95) reflects on the experiences to date with online public participation. The final section (pp. 95–96) concludes with a few thoughts on the future of participatory governance in the Information Society and similar challenges posed by online and offline forms of public participation.

Information Society as a technocratic ideology and tool for public participation

In his book *Citizens, Experts, and the Environment*, Frank Fischer argues that the Information Society, as an ideology, presents technological advance as social progress and that it conflates the concepts of information and knowledge (Fischer 2000). Reading between the lines, Fischer appears to recommend that we, as citizens, reflect upon whether we are really witnessing societal progress through the use of information technology (IT) or just the rapid development of some form of mass distraction. For example, the overwhelming flows of information on the Internet might present a danger of people becoming disengaged from existing political processes and, instead, viewing (and using) these tools only for their entertainment value.

Others have warned that the Internet is the greatest hegemonic device ever created by humankind and that it will lead increasingly to a globally monolithic, monocultural, and technocratic world (Bowers 2000). This homogeneous form of modern society would run counter to our environmental and cultural needs and to the preference – expressed by many, including (in the more radical forms) the "Deep Ecologists" and other environmental ideologies – to create small-scale and autonomous communities (Devall and Sessions 1986; Schumacher 1989; Katz, Light, and Rothenberg 2000).

I do not subscribe to the above world-view, nor do I have fears about the use and misuse of the Internet. Like many of the tools created by people in the past, the Internet will be operated in a responsible manner. There are questions, however, about the use of the Internet and about the implications that the Information Society would have for participatory forms of governance and the development of discursive institutions capable of rapidly responding to the stresses and pressures, particularly in the environmental arena. The potential positive and negative ramifications of the Internet on wider society have been extensively documented (Mitchell 2000; Slevin 2000; Toregas 2001). Nevertheless, it is only recently that researchers have focused on the possible implications for community engagement within the framework of emerging forms of digital or e-governance. Some kind of IT-led transformative process is under way, with the potential to alter the *modus operandi* of interactions between governmental bodies and the general citizenry. Nevertheless, little is known about the direction of current changes and their potential implications for the future forms of governance, although this is the subject of intensive debate in the ongoing preparations for the World Summit on the Information Society (WSIS), the first part of which has taken place in Geneva in December 2003 while the second part will be held in Tunisia in December 2005.

The Geneva Summit primarily focused on agenda setting, and the adopted Action Plan lists a series of issues grouped in nine main clusters (WSIS 2003). Access to information and knowledge is one of these key clusters. The Plan calls for the development of policy guidelines on the promotion of public-domain information as an important international instrument promoting public access to information. Governments are also encouraged to provide adequate access through various communication resources, notably the Internet, to public official information. This would involve establishing legislation on access to information and the preservation of public data, notably in the area of the new technologies. Substantive discussions also took place, and proposals were made, on other key areas related to participatory governance – including the ethical dimensions of the Information Society and also the issue of e-inclusion for remote areas, indigenous groups, and poor communities. These topics are bound to receive greater attention as we move toward the Tunisia Summit.

There is a wide range of thematic areas associated with the Information Society and its potential to influence development, including e-democracy, e-government, tele-education, e-commerce, tele-services, telework, digital divide, and social exclusion. Existing experience in these different areas has highlighted many significant barriers to the adoption of IT for public-participation purposes, but some clear ideas are emerging on modalities for their potential application to bridge the contemporary perception divide between governments and the communities that they serve. The term "perception divide" describes the situation whereby the administration (national and local politicians, officials, and experts) has a perception of an issue that differs from that of the broader lay community. This difference may be for a variety of reasons – such as a lack of communication, the tendency to form expert cliques, or arrogance that "we know best." The following sections examine some of the issues associated with the use of the Internet in public participation in more detail.

The Internet and inclusive governance

A review of recent literature reveals that an increasing number of policy makers and researchers around the world are working valiantly to link information and decision-making with the global trends and local needs. They are reflecting upon the pressing global problems facing modern communities and examining ways in which practical measures can contribute to understanding and amelioration of existing problems. This tendency is evident in recent efforts to deal with climate change, where bottom-up initiatives have been accompanied by complementary efforts to downscale both science and policy. In this increasingly globalized and interconnected world, Ulrich Beck has pointed out that no action or event is purely local, because "all inventions, victories and catastrophes affect the whole world" (Beck 1999). The opportunities (virtual and real) associated with this new electronic interdependence reflects Marshall Mc-Luhan's "global village" (McLuhan and Fiore 1968).

To put it simply, globality implies the coming-together of local cultures, a process that has become known as "glocalization." This is not an entirely neutral development, as Zygmund Bauman explains that both globalization and localization can be understood as expressions of new polarizations and stratification in society (Bauman 1998). Nowhere is this more apparent than with respect to the emergence of the Information Society and the Internet; this, therefore, presents the first challenge to those who equate the Information Society with progress. The contemporary "digital divide" at the global level is clear, but what remains uncertain is the potential impact on the distribution of power, wealth, privileges, and freedoms in all corners of the world that the Internet could bring. Social projects that seek to bridge the digital divide by providing greater community access to information technologies, while fundamentally important, must be scrutinized in the context of the motivations of the stakeholders involved in project promotion.

There are two potential implications of the widespread use of IT to support public participation. First, there is the information processing and dissemination element, whereby increasingly sophisticated environmental information in diverse forms (including via geographic information systems; GIS) is disseminated real-time through the Internet. Second, new forms of civic engagement are likely to emerge through websites that promote online interaction between citizens and government policy makers using a range of tools.

A study by the British Council published in 1999 supports this supposition (British Council 1999). This study looked at emerging practice with the application of Internet to public participation and indicated five possible benefits – namely, increased information accessibility, greater public involvement, public-awareness raising, promotion of enhanced communication, and stimulation of discussion on the merits of e-governance. Moreover, the British Council study argued that e-governance can be defined as encompassing the use of a variety of information technology tools by government in order to connect directly with citizens and to enhance service delivery, to provide for sustainable economic development, and to safeguard democracy.

Another recent review of experience and potential use of e-governance to support development across the globe outlined the main benefits of Internet use in terms of cost reductions, producing more for less, achieving results more rapidly to a higher quality, and doing so in new ways (Heeks 2001). Nevertheless, the same study identified six barriers hindering the degree of "e-readiness" of countries in different parts of the world. These are basically infrastructure problems associated with data systems (i.e.

the quality of data and its security), regulations, institutions, human capacities, technology, and leadership (i.e. the existence, or lack, of e-champions).

Looking at experience in the United States of America, a report on the development of local e-governance by the Center for Technology in Government highlighted four key lessons based on experience with online public participation from 1993 to 1999 (Dawes et al. 1999). These can be summarized as follows:

- IT projects need to be driven by programmatic goals, not by technology. If the outcome is to improve service performance or ensure more effective delivery of information, then this should remain central, and potential management and policy implications should be fully evaluated.
- Government-supported IT innovation for public participation should be approached from a learning perspective. Emphasis should be placed on the development of prototypes that can evolve, be evaluated, and eventually grow.
- Government complexity needs to be addressed. Successful IT projects require buy-in for different stakeholders within and outside local government.
- Professionalism and personal commitment are essential for success in online public-participation projects.

The report recommended that these lessons be addressed at the start of IT projects to ensure a culture in government that encourages innovation, fosters experimentation, and values thoughtful analysis.

The importance of considering local stakeholders in the development of local IT-based public-participation projects should not be underestimated. Local NGOs and communities face similar problems to those of the administration as they try to adapt to new demands related to the emergence of the Information Society. A 2001 study by the Surdna Foundation indicated that long-term IT-induced structural changes are just over the horizon for the non-profit sector, and that this IT-driven process will change how they work, reach their audience, deliver on their goals, and raise funds (Surdna Foundation 2001). Similar changes are taking place with online communities related to specific issues such as the environment. This initial experimentation is based primarily on geographic locations and existing (rather than virtual) communities, although this might not remain the case for long. A good example is the Seattle Community Network (SCN), established in 1995 by the local chapter of the Computer Professionals for Social Responsibility (Seattle Community Network 2003). The SCN provides local environmental organizations with access to a number of online interactive tools, including telnet login, Web-mail, calendars, mailing lists, and Web hosting. Another interesting example is the Minnesota E-Democracy group, a non-partisan citizen-based organization established in 1994, whose mission is to improve participatory democracy in Minnesota through the use of information networks (Minnesota E-Democracy 2003). It seeks to increase citizen participation in elections and promote public discourse on a range of is-sues through the use of IT.

Experience from these community-based and government-initiated activities suggests that the future of public participation is likely to be shaped by the forces promoting the digitization of governmental informa-tion, as well as service improvements, and by the traditionally counter-vailing civil-society forces promoting participation and citizen empower-ment. Significant progress has been made already with the development of the basic infrastructure, and interesting examples of environmental e-governance can already be found.

Reflecting upon existing experience with online public participation

When considering the experiences to date with online participation in the United States, Japan, and Europe, there are a number of interesting sim-ilarities. In all three, greater accessibility to information has been accom-panied by calls from many sectors for increased online interactivity and citizen participation. For example, in the United States, a 2000 study on environmental democracy and environmental governance at the state level evaluated the performance of local government environment web-sites against a set of criteria related to information access on the state of the environment and regulations, as well as interactivity in terms of citizen input, comment, and communication via the website (Beierle and Cahill 2000). The report concluded that few of the 50 states surveyed had quality opportunities for interactive electronic public involvement. In some instances, local officials expressed serious reservations about the possibilities of increased interaction for the following reasons:

- Online initiatives affect the internal organization of bureaucracies, re-quiring increased coordination and cooperation.
- Responding to the external demands of stakeholders forces agencies to be strategic in their use of resources for online efforts.
- These demands for internal prioritization create tensions between de-partments; as a result, engaging citizens online appears to be a consid-erably lower agency concern than streamlining the process aimed at the regulated community.

Similar studies are under way in Europe, including a major research project from 1998 onwards undertaken by the European branch of the International Council for Local Environmental Initiatives (ICLEI). The project, called ICTULA (Information and Communication Technology Use with Local Agenda 21), explores experience with the use of the Internet to support local environmental policy-making in five European cities – Amsterdam, Darmstadt, Hanover, Liverpool, and Turku. Initial findings from an associated survey of 52 European local authorities found that 58 per cent of the authorities were using the Internet to support their work, 21 per cent were using e-mail to support Local Agenda 21 networking, and 33 per cent were using Web pages to support Local Agenda 21 (ICTULA 1998). Looking specifically at experience in Darmstadt and Hanover, a number of risks associated with the use of IT were identified: these were the potential flood of information, the possible alienation of interpersonal contacts, the acceleration of all processes, and the rise of new dependencies (e.g. "if it can't be done without IT, it won't be done"). The benefits associated with Internet use in the context of Local Agenda 21 were highlighted as the potential for greater citizen involvement, opportunities for local authorities to share experiences rapidly, and new options for coordinating local activists.

In Japan, local authorities have a long record with telemetric environmental-monitoring systems, including automated systems linked through telephone lines to pollution-control centres. These are supported by GIS, remote sensing, modelling, and simulations, and they are integrated into comprehensive, local, environmental-information systems. In recent years, the environmental administration in Japan has driven a process to put as much information online as possible. Central to this effort is the national Environmental Information and Communication Network established in 1989 to provide extensive information on the natural environment, air, water, soil pollution, waste, and energy. It also provides access to resources (news, site links, databases and forums) and information related to eco-business and environmental-education activities. A good example at the local level is Kanagawa Prefecture, located south-west of Tokyo, which provides online real-time air-pollution data (at hourly intervals) and highlights breaches in local environmental standards.

In all these countries, considerable efforts are being made to expand the range of online public-participation activities. Important lessons have been learned, but there is still a long way to go before the full opportunities associated with the use of the Internet to promote enhanced forms of public participation become clear. In my view, the jury is still out.

Concluding remarks

There are many potentially positive impacts of Internet use to support public participation on environmental issues such as international water-resource management. Moreover, there could be an additional bonus, for example when the Internet is used to rapidly internationalize examples of good practice through online networks and the creation of associated Web-based epistemic communities. On the negative side, IT use is likely to bring advantages primarily to the digitally connected, and many governments, already strapped for funds, will struggle to expand accessibility for their citizens. Moreover, there is a real danger of the technological delivery system being viewed as more important than the "message," so that resources are heavily invested in IT instead of tackling environmental or educational problems directly. Let us hope that this will not be the case.

Online public participation is similar to offline versions. The same age-old problems have to be tackled, including how to develop trust and credibility. Moreover, there is the issue of how to reach those traditionally less active or the so-called "middle many," who could influence the process in a positive manner if they had the incentive to get involved. On top of this, there remains the need to explain complex information, especially in the environmental arena. As with all public participation, a clear communication strategy, responsive to local needs, is essential.

REFERENCES

Arnstein, S. 1969. "A Ladder of Citizen Participation." *Journal of the American Institute of Planners* 8(3):217–224.

Bauman, Z. 1998. *Globalization – The Human Consequences*. London: Polity Press.

Beck, U. 1999. *World Risk Society*. London: Blackwell Publishers.

Beierle, T., and S. Cahill. 2000. *Electronic Democracy and Environmental Governance: Survey of States*. Washington, DC: Resources for the Future.

Bowers, C.A. 2000. *Let Them Eat Data – How Computers Affect Education, Cultural Diversity and the Prospects of Ecological Sustainability*. Athens, Ga.: University of Georgia Press.

British Council. 1999. *Developments in Electronic Governance, Information Services Management*. Manchester: The British Council.

Dawes, S.S., P.A. Bloniarz, D.R. Connelly, K.L. Kelly, and T.A. Pardo. 1999. *Four Realities of IT Innovation in Government*. Albany, NY: Center for Technology in Government.

Devall, Bill, and George Sessions. 1986. *Deep Ecology*. Layton: Gibbs Smith Publisher.

Fischer, Frank. 2000. *Citizens, Experts, and the Environment: The Politics of Local Knowledge*. Durham: Duke University Press.

Heeks, R. 2001. *Understanding e-Governance for Development*. i-Government Working Paper No. 11. Manchester: Institute for Development Policy and Management.

ICTULA (Information and Communication Technology Use with Local Agenda 21). 1998. *User–Expert Dialogue in Darmstadt and Hanover*. Darmstadt: Institut fur Zielgruppenmarketing und Kommunikation.

Katz, Eric, Andrew Light, and David Rothenberg (eds). 2000. *Beneath the Surface: Critical Essays in the Philosophy of Deep Ecology*. Cambridge, Mass.: MIT Press.

McLuhan, M., and Q. Fiore. 1968. *War and Peace in the Global Village*. New York: Bantam.

Minnesota E-Democracy. 2003. Internet: ⟨http://www.e-democracy.org/discuss. html⟩ (visited 15 November 2003).

Mitchell, W.J. 2000. *E-Topia*. Cambridge, Mass.: MIT Press.

Reuters. 1996. *Dying for Information? An Investigation into the Effects of Information Overload in the USA and Worldwide*. Based on research conducted by Benchmark Research. London: Reuters Limited.

Schumacher, E.F. 1989. *Small is Beautiful: Economics as if People Mattered*. New York: Perennial.

Seattle Community Network. 2003. Internet: ⟨http://www.scn.org/⟩ (visited 15 November 2003).

Shenk, D. 1998. *Data Smog: Surviving the Information Glut*. Revised and updated edition. San Francisco: Harper.

Slevin, J. 2000. *The Internet and Society*. Cambridge: Polity Press.

Surdna Foundation, Inc. 2001. "More than Bit Players: How Information Technology Will Change the Way Nonprofits and Foundations Work and Thrive in the Information Age." Report by Andrew Blau to the Surdna Foundation. Internet: ⟨http://www.surdna.org/documents/morefinal.pdf⟩ (visited 15 November 2003).

Toregas, C. 2001. "The Politics of E-Gov: The Upcoming Struggle for Redefining Civic Engagement." *National Civic Review* 90(3):235–240.

World Summit on the Information Society (WSIS). 2003. "Plan of Action." Document WSIS-03/GENEVA/DOC/5-E. Internet: ⟨http://www.itu.int/dms_pub/ itu-s/md/03/wsis/doc/S03-WSIS-DOC-0005!!MSW-E.doc⟩ (visited 14 April 2004).

Part II

Experiences from international watersheds

5

Public participation in the management of the Danube River: Necessary but neglected

Ruth Greenspan Bell and Libor Jansky

Introduction

Public participation in environmental decision-making is best characterized as an evolving issue in the countries in economic and political transition of Central and Eastern Europe and, in particular, of the Danube
River Basin (see fig. 5.1). The watershed year from which progress is normally measured is 1989, the year when many countries in the Eastern
Bloc made the transition from communist governments. In the period
before 1989, although many laws – even constitutions – appeared to
welcome public involvement in principle, in practice this was a severely
flawed system that not only did not value public opinion but also, by
enacting laws and applying them selectively, acted in ways that had the
effect of eroding respect for law as an institution. The substitute arrangements that developed in the absence of a reliable rule of law society are
well known. People often developed informal arrangements, which were
vital to the functioning of the economy and the welfare of the people
(Wedel 1986).

The discrepancy between the official inclusion of a participatory system and a clean environment and the practical implementation of such
principles was clear. At some point, the environmental situation became
an example of the problems in the system. Concern about the environment became a vehicle for expressing more fundamental opinions about
how government made decisions and, over time, specific environmen-

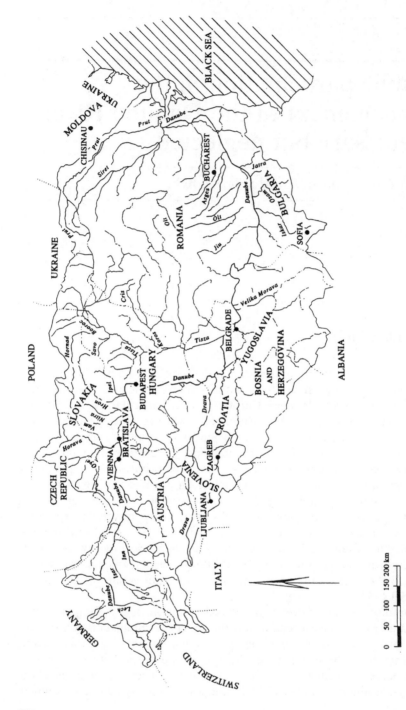

Figure 5.1 Map of the Danube River Basin (Source: Environmental Programme for the Danube River Basin 1994)

tal issues became a means for promoting the very idea of public participation.

Although there are a number of examples of this, this chapter focuses on the efforts to defeat the proposed Gabčíkovo–Nagymaros Dam. In this context, decisions about the Danube River were taken without public consultation, and the very failure to consult became a vehicle that mobilized public opinion and became a factor that led to the downfall of the communist regime in Hungary.

With the political and economic changes that began in 1989, the entire debate shifted. The driving issues became how to build a democracy, how to change long-standing habits, and how to encourage and incorporate public participation in public decision-making. Along with genuinely open elections and democratically elected legislatures, many countries began to build the formal infrastructure for open government, writing laws such as Freedom of Information Acts.

Another process that began in earnest was a joint effort among Danube River Basin countries to address their joint water-pollution problems. Cleaning up the Danube provides a practical example of how environmental decision-making is formulated and the role of public participation in practice with respect to a major, transboundary, environmental effort. Based on what has happened to date, one might argue that, although their goals are admirable, those administering the Danube River clean-up process have not yet fully internalized the principles of public participation (Bell, Stewart, and Nagy 2002; Bell and Fülöp 2003).

Lessons from history about shared water bodies

Background

Most efforts in Central Europe are best understood in the context of the considerable history of this region, and this is equally true for state-to-state efforts to manage water pollution.

Hungarian national identity was formed in the struggle for independence from the Habsburgs that subjugated the Kingdom of Hungary and relegated it to the status of a colony from 1526 to 1867, when a dual Austro-Hungarian monarchy was formed. At the same time, Hungarian domination since AD 907 over Slovak (among other Slav) territories gave rise to Slav nationalism. This ultimately sparked the flame of the First World War that engulfed the whole of Europe, led to the collapse of the Austro-Hungarian monarchy, and redrew the map of Central Europe.

The 1920 Treaty of Trianon that endorsed the establishment of

Czechoslovakia and Yugoslavia, and the expansion of Romania and Ukraine at the expense of Hungary, gave rise to new sources of resentment and conflict within and among the Danube River Basin states. Among other problems, this proved a hindrance for the joint management of the shared water resources of the Middle Danube. As a result of the treaty, Hungary lost two-thirds of its pre-1920 territory, two-thirds of its total population, one-third of its Hungarian population, and 94.5 per cent of its hydroelectric potential (Fitzmaurice 1998). At the same time, the political union between Czechs and Slovaks marked the beginning of an uneasy political partnership that eventually affected the historical development of the Gabčíkovo–Nagymaros project.

Political turmoil in Europe in the first half of the twentieth century and, in particular, in the Central European Danubian Basin, prevented the undertaking of joint water-management projects. The transition period during and between the two World Wars, involving the frequent transfer of political control over the Danubian lands between Hungary and Czechoslovakia, brought about conflicting political claims on these shared water resources (Lipschutz 1997). The political unification of the region under socialist rule, though, provided the political basis and the economic stimulus for joint water management between Hungary and Czechoslovakia, and the social impetus for cooperation lubricated by the long history of coexistence of Hungarians and Slovaks under the Habsburg and later the Austro-Hungarian monarchies.

Thus, joint planning for the modification of the middle reaches of the Danube between the two countries began in the 1950s, following a proposal by the Hungarian Academy of Sciences. The Hungarian proposal received the approval of the Council for Mutual Economic Cooperation (COMECON) among the socialist states in the early 1960s. However, domestic developments in the two countries prevented completion of the project until 1977 (Fitzmaurice 1998). In Hungary, these developments took the form of political turmoil followed by changing economic orientation and objectives; the Czechs and Slovaks in Czechoslovakia saw internal struggles over diverging interests. However, changes in domestic leadership in the two countries combined with external factors, such as the devastating floods and the sharp increase in world oil prices in the 1970s, brought about the consensus necessary for finalization of the planning stage of the project (Jansky, Murakami, and Pachova 2004).

Uninvited public participation

The history of the effort to defeat the Gabčíkovo–Nagymaros Dam is a classic example of an environmental protest movement reacting against a government decision and trying to change it. However, the story of the movement against the dam is also an example of how the expression of political dissent channelled into environmental concerns became a vehicle for the development of forms of public participation and budding institutions of civil society. Issues of environment, democracy, and dissent became intertwined, and the public found a way for expression of views about government action in a context that did not welcome such expressions of opinion.

To understand how it came about, it is important to understand the role of environmental dissent in the countries of Central and Eastern Europe before 1989. In the 1960s and 1970s, the environment became a "safety valve" subject, in which dissenting opinion could, gradually and cautiously, be expressed. In some countries, there was a slow relaxation of the government's monopoly on information.

Hungary is an example of a country where, at that time, scientists and experts were invited to offer advice to the government on a number of issues that were considered non-sensitive, including the state of the environment. In return, these experts received privileged information and data, as well as some level of financial and technical support not accessible by the general public – a form of exchange in which the State apparently hoped to co-opt the intellectual opposition. However, this discourse was not an invitation to openly criticize the political system; instead, it was a restrained exploration of "grey" issues. Explicit discussions among those who had doubts about the political and economic system continued to be held strictly in private.

In any case, independent of the influence of politics and political systems, debates about how to manage environmental issues and which environmental issues to address were the preserve of technical and scientific experts in Central Europe. Halina Brown and her colleagues have explained this phenomenon in Poland, referring to the "bureaucracy's deeply entrenched administrative resistance to external scrutiny and its disdain for the value of lay persons' contribution to data analysis and policy making" (Brown, Angel, and Derr 2000). They also highlighted the degree to which prevailing cultural mores favoured delegating problems to experts who would solve such problems in closed meetings. Therefore, the initial opposition to the dam project was consistent with this, as it was led by "experts." Addressing Poland in the mid-1990s, Brown and her colleagues also noted that "the independent ecological organizations have no traditions of participative legal process and are too fragmented to mobilize their limited resources necessary for such participation" (Brown, Angel, and Derr 2000), and enterprises continue to be recipients of regulations rather than participants in their formulation. However, there is reason to believe, as noted later, that this situation is slowly changing.

In the same time period, countries of the region were influenced by international environmental movements of the 1970s and undertook formal activities such as writing environmental laws consistent with this movement. Hungary is, again, a good example. Following the 1972 Stockholm United Nations Conference, the Hungarian Academy of Science set up the first major environmental conference in the country (Berg 1999), and the government revised the constitution to include a right to a healthy environment (Enyedi and Szirmai 1998).

The subject had such resonance that authorized political parties began to use environmental issues to revitalize themselves. In the mid-1970s, the National Patriotic Front (NPF), an offshoot of the Communist Party created by Janos Kadar, sought to involve non-communist members of the intelligentsia in the task of governing by providing a forum for discussion. NPF's interest in the Green agenda resulted in the 1976 Environmental Framework Act, the first comprehensive Hungarian legislation that treated environmental protection as a separate issue and established an institutional system for environmental management.

When the mounting financial and economic problems of the 1980s contributed to the waning of the government's attention to environmental issues and the environment continued to deteriorate, environmental advocates felt that the government was unresponsive to their concerns and the recommendations of the scientific community (Enyedi and Szirmai 1998). In addition to discussing technical aspects of environmental protection, they began to question the generally undemocratic character of the system.

The plan to build the Gabčíkovo–Nagymaros Dam became the focal point for much of this anger. The construction of the dam was initiated as a joint project between Czechoslovakia and Hungary in the late 1970s. The ostensible purpose of the dam was seemingly environment favourable, representing an attempt to reduce dependence on foreign sources of energy following the oil crisis.

Shortly after its conception, a number of scientists and experts raised doubts about the benefits of the project. Professionals engaged in an intense dialogue about the subject and attempted to influence government policy. Their inability to influence changes in government policy motivated frustrated scientists to expand their audience. They began to leak information, circulate petitions, and issue newsletters, which (in addition to scientific conclusions) began to examine the social context of the issue. The number of groups focused on this issue grew. One of the most prominent of the groups that sprang up around the dam issue, the Danube Circle, managed to cooperate with the Austrian Greens to stop Austria's plan to provide aid to finance the dam project. For its efforts, the Danube Circle received the Right Livelihood Award (characterized as an "alternative Nobel Prize") in 1985. The Danube Circle remains the most widely recognized environmental group in Hungary today.

Through much of the 1980s, the Hungarian government was unresponsive to the demands of the environmental groups with respect to the dam project. Indeed, at one point, the government tried to gain more control over the movement by forming its own government-sponsored organizations (Waller 1998). The Society for Environmental Protection, founded in 1988, was intended to "provide support for state policy and to channel popular concern into acceptable activities" (Berg 1999).

Although their progress was uneven, the protestors' efforts contributed to the development of a civil sector and, for the first time, some involvement of the public in state affairs. Increasingly through this period, environmental groups were allowed to exist. Although their most outspoken leaders were periodically harassed, they were not arrested. Environmental groups became a place for scientists and for cautious dissidents to come together to challenge, albeit indirectly, the socio-economic structures that produced pollution and shaped environmental policy.

By the end of 1985, a number of independent environmental groups existed, each with its own goals and tactics. Some, like the Danube Circle, the Blues, or the Foundation for the Danube, concentrated on the construction of the Gabčíkovo–Nagymaros Dam and directly confronted the one-party system. Others, formed around universities and regional issues, sought to solve local environmental problems and engaged in collecting and disseminating information (REC 1997).

By 1988, the Communist Party came to be dominated by the reformist faction, which voted to stop the Gabčíkovo–Nagymaros Dam project in response to mass demonstrations. This was a significant victory for the environment movement. However, it also was but one signal of the beginning of a different, more democratic era. Environmental activists had invented an effective form of public participation and had used that both to defeat the dam project and to undermine a government that ruled in defiance of public will, not because of it.

In comparison, the situation in Czechoslovakia was very different, including the process of democratic transition. The different political context in each country meant that ecological movements had different roles. As previously noted, a Hungarian non-governmental organization (NGO), the Danube Circle, played an increasingly public role in voicing awareness about the environmental consequences of the dam and began campaigning against construction in the early 1980s. However, in Czechoslovakia it was more difficult for such a movement to enter the political arena. Slovak environmentalists were isolated and no other group had sufficient information regarding the situation (Šnajdr 1999). Indeed, the very concepts of public participation, civil society, or what is today called a non-profit sector were in their infancy. Moreover, in Slovakia, the way in which the dam was seen differed from that in Hungary: it became a symbol of stronger independence from historical Hungarian influence and of the glories of engineering performance (Sukosd 1997).

Structured public participation and the post-1989 Danube clean-up process

It is useful to compare the mobilization that occurred around the dam project and public efforts to stop it with more recent efforts to clean up the Danube River. There are some distinct differences between the two efforts, and a few odd similarities.

In the mid-1990s, the countries of the Danube River Basin came together to mount a concerted regional effort to address the continued deterioration of Danube River water quality. Many years of human activity and polluted effluents had produced high loads of nutrients and toxics that, in turn, contributed to eutrophication of the Danube and the Black Sea. The countries of the Danube Basin contribute significant amounts of untreated effluent, including faecal coliform bacteria (often raw sewage), organic compounds, and heavy metals. Many of the countries in economic and political transition that affect the Danube watershed have either primitive wastewater-treatment plants in their large cities or none at all, and some have only begun the process of making industry comply with environmental requirements. Mining accidents such as the Baja Mare incident in Romania create additional ecological stress, as does non-point pollution from agriculture and livestock operations.

Efforts to improve the Danubian environment have been organized through regional agreements that establish a large and somewhat cumbersome bureaucratic structure. One of these is the Convention on Cooperation for the Protection and Sustainable Use of the River Danube ("Danube River Protection Convention" or "DRPC"). It was signed in Sofia, Bulgaria, on 29 June 1994 by 11 Danube Riparian States – Austria, Bulgaria, Croatia, Czech Republic, Germany, Hungary, Moldova, Romania, Slovak Republic, Slovenia, and Ukraine – and the European Union (EU). As the funding arms of the EU, the Phare and Tacis programmes have been important supporters of this effort. Through the DRPC, the signatories have agreed on the conservation, improvement, and rational use of surface waters and groundwater in the catchment area; on control of the hazards originating from accidents involving substances hazardous to water, floods, and ice hazards; and on contributing to reducing the pollution loads of the Black Sea from sources in the catchment area (Article 2.1). Through Article 2.2, the states committed to taking "all appropriate legal, administrative and technical measures to at least maintain and improve the current environment and water quality conditions of the Danube river and of the waters in its catchment area and to prevent and reduce as far as possible adverse impacts and changes occurring or likely to be caused."

To coordinate the efforts to achieve sustainable and equitable water management in the Danube Basin, the Convention establishes a number of functional bodies. The Conference of the Parties is the highest-level body, tasked with providing the overall policy for the work under the Convention; it convenes every few years. The International Commission for the Protection of the Danube River (ICPDR) is the Convention's main decision-making body. The Convention also establishes an International Commission Permanent Secretariat, various expert and ad hoc groups, and a supporting body – the Programme Management Task Force (PMTF) – which includes senior representatives of the riparian countries, international organizations, governments, and NGOs.

A few months after the June signing of the DRPC (on 6 December 1994 in Bucharest, Romania), the environment ministers for the Danube countries and the European Commissioner for the Environment adopted the Danube Strategic Action Plan (SAP). This is the road-map for achieving the goals of regional integrated water management and riverine environmental management expressed in the Convention. The SAP sets short-, medium-, and long-term targets for the period 1995–2005, and it defines a series of tasks to meet them, including sector-specific tasks for public authorities at central, district, and local levels; municipal water companies and utilities; industrial enterprises; the general public and NGOs; agricultural enterprises; and the farming community. The activities contemplated by the SAP include capacity building, policy development, and pilot programmes.

The process used by the Danube bodies to clean up the Danube has similarities to problem-solving in the pre-1989 period. The entire effort tended to work from the position that reducing the considerable Danube-pollution load is a technical problem with technical solutions: once the problems are identified and priorities and solutions found, it is enough to identify who (including the NGOs and citizen groups) must undertake which activity to reach a solution. At least in the early stages of the Danube effort, the role of the public was seen from the perspective of generating support for already identified solutions.

In a typical example, the ICPDR issued recommendations on Best Available Techniques that are to be used in priority industrial sectors, as well as implementation timetables. The implementing step is a statement to the effect that recommendations on Best Available Techniques for these important industrial sectors should be made available on a large scale to the administrative authorities, the industry, and the interested public, who should "translate" the recommendations into the "different administrative languages" used in the Danube River Basin. The plan appears to see law creation as a relatively ministerial act, rather than one that reflects democratic processes and all the compromises and delays that democratic procedures inevitably present.

In another example, the ICPDR commissioned a study on cleaner production in Danube countries (Environmental Programme for the Danube River Basin 1994). The study identified responsible sectors including cities, rural towns and villages, industry, energy production and transport, and agriculture, and a number of stakeholders including public authorities, public and private enterprises, the general public both as citizens and consumers, as well as NGOs, "precisely" defining their roles in the Action Plan.

In this sense, the ICPDR has done a good job of identifying priorities, providing a forum for governments to act together, conducting technical studies, and making technical recommendations. However, in this type of approach, public participation is treated as another element that is simply activated by the planning effort. Thus, the ICPDR allocates funding for small grants to NGOs as part of the overall process for awareness raising, which is supported as part of an already formulated programme. One significant exception to this is an ongoing project, funded by the GEF, to develop institutions and procedures for countries to operationalize access to environmental information regimes.

There is a distinct irony in the way in which this approach echoes the pre-1989 decision process. Like the old regime, the overall Danube plan was formulated with little meaningful public participation and reflects a technocratic way of managing pollution. Experts "solve" these problems in closed meetings, not in open democratic processes. Indeed, like the Polish example provided by Brown and colleagues (Brown, Angel, and Derr 2000), one possible interpretation of the plans and their formulation might be said to reflect "resistance to external scrutiny" and "disdain for the value of lay persons" contributing to data analysis and policy-making. Public participation in some ways is an add-on. In this respect it is more similar to the pre-1989 Hungarian effort to generate organizations that seemed to be supporting government's efforts and to enlist representatives of the public to play designated roles in the plan devised for the Danube, rather than a genuinely consultative or participatory process.

Where this differs from pre-1989 approaches is that the Danube organizations are trying to generate popular support for an activity that appears genuinely to be battling against environmental degradation; they are not trying to use their clout to undercut public opposition to an unpopular government action. However, whether this represents an effort to include the public in developing and implementing solutions remains unclear.

Public-participation considerations in the Danube clean-up process

Why should how public participation takes place make a difference?

To understand the nature of the concerns expressed in this chapter, it is necessary to think about the purpose and role of public participation in achieving an environmental agenda. Fundamentally, environmental laws generally require a high level of public engagement and mutual responsibility for regulation to be effective. Far from requiring only a few polluting factories to install control technology, achieving a cleaner Danube River will demand specific (sometimes inconvenient and often costly) behavioural changes from a diverse group of people throughout the river basin. Poor water quality is the responsibility of numerous non-point sources (including farmers, gardeners, and urban residents) as well as industrial point sources. This need for public engagement is particularly important, as one stated goal is to reduce pollution from diffuse and widely distributed non-point sources.

To achieve this requires widespread knowledge, commitment, and mobilization. There is reason to believe that this type of outcome is perhaps more likely to succeed when the large numbers of people who must undertake these activities have respect for, and confidence in, the decision-making system and are willing to follow the law. Sociological studies in the United States suggest that confidence is built when the process of setting the rules is perceived as fair and the public feels that its views have been heard (Davies and Mazurek 1998). Indeed, there is some reason to believe that, under such conditions, people who disagree with the final decision are more likely to go along with it. Conversely, "when legitimacy diminishes, so does the ability of legal and political authorities to influence public behavior and function effectively" (Tyler 1990). The history of mandated laws may be one reason why there was little compliance with environmental laws of previous regimes in Central and Eastern Europe.

In addition to developing a belief that the laws fairly represent shared concerns, good environmental rules benefit from a healthy flow of information to and from government as they are formulated. Government lawmakers and environmental-protection officials are rarely omnipotent, and the problems they face are complex. When they obtain data, experience, and opinions from industry and the affected public, they can develop more realistic and achievable requirements. However, to engage in this dialogue, the government must be willing to communicate its decision-making process, what data it is relying on, and what it wants to achieve. It must listen, and this exchange should take place while rules are being formulated, not afterwards.

There are other reasons to engage the public at large. Despite the emphasis of the Danube process on the development of expert opinions and technical solutions and of the presentation of these to the public for affirmation, environmental decisions involve a great deal more than good science. This means that it is not enough to simply engage experts. Environmental protection is, in part, a process of determining what level of risk that society is willing to accept or tolerate, and what it is willing to spend to reduce those risks. Families concerned about their drinking-water, and mothers with asthmatic children who breathe polluted air, contribute important intuitions about the human context and tolerance for risk (Fiorino 1989). Even technical tools of environmental decision-making, such as risk assessment and cost–benefit analysis, include significant subjective judgments that are best made with explicit attention to public values in consultation with the public (Fiorino 1989).

These are the arguments for involving the public at an early stage in environmental decision-making and for open, transparent processes. Reversing the order – and engaging in consultation after a decision has already been made – may produce "cleaner" or more scientifically "correct" decisions. In such a system, the decisions would not be complicated by popular concerns or delayed by democratic processes, which can be time-consuming and expensive. However, formulating policy in this way runs the great risk of eroding public trust and belief in the legitimacy of the decision-making process. And "buy-in" (i.e. obtaining whole-hearted general public agreement/acceptance) is essential because, in the end, the rules must have advocates and workers if they are to be implemented and, in short, to succeed. Disputes regarding policy and science inevitably will be resolved by compromises, and few of the participants in the process are likely to perceive themselves as outright winners or losers. Nevertheless, even those who disagree with the final result might be persuaded to work together on implementation and not to ignore or sabotage the decision.

The danger for the Danube process is that the best-formulated plans may never be implemented. Much like the pre-1989 laws, they could sit on the books largely unused. It is to their credit that the Danube planners have recognized this challenge and built a public-participation component into the plan, intended to mobilize domestic will and enthusiasm. Their challenge, however, will be to make sure that they have not chosen a flawed approach.

Why resist public participation?

Why would officials resist public participation? Why would decision makers prefer public relations to public participation? The answers to these questions cannot be entirely attributed to pre-1989 practice. Even in long-standing democracies, the challenge to include the public in decision-making is ongoing. Merely writing laws is not sufficient. In the United States, environmental advocacy groups have had to be vigilant: they bring lawsuits to ensure that laws are implemented in the ways envisaged by the law drafters, and inform the media to help enforce (albeit informally) their right to make their views known in a timely manner.

Some decision makers see public participation as inefficient and a nuisance, as steps to involve the public can delay the decision-making process and increase expenses. It takes time to be consultative; meanwhile, the environment is deteriorating. Moreover, public opinion can be frustrating for environmental experts. The public often demands action on environmental problems that experts might rate as a lower environmental priority. Often, as a result, public funds are spent on environmental problems that do not present the greatest hazards.

On the other hand, environmental protection everywhere works at a seemingly glacier-like pace, and the experience with "efficiently" derived environment requirements has not necessarily resulted in better or faster clean-up than democratically derived standards – indeed, quite the contrary: many countries have well-drafted and comprehensive environmental laws but little implementation. Experience in democracies seems to suggest that, in the long run, public involvement can be an important factor in helping move the environmental agenda forward and giving life to laws.

The international funding agencies also recognize this, but are experiencing some difficulties in implementing solutions. The GEF Operational Strategy on international waters recommends bottom-up participation with NGOs and communities. However, apparently, the most recent Project Implementation Review (PIR) authored by GEF's Monitoring and Evaluation (M & E) Unit contains strong criticism of GEF International Waters projects for not engaging citizens and NGOs in bottom-up processes.

A better process? Could genuine public engagement in the Danube process lead to a cleaner Danube River?

Engaging the public *ex post facto* is better than not at all. Nevertheless, such practice more resembles public-relations efforts to shape public opinion than it does a democratic process in which the public plays a role in selecting which environmental challenges will be the subject of government attention, the level at which the environmental challenges will be addressed, and how much is to be spent. Clearly, there is a balance to be struck between a tidy process to find solutions to complex environmental problems, and a wide-open public process. At minimum, however, there must be some evidence that a sufficient number of members of the public are willing to undertake the expensive and inconvenient chores of environmental protection.

In general, bodies such as the ICPDR that implement programmes (for example, cleaning up the Danube River) would be well advised to consider a broader definition of public participation and to give a stronger charge to the governments with which they interact, to figure out constructive ways to engage the public. If it did so, the ICPDR would not be acting alone. Several of the countries of the Danube region, such as Hungary and Slovenia, are EU accession countries. They are already formally committed to various forms of meaningful citizen participation in carrying out environmental and pollution-reduction goals, chief among them the right of their citizens and others to obtain environmental information on request. This kind of information can fuel non-governmental actors. An effective NGO community with access to relevant environmental information can be a strong catalyst for environmental change, and can help to mobilize the large number of non-governmental actors who must act in order to reach the goals set out in the Danube Strategic Action Plan.

Conclusions

Prospectively, the Danube process is giving greater weight to the values of public participation in environmental decision-making. Indeed, a considerable percentage of the funding of the second (and, presumably, last) tranche of funding for this effort will be spent on efforts to increase a public voice in the Danube clean-up. The newly appointed Executive Director of the ICPDR, Philip Weller, came from an NGO background. Earlier, he had directed a US–Canadian effort (Great Lakes United), which was a unique binational coalition of interest groups that included local municipalities, research organizations, businesses, and NGOs in both Canada and the United States, and which focused on cleaning up the North American Great Lakes (see chap. 6, this volume). Immediately prior to coming to the ICPDR, he was Director of the Worldwide Fund for Nature's Danube Carpathian Programme, 1995–2002. This is all to the good. There is every reason to believe that he will show great sensitivity to the need to include a public voice in Danube decision-making.

The most recent chapter of the Danube River case is also an interesting example of conflict resolution in international water systems. Following the governmental changes in 1989, Hungary and Czechoslovakia appealed to the International Court of Justice (ICJ) in 1993 to resolve their dispute (ICJ 1997). The resulting decision demonstrated that the ICJ could be instrumental in resolving conflicts among riparian states of an international water system, including issues of the environmental aspects of a project, rather than just conflicts about the sharing of water resources among riparian states. Significantly, during the ICJ's deliberation of the Gabčíkovo–Nagymaros case, the ICJ for the first time accepted *amicus curiae* (friend of the court) briefs from interested NGOs. It is also interesting to note that the judgement by the ICJ in many parts referred to the then-pending 1997 UN Convention on Non-Navigational Use of Water Resources in International Water Systems (Nakayama 1998).

Whether this represents a trend for resolution of inter-State environmental and water disputes is not clear. In the period since the ICJ's decision, there has been no visible movement to submit another similar case to the ICJ. The Danube River case may be unique in that the two basin countries agreed in a relatively short time to have the issue resolved in this way. In part, this motivation arose from their desire to apply for membership of the EU, which stipulated that there should be no outstanding conflicts between the two nations. The EU's political leverage and the procedures of the ICJ filled the post-socialist institutional vacuum in which the two countries found themselves, following the disintegration of their former structures for regional political security and economic cooperation. It is safe to assume that other cases will require a similarly strong motivation to resolve such conflicts quickly.

More recently, both sides agreed that the decision regarding whether to complete the final stage of this originally joint investment is not a political but a technical issue. The ICJ did not allow environmental arguments in 1997, but it gave both countries the opportunity to negotiate in good faith in the light of the current facts and taking all necessary measures to ensure the achievements of the objectives of the Treaty of 1977. Ultimately, the EU and the ICJ left the water-management issues and their actual and potential environmental threats for Hungary and Slovakia to resolve. Accordingly, the only criteria for solution seem to be the economic considerations of the sides involved.

Another significant aspect of the Danube River case is that the 1997 UN Convention may take the position of a *de facto* code of conduct for riparian states in an international water system, despite the fact that the Convention was not adopted until late 1997, after the ICJ's judgement.

REFERENCES

Bell, Ruth Greenspan, and Sándor Fülöp. 2003. "Like Minds? Two Perspectives on International Environmental Joint Efforts." *Environmental Law Reporter* 33:10344; also Resources for the Future (RFF) discussion paper on Internet: ⟨http://www.rff.org/Documents/RFF-DP-03-02.pdf⟩ (visited 14 October 2003).

Bell, Ruth Greenspan, Jane Bloom Stewart, and Magda Toth Nagy. 2002. "Fostering a Culture of Environmental Compliance Through Greater Public Involvement." *Environment* 44(8):34–44. October.

Berg, Marni M. 1999. "Environmental Protection and the Hungarian Transition." *Social Science Journal* 36(2):227.

Brown, Halina Szejnwald, David Angel, and Patrick G. Derr. 2000. *Effective Environmental Regulation: Learning From Poland's Experience*. Westport, Conn.: Praeger.

Davies, J. Clarence, and Jan Mazurek. 1998. *Pollution Control in the United States: Evaluating the System*. Washington, DC: Resources for the Future.

Environmental Programme for the Danube River Basin, Tasks Force for the Programme. 1994. *Strategic Action Plan for the Danube River Basin 1995–2005*. Brussels: Environmental Programme for the Danube River Basin.

Enyedi, György, and Viktória Szirmai. 1998. "Environmental Movements and Civil Society in Hungary." In Andrew Tickle and Ian Welsh (eds). *Environment and Society in Eastern Europe*. New York: Longman. 146–155.

Fiorino, Daniel J. 1989. "Technical and Democratic Values in Risk Analysis." *Risk Analysis* 9(3):293–299.

Fitzmaurice, J. 1998. *Damming the Danube: Gabčíkovo and Post-Communist Politics in Europe*. Boulder, Colo.: Westview Press.

International Court of Justice. 1997. Gabčíkovo–Nagymaros Project. (Hungary/ Slovakia). 1997 ICJ 6. 25 September. Internet: ⟨http://www.icj-cij.org⟩ (visited 21 July 2003).

Jansky, L., M. Murakami, and N.I. Pachova. 2004. *The Danube: Environmental Monitoring of an International River*. Tokyo: UNU Press.

Lipschutz, Ronnie. 1997. "Damming Troubled Waters: Conflict over the Danube, 1950–2000," *Intermarium* 1(2). Internet: ⟨http://www.columbia.edu/cu/ sipa/REGIONAL/ECE/dam.html⟩ (visited 12 October 2003).

Nakayama, M. 1998. "Possible Role of International Organizations in Management of Water Resources and Abatement of Conflict in International Water Systems." *Journal of Japan Society of Hydrology and Water Resources* 11(7):723–731 [in Japanese].

REC (Regional Environmental Center for Central and Eastern Europe). 1997. *Problems, Progress and Possibilities: A Needs Assessment of the Environmental NGOs in Central and Eastern Europe*. Szentendre, Hungary: REC.

Šnajdr, Edvard. 1999. "Green Intellectuals in Slovakia." In Bozoki Andras (ed.). *Intellectuals and Politics in Central Europe*. Budapest: Central European University Press, 207–224.

Sukosd, Miklos. 1998. "The Gabčíkovo–Nagymaros Dam: Social, Political and

Cultural Conflicts." In A. Vlavianos-Arvanitis and J. Morovic (eds). *Biopolitics: The Bio-Environment*. Bratislava: Biopolitics International Organization.

Tyler, Tom R. 1990. *Why People Obey the Law*. New Haven: Yale University Press.

Waller, Michael. 1998. "Geopolitics and the Environment in Eastern Europe." *Environmental Politics* 7(1):29–53.

Wedel, Janine. 1986. *The Private Poland: An Anthropologist's Look at Everyday Life*. New York, NY: Facts on File.

6

Citizens working across national borders: The experience in the North American Great Lakes

John Jackson

Introduction

The ever-increasing number of agreements among governments for the protection and enhancement of the environment in shared waterbodies that cross international boundaries shows the wide recognition of the need to work cooperatively internationally on environmental issues. Many treaties and international agreements lay out principles, goals, action plans, and monitoring mechanisms for shared waterbodies. These agreements usually include international governmental institutional arrangements.

Much more than international governmental mechanisms are needed, however. Over and over again, it has been shown that citizen action is critical to the protection of our waterbodies. Those living around these bodies of water are the ones who are most directly affected by them, who share in their use and enjoyment, and who value these waters for their multitude of essential and delightful facets. They bring the most passion, determination, and creativity to the search for solutions to the problems.

Citizens' groups usually arise around specific issues in specific locales. Nevertheless, citizen action is also needed on an ecosystem-wide basis. This means that when a waterbody crosses international boundaries, mechanisms are needed that support basin-wide international citizens' actions. For it is only by bringing pressure to bear in a united way on all responsible government jurisdictions simultaneously that there is any hope of protecting shared water systems. Mechanisms are needed to encourage citizens to get to know each other, to discover and define their shared goals, and to speak loudly and clearly, in a non-national way, to responsible government authorities and industry when they are being negligent or rapacious.

Great Lakes United is an example of an organization that helps citizens to achieve their potential in fulfilling this kind of role. This chapter outlines the formation and work of Great Lakes United. It also discusses the problems encountered by organizations that try to play this kind of international role and the ways in which Great Lakes United has tried to address these problems. This chapter draws upon the author's experience as a Board member of Great Lakes United throughout all of its 21 years, including serving as President for six years (he is currently Director Emeritus).

The North American Great Lakes

Almost 20 per cent of the world's fresh water is in the North American Great Lakes (see fig. 6.1), the largest system of fresh surface water on the globe. In total, the Great Lakes hold a volume of about 23,000 cubic kilometres (5,500 cubic miles) of water.

The five Great Lakes and their connecting channels and the St Lawrence River create one integrated ecosystem stretching 4,000 km (2,500 miles) from the heart of the North American continent to the Atlantic Ocean. The area drained by the Great Lakes and their connecting rivers is approximately 520,000 square kilometres (201,000 square miles) (Fuller and Shear 1995).

This vast basin contains a wide variety of natural habitats and is home to a rich diversity of wildlife and plants. The Great Lakes are also home to over 33 million people: one-quarter of Canada's population and approximately 10 per cent of the United States population lives within the Great Lakes Basin. An additional 4.5 million Canadians live near the St Lawrence River. Approximately 350,000 of the people in the Great Lakes–St Lawrence River Basin are descendants of the first peoples of the Great Lakes Basin. The Great Lakes–St Lawrence River system covers many government jurisdictions: these include 2 national governments (Canada and the United States); 10 provinces and states (Illinois, Indiana, Michigan, Minnesota, New York, Ohio, Ontario, Pennsylvania, Quebec, and Wisconsin); 110 First Nation and Tribal governments; and hundreds of municipalities.

The Great Lakes and St Lawrence River system is an ecosystem suffer-

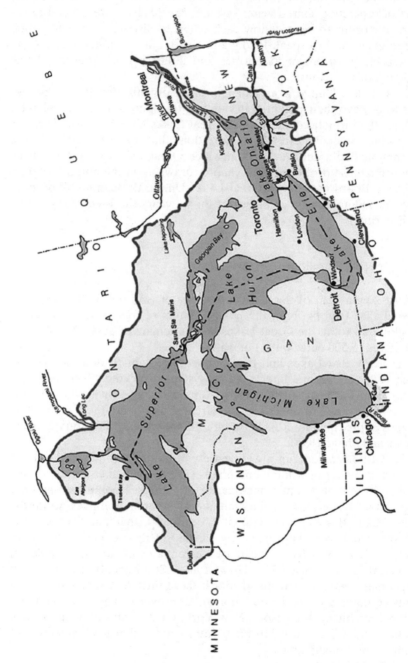

Figure 6.1 Map of the North American Great Lakes (Source: International Joint Commission 2003)

ing from many stresses. In the 1950s, 1960s, and 1970s, awareness of serious problems began to escalate. The lake trout were being devastated by the eel-like sea lamprey (*Petromyzon marinus*), which had made its way into the Great Lakes as a result of the opening of the St Lawrence Seaway. Dead fish were being washed up on the shores of Lake Erie as a result of excess algal growth in the lake. The Cuyahoga River in Cleveland caught fire because of contaminants floating on its surface. Nine hundred families were moved from Love Canal in Niagara Falls, New York, because of a leaking hazardous-waste dump. Other dumps along the Niagara River were found to be leaking dioxins into the river and from there into Lake Ontario. Eagle populations were plummeting, and cormorants with seriously deformed bills were being found. Their health problems were the result of persistent, toxic, bioaccumulative contaminants pouring into the lakes. People in distant water-starved parts of the North American continent wanted to divert water from the Great Lakes to quench their ever-growing thirst. Wetlands were being drained and built upon, reducing the total extent of wetlands by four-fifths of that 150 years previously. The image of the Great Lakes was suddenly becoming a negative one, perhaps best epitomized by William Ashworth's book released in 1986 entitled *The Late, Great Lakes*.

In response to this growing awareness of problems, governments began working together across international borders. The Great Lakes Fishery Commission was formed in 1955 to try to suppress the sea lamprey populations. The International Joint Commission (IJC), which had been formed in 1909 when the Boundary Waters Treaty was signed, was given new responsibilities under the Great Lakes Water Quality Agreement (GLWQA) signed in 1972 by the US and Canadian governments. At first, its role focused on eliminating the pollutants that were creating excess algal growth in such places as Lake Erie. The Agreement was revised in 1978 to include a focus on virtually eliminating pollution by persistent toxic substances. The Great Lakes states and provinces committed themselves to work together to control diversions of water from the Great Lakes Basin and to reduce water use within the basin when they signed the Great Lakes Charter in 1985.

Although significant progress has been made as a result of this increased binational attention, serious environmental crises remain in the Great Lakes–St Lawrence River Basin at the beginning of the twenty-first century. The United States General Accounting Office concluded in 2003 that "despite early successes in cleaning up the nation's water, the Great Lakes Basin continues to face significant environmental challenges" (USGAO 2003). In an assessment of the state of the Great Lakes in 2002, government scientists concluded that drinking-water quality was good and that it is becoming safer to eat fish caught in the Great Lakes.

However, they found that conditions for 70 per cent of their indicators of ecosystem health were "mixed," "mixed–deteriorating," or "poor." Among other items, mixed or poor conditions were noted for air quality, and for the condition of some wildlife and fish such as lake trout, salmon, preyfish, amphibians, wetland birds, and native mussels (Environment Canada and US Environmental Protection Agency 2002).

In addition, new challenges are becoming evident in the Great Lakes–St Lawrence River Basin. Dead zones have once again appeared in Lake Erie. Water levels are falling, which may be linked to climate change. New types of substances such as hormone mimics, including substances such as flame retardants and pharmaceuticals, are harming wildlife and human health. New invasive species of animals, plants, and microorganisms have entered the Great Lakes and are disrupting the ecosystem; dozens of new invasive species are identified as being on their way. Increased urban and developmental sprawl is destroying more of the shorelines and the precious habitat.

The need for concerted effort across the Canadian–United States border has by no means diminished.

The formation of Great Lakes United

During the 1960s and 1970s, numerous citizens' and environmental groups arose around the Great Lakes. Increasingly, these groups became involved in Great Lakes issues as the profile of these issues rose. In some cases, it led to groups collaborating with environmental groups on the other side of the international border. This was most true in the Niagara River area, as groups tried to confront the massive quantities of toxic wastes leaking out of the dumpsites on the US side of the river.

In 1981, one of the largest groups in the Great Lakes Basin – the Michigan United Conservation Clubs with a membership of approximately 200,000 in Michigan – obtained funding from the Joyce Foundation to explore the creation of a binational, basin-wide organization. The talks that took place through this project resulted in a founding meeting in May 1982 on Mackinac Island, Michigan, near the centre of the Great Lakes Basin. Fifty-five citizen activist delegates from Ontario, Quebec, and the eight Great Lakes' states attended that first meeting. At this meeting and a follow-up meeting attended by 110 people six months later in Windsor, Ontario, the details of a new basin-wide organization were hammered out.

Whereas everyone agreed with the need for a basin-wide citizens' organization, there was heated controversy about the nature of the new organization. Two speakers at the Mackinac meeting starkly laid out the differences. Jay Reed, of the National Audobon Society, maintained: "The need here is not one of creating another advocacy group, but one of supplying information and strengthening existing advocacy groups." On the other hand, Barry Freed (a.k.a. Abbie Hoffman), of Save the River, declared: "Information is the power, but we have to go beyond the information and start getting into advocacy. I don't want to leave here with just a box of fudge and a newsletter!" (Great Lakes United 1992). The differences articulated themselves particularly in the debates over whether the new organization should have a strong executive director and board independent of the member groups, and over concerns about potential domination of the new organization by a few large membership groups on the US side of the basin.

Jack Manno, Director of the Great Lakes Research Consortium and long-time follower and participant in Great Lakes issues, described the impetus for the intense feelings around the debate over these issues in the following way:

[Some] argued for the formation of a strong regional organization that could advocate positions with a single voice representing the scores of groups with environmental portfolios. Many others saw a need for a central information clearinghouse and networking node for existing groups, but feared a new organization would compete with them for influence, funding, and members. The issues of organizational structure were mirrored in leadership styles. [Some] worked in, or were used to, organizations with top-heavy, authoritarian decision-making styles, whereas many of the environmental organizations involved early in the Great Lakes coalition building promoted a more egalitarian, participatory style. (Manno 1993)

The intense debate ultimately resulted in the formation of an organization more focused on coordination and support of the voice of other groups than on being a strong independent advocacy group. A decentralized structure was created in which the direction of the organization was to be set by policy resolutions passed by the member groups at annual meetings, and in which the power focused on a part-time, non-paid president rather than the executive director. In addition, a task-force structure was set up on issue areas. These task forces were to be a mechanism through which members could work together to develop policies and programmes. Today, task forces are organized on the topics of clean production, green energy, healthy communities, sustainable waters, and biodiversity and habitat protection.

To avoid the possible domination of a few large groups or one part of the Great Lakes Basin, each member group was given one vote at annual meetings, regardless of the size of the group's membership or budget.

The by-laws also specified that there had to be board members from each lake and the St Lawrence River. The by-laws designated five seats for Canadians, including a requirement for at least one of these to be from Quebec, and five from the United States. Ten years later, the by-laws were revised to designate two seats for members of First Nations or Tribal organizations.

After all this had been worked out, the issue of a name for the new organization was left to be decided. Bob Boyce, of the New York State Conservation Council (who would later be elected as the organization's first President) made the following proposal: "I suggest we call ourselves 'Great Lakes United.' The name says we each maintain our autonomy, but we're working together for a common cause. And its acronym is 'GLU' – the group that holds the lakes together." (Great Lakes United 1992).

And so a new organization had been formed with a membership spanning the vast expanse and diversity of the Great Lakes–St Lawrence River Basin. The tensions around its formation were to be questions that repeatedly arose during its history.

Over the past 22 years, Great Lakes United has developed a diverse membership of approximately 170 community-based and regional organizations from the United States, Canada, and the First Nations and Tribes. This includes a wide range of environmental, labour, conservation, and community groups.

The work of Great Lakes United

Over the past 22 years, Great Lakes United in cooperation with its member groups has worked on numerous issues of concern to citizens throughout the region. Some examples of these issues are listed below:

- Stopped US Army Corps of Engineers' proposals to extend the shipping season into the winter months, which would have increased environmental damage on shorelines and increased the risk of spills.
- Led public opposition to proposals for diversions of Great Lakes water to Pleasant Prairie, Kenosha, and the Crandon Mine (Wisconsin); and Akron (Ohio).
- Led citizen input into the development of the Great Lakes Charter among the Great Lakes provinces and states and, 15 years later, into the development of Annex 2001 to this Charter. This Charter places certain requirements on proposals to divert water out of the Great Lakes and around major new consumptive uses of water within the basin, and the Annex developed a new regime for making decisions regarding water use.

- Led citizen input into the implementation and review of the GLWQA between Canada and the United States. This includes leading citizen involvement in the activities of the IJC and the publication of *A Citizens' Guide to the Great Lakes Water Quality Agreement*.
- Held workshops for citizens working on the clean-up of the designated "areas of concern" or toxic hotspots around the Great Lakes and St Lawrence River. These were aimed at helping citizens to develop a strategy on how to best address the problems that they were encountering.
- Co-led a basin-wide zero-discharge campaign focused on a call for the phasing out of the use of chlorine as a feedstock.
- Led a successful basin-wide effort to defeat proposals for increased control structures in channels connecting the Great Lakes.
- Participated in the development of the Great Lakes Wetlands Conservation Plan.
- Led input by Canadian environmental groups into Canada's Strategic Options Process to develop regulations on 25 chemicals.
- Held a workshop on the stocking of fish in the Great Lakes, followed by the release of a publication on this topic.
- Coordinated citizen input into the development of legislation and strategies to prevent the introduction of invasive species into the Great Lakes Basin.
- Co-wrote and published a report on the threats to Great Lakes water quantities and flows entitled *The Fate of the Great Lakes: Sustaining or Draining the Sweetwater Seas* (Farid, Jackson, and Clark 1997).
- Held public hearings around the basin on Great Lakes issues to facilitate citizen input.
- Worked to prevent the building of more pipelines across the lakes and to prevent the drilling of more oil and gas wells under the lakes.
- Organized educational days in Ottawa and Washington, DC – the two federal capitals – to make legislatures more aware of the needs of the Great Lakes Basin.
- Participated in the development of the Stockholm Convention, the international treaty on persistent organic pollutants.
- Developed a campaign to persuade the automobile industry to address the problems created by their use of mercury switches.
- Coordinated the development of *The Great Lakes Green Book: A Citizens' Action Agenda for Restoring the Great Lakes–St Lawrence River Ecosystem* (Great Lakes United 2003).
- For 15 years, has been publishing a quarterly newsletter, which highlights the activities of citizens' groups around the Great Lakes Basin.

To give a more in-depth understanding of Great Lakes United's work, two of these activities are discussed in the next two sections. The first is Great Lakes United's campaign around the renegotiation of the GLWQA in 1987; the second is the development of the Citizens' Action Agenda during 2002 and 2003.

The renegotiation of the Great Lakes Water Quality Agreement

The renegotiation of the GLWQA in 1987 was a prime example of Great Lakes United's success in pulling together the basin's residents to set the agenda. The GLWQA between Canada and the United States was negotiated in 1972 and 1978 under a veil of diplomatic secrecy. In contrast, the citizens in the Great Lakes and St Lawrence River Basin played a major role in the development of the 1987 changes to the Agreement. This expanded public role developed as a result of Great Lakes United's initiatives (Jackson and Eder 1991).

The terms of the 1978 GLWQA required that the Agreement be reviewed in 1987. Because Great Lakes United believed that those most capable of judging the successes and failures of the Agreement were the residents of the basin, they set up the Citizens' Hearings on Great Lakes Water Pollution. These hearings were designed to give members of the public an opportunity to express their concerns and to present proposals for improving the GLWQA. Great Lakes United committed itself to convey these concerns to the government bodies responsible for reviewing progress in implementing the Agreement.

Great Lakes United organized 19 hearings in locations scattered across the wide expanse of the lakes between Duluth, Minnesota, and Montreal, Quebec. Over 1,200 people attended the hearings; 381 made presentations or presented statements by mail. The presenters came from a wide range of organizations and backgrounds, including citizens' groups, aboriginal groups, environmental groups, long-term residents, fishing and hunting associations, schoolchildren, wildlife groups, labour, industry, chambers of commerce, clergy, academics, political parties, employees of the IJC, and federal, provincial, state, and municipal elected officials and civil servants.

The message conveyed by these people was almost unanimous. The lakes' residents saw zero discharge of persistent toxic substances as an imperative for their future and the future of the lakes. They condemned the governments for failing to live up to the objectives in the GLWQA and for (in some instances) not even enforcing their own laws and regulations. They called on the governments to be more aggressive in protecting the lakes. They also insisted on being more directly involved in decision-making on issues that affect the quality of the lakes and the quality of their lives.

As a result of the hearings, Great Lakes United concluded that the root of the problem was a lack of political will. Intense, ongoing pressure is the mechanism that produces political will. They concluded that lack of information and lack of mechanisms for holding the governments accountable to the public have militated against the generation of such public pressure.

Great Lakes United documented the findings from its hearings in a report entitled *Unfulfilled Promises: A Citizens' Review of the International Great Lakes Water Quality Agreement* (Great Lakes United 1987). This report conveyed the concerns and hopes of the basin's residents. *Unfulfilled Promises* also detailed their ideas for cleaning-up and protecting the Great Lakes.

This report was strategically timed to come out just as the Canadian and US governments were beginning to review the Agreement. In this way, the public helped to take the lead in setting the review agenda.

Several meetings were held between Great Lakes United and Canadian and US government officials to review the public's concerns as expressed in *Unfulfilled Promises* and to discuss the governments' plans for review. As a result of Great Lakes United's persistent articulation of the public's voice developed during the tour of the lakes, copies of preliminary government proposals for amendments to the GLWQA were released to Great Lakes United for comment. The governments also gave the public a chance to review drafts at seven public hearings (no such hearings had led up to the 1972 and 1978 Agreements).

Finally, five representatives of environmental groups, three of which were from Great Lakes United, were granted observer status in the negotiations between the Canadian and US governments. This meant that members of the public were part of each negotiating team, helping to develop positions, strategies, and language for amendments. They also sat at the negotiating table and were called on for comment and input throughout the negotiations.

This high degree of public input into usually secretive international negotiations was virtually unprecedented. The uniqueness of this situation was emphasized in a letter from Canada's Minister of the Environment to Canada's Secretary of State for External Affairs. He wrote, "Although I realize that it is unusual to involve the public directly in government-to-government consultative sessions, I believe that the presence of GLU would be useful" (McMillan 1987).

The unique cross-border nature of Great Lakes United meant that they played a special role in those negotiations. One observer described it as follows:

Their [Great Lakes United's] very presence, on both sides of the negotiating table, affirmed the cross-boundary nature of the issues and challenged the presumption of separate national interests built into the structure of binational negotiations ... On a more practical level, their knowledge of both sides' positions, their familiarity with most of the negotiators, and their appreciation for the inter- and intra-agency politics on both sides of the border, gave them a more heightened understanding of the issues than most members of either delegation. (Manno 1993)

Another outcome of Great Lakes United's public hearings and participation in the renegotiation of the GLWQA was that activists throughout the Great Lakes–St Lawrence River basin became more aware of the importance and potential of the Agreement and became advocates for its implementation. They, in effect, adopted this document as their Bible, which guided them as they pushed for the clean-up and protection of the basin: "The existence of the community continually advocates and holds legitimate the goals of the Agreement" (Botts and Muldoon 1997).

The Citizens' Action Agenda

In 2000, the US Congress allocated US$7.8 billion over 10 years for a large-scale effort to restore the Florida Everglades. This happened at a time when government funding for Great Lakes programmes had been undergoing substantial cuts for several years. Envy of the money allocated to the Everglades escalated efforts to bring a larger focus and attached financial resources to the Great Lakes.

In 2002, under the Great Lakes Legacy Act, Congress authorized US$250 million over five years for clean-up of contaminated sediments in the Great Lakes. However, the governments estimate that US$7.4 billion is needed for the clean-up of contaminated sediments and wastewater infrastructure on the US side of the Great Lakes (IJC 2003). For the fiscal year 2004, the US President recommended a budget that includes US$15 million of this authorized money. At that rate, over five years the total actual expenditures would be only $75 million of the $250 million authorized.

Funding for Great Lakes programmes on the Canadian side of the Great Lakes also underwent cuts during the 1990s. After an assessment of the Canadian Federal Government's Great Lakes programmes, Canada's Commissioner of the Environment and Sustainable Development concluded:

Important matters are adrift. Declining and unstable funding to federal departments has significantly impaired their ability to achieve their environmental objectives and meet Canada's international commitments. (Gelinas 2001)

Cuts during the same period by the Ontario provincial government were just as damaging.

Some funding restoration has occurred in Canada in the last few years. In 2000, the Canadian Federal Government allocated CAD$30 million over five years for Great Lakes work. In 2002, Ontario allocated CAD$50 million over five years for the Great Lakes. This compares with the governments' estimate that it will take CAD$1.9 billion for clean-up of contaminated sediments on the Canadian side of the Great Lakes (IJC 2003).

On the US side of the Great Lakes, there are currently several efforts to try to bring greater attention to Great Lakes programmes and to obtain focus and support equivalent to that of the Florida Everglades. For example, the Great Lakes Commission, an organization of the eight Great Lakes states, has developed the Great Lakes Program to Ensure Environmental and Economic Prosperity to try to influence the federal agenda and funding. Likewise, the Great Lakes Task Force, which is made up of the members of the US Congress who were elected from the Great Lakes region, is working to develop such initiatives.

On the Canadian side, Canada and Ontario signed the Canada–Ontario Agreement Respecting the Great Lakes Basin Ecosystem in 2002. They are now developing action plans on how to implement this Agreement.

Because Great Lakes United was concerned that these initiatives were being developed without adequate opportunity for input from citizen activists in the Great Lakes Basin, the organization decided that the activists in the Great Lakes–St Lawrence River Basin needed to set their own agenda for the Great Lakes:

Knowing that some of our government leaders are now considering a major investment in Great Lakes restoration, Great Lakes environmental, conservation, and labor groups developed the action agenda to help guide those efforts from a citizen point of view. Together these individuals and groups have developed a set of goals, targets and strategies for addressing the many challenges facing the Great Lakes–St Lawrence River ecosystem in the twenty-first century. (Great Lakes United 2003)

Great Lakes United spent two years pulling together *The Great Lakes Green Book: A Citizens' Action Agenda for Restoring the Great Lakes–St Lawrence River Ecosystem* (Great Lakes United 2003). It was a classic example of how a basin-wide organization is uniquely able to pull together a voice for an entire ecosystem.

Numerous groups in the Great Lakes–St Lawrence River Basin co-operatively developed the Citizens' Action Agenda under the co-ordination of Great Lakes United. Great Lakes United used several methods to develop this Agenda.

First, Great Lakes United surveyed its membership to discover which issues the groups felt it most important to focus on. It also came to a decision on issue areas by reviewing the documentation from 10 public hearings held by Great Lakes United around the basin in 1998. The testimony of the (approximately 325) people who spoke at these hearings was summarized in *Citizens Speak* (Great Lakes United 1998).

Great Lakes United then pulled together materials that various citizens' groups had already created on these issue areas and formed working groups of people already working on each issue to develop drafts of the Citizens' Action Agenda. Approximately 40 people from different groups were involved in the actual writing of the document. In addition, many more people provided input by reviewing drafts. Great Lakes United dedicated most of its 2002 annual meeting in Chicago to discussion of an early draft of the agenda. Throughout the following year, many more people reviewed drafts.

The Citizens' Action Agenda was then released at Great Lakes United's 2003 Annual Meeting in Sault Ste Marie, Ontario. The Citizens' Action Agenda contains approximately 150 recommendations to federal, provincial, state, tribal, first-nation, and municipal governments. The areas covered in the recommendations are toxic clean-up, clean production, green energy, sustaining and restoring water quantities and flows, protecting and restoring species, protecting and restoring habitat, and water- and air-quality regulations. Examples of some of the recommendations in the Citizens' Action Agenda are as follows:

- Complete clean-up and restoration activities in all 43 Great Lakes toxic hotspots or "Areas of Concern" by 2015.
- Adopt "extended producer responsibility" legislation requiring manufacturers to be fully responsible for the recovery and safe disposal of high-risk waste associated with their products, including automobiles, electronics, and packaging products.
- Increase the amount of electricity that must be generated by new, clean, renewable sources (i.e. wind and solar power) to 20 per cent by 2020, accompanied by a phase-out of coal and nuclear power plants.
- Adopt by 2004 a binding agreement for regulating the withdrawal of water from the Great Lakes system that is based on sound science for protecting the ecosystem.
- Phase out shipping and navigation practices (including the indiscriminate dumping of ballast water) that allow for the continued introduction of invasive species that are threatening the survival of native species in the Great Lakes.

- Set strict urban boundaries that remain fixed for at least a 20- to 30-year period to stop low-density urban sprawl.
- Increase the amount of protected wetlands by a million acres by 2025.
- Support conservation initiatives that maintain or restore interconnected habitats for Great Lakes wildlife.

Now that the Citizens' Action Agenda has been pulled together, Great Lakes United will be coordinating the development and implementation of a campaign to persuade governments to adopt its components.

Challenges of a cross-border citizens' organization

Developing and operating a citizens' organization that stretches across such a wide geographic area as the Great Lakes–St Lawrence River Basin and into two countries creates a set of difficulties that differ from those usually encountered by citizens' groups. Three of the most challenging of these are discussed here – cost of communication, involvement of a range of various types of groups, and developing and maintaining a multinational nature.

Cost of communication

Coordinating work across such a vast space as the 4,000 km (2,500 miles) from the heart of the North American continent to the Atlantic Ocean places major burdens on any organization, but even more so on a citizens' coalition that depends on volunteer workers and operates with a limited budget. Getting together to work and make decisions can be very challenging.

The costs and time involved in travelling to meetings can be prohibitive. As a result, efforts are made to limit the number of such meetings, and to replace face-to-face meetings with numerous phone calls. Nevertheless, these are not without costs. Although efforts are also made to communicate by cheaper e-mail, Great Lakes United's experience shows that the extent to which e-mail can be used to communicate is limited: it does not prove very effective as a means of drawing discussions to actual decisions – especially if there is any controversy around the matter. E-mail easily leads to misunderstandings and can actually escalate problems. For this reason, Great Lakes United spends a higher percentage of its budget on travel and phone than would be expected in most organizations: approximately 15 per cent of Great Lakes United's budget goes to travel and telephone.

Charitable foundations focused on supporting Great Lakes work by environmental groups have been essential to the viability of Great Lakes

United and the environmental movement across the basin. The Joyce Foundation (based in Illinois) provided the funding for the creation and start-up of Great Lakes United and has been a consistent funder ever since. Likewise, the Mott Foundation (in Michigan) and the Gund Foundation (in Ohio) have been important ongoing financial supporters. The ongoing and not always project-specific funding from some foundations has been critical to the viability of Great Lakes United.

Governments have also helped make basin-wide work possible by contributing travel costs to enable environmental group members to attend their meetings. This is especially true in Canada; little such funding occurs in the United States. In addition to being a time for citizens' groups to make input to governments, these meetings become essential opportunities for citizens' groups to network and strategize among themselves.

Involvement of range of types of groups

As discussed above, a major part of the debate at the founding of Great Lakes United centred on the respective roles of the grass-roots groups and the larger groups. Great Lakes United has always believed that the primary power of the organization derives from its ability to reflect the views and support the work of that vast range of citizens working on the issues in their local communities – those people who directly feel the negative effects of inappropriate human actions within the basin. At the same time, however, Great Lakes United has felt that it is essential to have the larger regional or national groups involved because of the skills, expertise, and power that they bring to the organization.

The major difficulty for the grass-roots groups in operating within such a coalition is that they are almost all without any paid staff, are completely dependent on dedicated volunteers, and have little (if any) money to spend on participating in a group outside their local community; their limited resources and time are all needed to deal with the local crises that confront them. In addition, they often are suspicious of the motivations of the larger groups and feel that their own energy and resources can be drained by them.

Great Lakes United has used several strategies to keep grass-roots groups heavily involved in the organization and to ensure that these groups play a major role in setting Great Lakes United's direction. These strategies include the following:

- *Ensuring strong grass-roots representation on the Board*: Currently, approximately 60 per cent of Great Lakes United's Board is made up of people from local grass-roots groups scattered across the basin; the other 40 per cent is divided equally between regional and national groups. The pattern has been similar throughout Great Lakes United's history. Great Lakes United pays the financial costs for its Board members to be involved in the Board.

- *Holding meetings in different communities around the basin*: To make it easier for the local groups to be involved and to provide input to Great Lakes United, the organization holds meetings around the basin. The annual meetings are moved to locations in all parts of the basin to make it easier for different groups to attend. The Board also holds its meetings in different locations and organizes a part of these meetings as an opportunity to meet members in those communities.

- *Setting policy at its annual meetings*: Great Lakes United's policies are set at its annual meetings. At these meetings, each group has one vote, regardless of the size of the group. This ensures that the grass-roots groups have a strong voice in policy-setting.

- *Having staff work in the field*: Instead of just working from their offices, Great Lakes United staff go to communities to meet and work with local groups.

- *Holding formal public meetings around the basin*: Periodically, Great Lakes United holds formal public meetings around the basin to hear from the grass roots. Two examples of this have been described earlier in this chapter – the hearings around the renegotiation of the GLWQA and the meetings leading up to the development of the Citizens' Action Agenda.

- *Funding groups to participate*: Great Lakes United writes funding into its project budgets to cover travel costs for grass-roots representatives to attend meetings and participate in Great Lakes United activities.

- *Funding local groups*: In some projects, Great Lakes United has included funding to give to local groups to carry out parts of the work that Great Lakes United needs to do. For example, when holding public meetings around the basin, Great Lakes United makes a small grant to a local group in each community for them to help organize the event. The local groups are also always included as co-sponsors of meetings held in their communities.

- *Keeping local groups informed of, and involved in, region-wide issues*: It can be challenging for local groups to dedicate the resources needed to keep informed of, and involved in, Great Lakes–St Lawrence River Basin-wide issues and opportunities to affect policies. Great Lakes United sees help for groups to do this as one of its primary roles. As the only truly basin-wide organization, Great Lakes United monitors and coordinates input on basin-wide issues. For example, over the past few years, the Great Lakes states and provinces have been working together to develop a binding, consistent, new regime for water takings and use throughout the basin. This is the Annex 2001 process developed under the Great Lakes Charter. Great Lakes United has played the lead role in coordinating the development of proposals for the content of this water-use regime. It has reached out to its members through action-alert e-mails and mailings and through public meetings for their input and to advise them of how to become involved. Great Lakes United has repeatedly played a similar role around issues concerning the GLWQA.

Developing and maintaining a multinational nature

One of the major challenges that Great Lakes United has had to confront throughout its history is to develop Great Lakes United as an organization that truly reflects the different nations within the Great Lakes–St Lawrence River Basin. This difficult task has three elements to it: (1) Canada and the United States; (2) inclusion of the First Nations and Tribes; and (3) inclusion of Quebec.

The primary understanding upon which Great Lakes United operates is that the Great Lakes and St Lawrence ecosystem transcends national boundaries and, therefore, Great Lakes United must operate beyond political boundaries, while respecting the differences.

Canada and the United States

Working binationally across the border is always a challenge. The power differences and cultural differences between Canada and the United States are ones that Great Lakes United must always be vigilant to balance.

Even though the differences between Canada and the United States may be less than the differences among those living around many international waterbodies, such differences are still significant. Language differences have been relatively minor – with the exception of Quebec, which is discussed later. Nevertheless, the use of English does vary between the two countries and can lead to unexpected misunderstandings: for example, "tabling a motion" means the exact opposite in each country. Thus, in the United States, "tabling" implies "cancelling," whereas in Canada to table a motion is to put it forward for later discussion.

The more striking differences are in the political systems and government decision-making processes in the two countries. How often have Canadian board members sat at meetings in bewilderment as US board members discussed who is the most important senator or house representative to lobby on a particular bill? What are those critical "conferees" that they talk about? And why do they not understand why a guide on legislator's voting records makes no sense in Canada? Likewise, US board members are frustrated by Canadian board members' strong focus on how to affect the Prime Minister or the Premiers. One way in which

Great Lakes United has tried to overcome this lack of understanding is by including people from both countries when the organization goes to Ottawa or Washington, DC, to discuss Great Lakes issues.

The most important difficulties are those that arise from the juxtaposition of a superpower with a middle power. These political realities operate not just at the level of government-to-government relations but also at the citizen level. Whereas Canadians fear that their interests and agenda will be submerged by the more aggressive style of the residents of the superpower, the residents of the United States have substantial difficulty in understanding that there may be differences in interests, needs, and approaches, and they may find it hard to respect and take those differences into account.

The by-laws of Great Lakes United were structured to try to ensure a balance of power between the Canadian and US memberships of the organization. Five seats were specifically reserved for Canadian and US members; the other seats on the board were determined by region and could be held by either a Canadian or a US resident. Two years later, the by-laws were changed to have two treasurers – one from Canada and one from the United States. After a particularly heated Annual Meeting in Cleveland, Ohio, in 1988, at which the Canadian candidate for vice-president lost to one from the United States, the by-laws were amended to require that the president and vice-president be from opposite countries. This was done to ensure more balance on the executive of Great Lakes United. Great Lakes United has also tried to achieve balance by locating an office in Canada as well as in the United States. In addition, staff positions are advertised in both Canada and the United States.

Structural provisions alone cannot ensure balance in the operation of an organization. The Board, staff, and membership of Great Lakes United always has to be vigilant to ensure that one nation does not dominate the other in deliberations and programmes. These challenges require time, patience, and the willingness – indeed, the desire – to learn about the other and to be sensitive to, and accepting of, the differences. It also requires appreciating and valuing those differences. These differences are one of the joys of working within Great Lakes United.

The First Nations and Tribes

There are approximately 350,000 aboriginal people and 110 First Nation and Tribal governments in the Great Lakes–St Lawrence River Basin. How best to recognize and take account of the special rights of the indigenous peoples as the first human inhabitants of that basin has sometimes been a source of conflict.

Great Lakes United has always believed that it is essential to recognize the special roles, rights, and contributions of these first inhabitants of the basin. To ensure a presence of First Nations and Tribal perspectives in its deliberations, Great Lakes United amended its by-laws in 1991 to require that one seat on the Board be filled by someone who is a member of a First Nations or Tribal organization. Eight years later, the by-laws were again amended to increase the number of seats reserved for representatives from such nations or organisations to two seats. (There could, in fact, be more than two representatives from these groups, because they could run for any other seat on the board.)

Great Lakes United has always ensured that representatives of the first peoples of the basin are invited to its public meetings. It also has played a role in getting recognition from other groups for the special role of the First Nations and Tribes. For example, when the IJC allocated time at its biennial meeting for Great Lakes United to give a presentation in 1991, Great Lakes United asked the IJC to also provide time for the First Nations and Tribes to give a presentation to the plenary. When the IJC refused to allocate such time, Great Lakes United turned over part of its time to the First Nations and Tribes. This set a precedent, which meant that, in future years, the IJC itself allocated time to the First Nations and Tribes.

Great Lakes United has developed programmes specifically oriented towards native people. For example, Great Lakes United has a staff person who developed and is now coordinating the Indigenous Peoples Hub of the Great Lakes Aquatic Habitat Network and Fund. Through this, Great Lakes United promotes networking and information exchange among indigenous people, and provides access to funding. Great Lakes United also works on educating First Nations and Tribal fish consumers about the current risk of eating fish, and on how to reduce exposure to contaminants in Great Lakes fish.

As well as these efforts, however, Great Lakes United must continually strive to learn from (and to encourage and support participation by) the First Nations and Tribes. The style of operation and the perspectives of the indigenous peoples of the Great Lakes–St Lawrence River Basin differ greatly from those of the immigrants after 1497. It is all too easy for organizations such as Great Lakes United to fall into token recognition, without enabling the views and needs of the First Nations and Tribes to truly affect how the organization operates and what it does.

Quebec

The people along the St Lawrence River in Quebec live with the consequences of the activities of those who live upstream throughout the rest of the Great Lakes system: one environmentalist in Quebec has referred to them as living along the sewer for the Great Lakes. Nevertheless, that part of the St Lawrence River in Quebec is not included in the GLWQA.

Great Lakes United has always pushed governments in Canada and the United States to recognize that the St Lawrence River is an integral part of the Great Lakes ecosystem and to include Quebec in Great Lakes programmes.

The culture of Quebec differs from that of the rest of Canada, and most of Quebec's inhabitants speak French; this brings a whole new set of obstacles to the achievement of full inclusion. Great Lakes United's by-laws require that at least one of the five at-large positions for Canadians be held by a resident of Quebec. Great Lakes United has also located an office in Montreal, Quebec, to improve contact with Quebec citizens' groups and to increase Great Lakes United's involvement in issues in the St Lawrence River and Quebec.

Nevertheless, Great Lakes United still faces difficulties in fully involving a broad Quebec membership. The Board conducts its meetings in English, which means that Board members from Quebec cannot be effective participants if they do not speak English. Likewise, almost all of Great Lakes United's publications are in English, which is a major barrier to developing and working with membership in Quebec. Operating as a truly bilingual organization is much more expensive than most not-for-profit organizations can afford.

Conclusions

The 21-year history of Great Lakes United has shown the value of having an environmental non-governmental organization that crosses political boundaries. Ecosystems do not recognize political boundaries; therefore, if we are to adequately address problems in watersheds that cross political boundaries, we must work across these boundaries. In the context of the North American Great Lakes, basin-wide citizens' groups have been shown to be essential to push governments to break down the artificial barriers created by political systems.

The other major lesson from Great Lakes United's experience is that it is critical to have this basin-wide organization driven by grass-roots citizen activists. These are the people who are the most effective advocates on behalf of the waters that are so critical to their lives, because they most fully understand the impacts of inappropriate behaviour and push towards true long-term solutions to the problems.

REFERENCES

Ashworth, William. 1986. *The Late, Great Lakes: An Environmental History.* Toronto: Collins Publishers.

Botts, Lee, and Paul Muldoon. 1997. *The Great Lakes Water Quality Agreement: Its Past Successes and Uncertain Future*. Hanover, New Hampshire: Institute on International Environmental Governance.

Environment Canada and US Environmental Protection Agency. 2002. "SOLEC 2002: Implementing Indicators." Draft for Discussion at SOLEC 2002. October.

Farid, Claire, John Jackson, and Karen Clark. 1997. *The Fate of the Great Lakes: Sustaining or Draining the Sweetwater Seas*. Buffalo, NY: Canadian Environmental Law Association and Great Lakes United.

Fuller, Kent, and Harvey Shear (eds). 1995. *The Great Lakes: An Environmental Atlas and Resource Book*. 3rd edn. Toronto and Chicago: Government of Canada and United States Environmental Protection Agency.

Gelinas, Johanne. 2001. *A Legacy Worth Protecting: Charting a Sustainable Course in the Great Lakes–St Lawrence River Basin*. October. Ottawa: Office of the Auditor General.

Great Lakes United. 1987. *Unfulfilled Promises: A Citizens' Review of the International Great Lakes Water Quality Agreement*. Internet: ⟨http://www.glu.org⟩ (visited 7 February 2005).

Great Lakes United. 1992. *Ten Years of Action: Ten Years of Celebration*. Buffalo, NY: Great Lakes United. Internet: ⟨http://www.glu.org⟩ (visited 7 February 2005).

Great Lakes United. 1998. *Citizens Speak: Great Lakes United's 1998 Hearings on the State of the Great Lakes*. Internet: ⟨http://www.glu.org⟩ (visited 7 February 2005).

Great Lakes United. 2003. *The Great Lakes Green Book: A Citizens' Action Agenda for Restoring the Great Lakes–St Lawrence River Ecosystem*. Internet: ⟨http://www.glu.org⟩ (visited 7 February 2005).

International Joint Commission (IJC). 2003. Status of Restoration Activities in Great Lakes Areas of Concern: A Special Report. April. Washington, DC, and Ottawa: IJC. Internet: ⟨http://www.ijc.org⟩.

Jackson, John, and Tim Eder. 1991. "The Public's Role in Lake Management: The Experience in the Great Lakes." In M. Hashimoto and B.F.D. Barrett (eds). *Socio-economic Aspects of Lake Reservoir Management. Guidelines of Lake Management* 2:31–46. Otsu, Japan: United Nations Environment Programme.

Manno, Jack. 1993. "Advocacy and Diplomacy in the Great Lakes: A Case History of Non-Governmental Organization Participation in Negotiating the Great Lakes Water Quality Agreement." *Buffalo Environmental Law Journal* 1(1).

McMillan, Tom. 1987. Letter from Canada's Minister of the Environment to Joe Clark, Canada's Secretary of State for External Affairs, August 24.

USGAO (United States General Accounting Office). 2003. Great Lakes: An Overall Strategy and Indicators for Measuring Progress Are Needed to Better Achieve Restoration Goals. April. Washington, DC: United States General Accounting Office.

7

Public participation in watershed management in theory and practice: A Mekong River Basin perspective

Prachoom Chomchai

Introduction

With an average annual discharge of 500 billion cubic metres (BCM), a length of 4,800 km, and a basin area of 795,000 km^2, the Mekong River constitutes one of Asia's most substantial resources. In terms of energy, it is equivalent to an oil well producing approximately 1.5 million barrels of crude petroleum per day, but renewable and without the concomitant pollution. The basin's population of about 70 million, with an annual per capita income of less than US$400, however, is impoverished. Thus, there is poverty amidst plenty.

The current and future livelihoods of much of the basin's inhabitants hinge on the sustainable development of the basin's resources. Of the basin's inhabitants, 80 per cent are farmers and fisherfolk who depend on the river for irrigation water and the possible catch of more than 1,000 species of fish living in it. Apart from energy and food, the Mekong also provides a relatively cheap means of communication, although it is not navigable throughout its length. In addition, the river is potentially suitable for ecotourism and flood-control development.

Public participation in watershed management can be treated either as an end in itself or as a means to an end. To some, it may be desirable as an end *per se* because of its democratic nature; to others, it is a means to improve governance of the resource. Fortunately, public participation has been a traditional feature of the Mekong River Basin; whereas there is no necessary connection between public participation and good governance, they frequently support and reinforce one another.

This chapter examines the history and continuing evolution of public participation in theory and practice in the Mekong River Basin. It considers local, national, and basin-wide approaches, starting with the traditional approaches within the basin. It then considers some of the challenges in promoting public involvement in watercourse governance, concluding by examining recent initiatives to provide a regional framework for enhancing transparency and public participation in decision-making.

Traditional approaches and principles

The inhabitants of the Mekong River Basin are no strangers to participatory approaches and principles, which have been observed in local communities since the distant past. This has fortunately existed in tandem with a "green" ideology derived from the Hindu and Buddhist principles of non-violence toward nature. Together, they have helped to maintain a sustainable ecology until the advent of contemporary development, which has more exacting demands on resources.

In fact, scrutiny of a handful of local communities confirms the deep-rooted nature of participatory principles that have evolved in the context of communal subsistence and cohesion. This holds true for both the wet and the arid parts of the Mekong River Basin. For example, a social impact assessment (SIA) of the planned Kaeng Sua Ten Dam in Thailand's northern Phrae Province on the edge of the Mekong watershed reveals invaluable approaches within the indigenous system of natural resources management. However, the proposed dam could mean the permanent loss of the villagers' traditional knowledge about the forest and its biodiversity (Bangkok Post 2000).

For centuries, the mountainous area of northern Thailand has been dotted with small irrigation systems (*muang faai*) built and managed by farmers (Sluiter 1992). A similar system exists in Luang Prabang, the former capital of Laos on the other bank of the Mekong (Sluiter 1992). The *muang faai* system has always been accompanied by a strict set of rules maintained by *muang faai* leaders to ensure that the surrounding forest is safeguarded and the water distributed fairly to all members of the irrigation group. Recent changes brought about by imposed development projects such as large-scale logging, however, have threatened the viability of the traditional *muang faai* system. Arid regions of the basin have comparable experiences: traditional structures known as *thamnob*, counterparts to the *muang faai*, store irrigation water and have been maintained by farmers with full public participation and have helped to ensure sustainable development (Ekachai 2003).

As an alternative to large-scale developments that have proved to be problematic, Care Thailand, funded by the new Danish Cooperation for Environment and Development (DANCED) has launched the Integrated Natural Resources Conservation (INRC) project to broaden community planning by bridging the gap between villagers and government officers (Kungsawanich 2001). The project adopts a bottom-up approach of reinforcing traditional community participation in natural resource management, whereby efforts are made to settle conflicts over the use of natural resources between ethnic groups and state agencies. In doing so, Care has worked closely with *tambon* (sub-district) administrators in the project area. In retrospect, mistakes of past management by international aid agencies could be pinpointed. Contrary to previous experience, forest encroachment in the area occurred when villagers were dominated by profit-driven, cash-crop plantation activities. Moreover, as monocrop plantations consumed huge amounts of water, water wars between highlanders and lowlanders ensued. Instead of imposing a set of solutions on the communities, this renewed bottom-up approach has established village committees and mini-watershed networks to work out rules and activities for forest conservation. Although Care's approach is leading to the slow recovery of forest areas, the threat of future deforestation remains and the constant challenge is to find a proper balance between private economic gain and collective ecological well-being.

It is to be noted that the traditional participatory principle, in contrast to its modern counterpart, is essentially non-aggressive, non-assertive, inward-looking, and non-confrontational (*Bangkok Post* 2001). In particular, it evolved in a context of deference to authority, where the ruler was believed to be benevolent. To the extent that such a principle still is observed in local communities, it may be said to be a relic of the past.

Indeed, traditional public participation in water management has been more prevalent than may appear at first sight. A study by the Thailand Development Research Institute (TDRI) found that US$1.6 billion in Thai government funds for 550,000 small water sources across the country were wasted over the past two decades because these sources have been neither fully used nor properly maintained (Ruangdit and Theparat 2003). By contrast, most water sources managed by local inhabitants are in good condition, providing clean drinking-water year round. In view of this, it is likely that the government will soon transfer the power to manage small water sources to local administrative organizations as part of the general programme for decentralization and devolution.

The TDRI study of sites in the Mekong River Basin shows that, in an open-access system of water being drawn for collective use, there has been indigenous public participation in watershed management throughout. Thus, parts of the basin, wet and arid alike, had been dotted with traditional *muang faai* and *thamnob* irrigation structures that helped the basin inhabitants to achieve, through participatory management practices, ecological balance and sustainable development until the advent of large-scale, public-sector dams.

It is tempting to argue that the coexistence of participatory principles with the "green" ideology constitutes a strong case for requiring public participation in watershed management. There is, however, no assurance that the indigenous green ideology is sufficiently robust always to ensure sustainable ecology, particularly when faced with the prospect of private economic gain, even when short-term in nature. The situation is particularly precarious when there is a need for communities not only to balance collective ecological well-being against private economic gain but also to avert conflicts over the distribution of such gain, which could destroy traditional communal cohesion.

The Mekong traditional participatory principle is by no means unique, having counterparts elsewhere: for example, it is similar to collective-management approaches adopted by the pre-Columbian Kogi Indians (Delannoy 2001).

That the indigenous Mekong participatory principle should favour governance is intriguing from a public-finance analytical standpoint. For one thing, good governance is a "public" good from which a potentially infinite number of people could benefit simultaneously, but which, because of "market failure," could not be left to the market to provide on its own. Of course, the freeriders' quality of life benefits from good governance, though the freeriders may continue to ravage key elements of the environment for personal gain. For another thing, good governance, like insurance, may also be seen to be a "merit" good, to which people tend to attribute insufficient merit. It may, however, represent a fresh breed of merit goods, since in contradistinction to such classical cases as housing, the "merit" want it meets is imposed not from above but from below – the very livelihood of people threatened by an absence of good or effective governance, especially in public-sector projects, being all too common. To the extent that the government lacks the political will to address environmental deterioration, and the workings of its machinery are thwarted by "government failure," the person in the street may be said to be playing an avant-gardist role in environmental governance.

Advent of top-down development

The State's management of natural resources has relied too heavily on an open-access regime. As noted above, in certain instances the imposition of collective self-discipline has curbed the worst excesses, whereas elsewhere the permissiveness of the regime has resulted in significant environmental harm. The main difficulty with the regime is that it provides the impetus for the abuse and overuse of water, forest, and fishery resources (Kaosa-ard and Wijukprasert 2000), ills of the tragedy of the commons (Hardin 1968). The widespread presumption of ownership leads the bulk of the rural people to build their lives (and even their communities) around the use of these resources and gives people no incentive whatsoever to exert themselves by, for instance, keeping their own fish-ponds or cages.

The open-access system being the order of the day, the indigenous green ideology has unfortunately been unable to withstand the impacts of globalization, population growth, and economic growth. In fact, in the past 150 years or so, export-led growth; population growth; and increased mobility, industrialization, and urbanization have wreaked havoc on Thailand's apparently robust environment, as well as elsewhere in the Mekong River Basin.

Rapid population growth is tied to Thailand's (and the region's) economic growth, particularly between 1988 and 1997, when double-digit growth placed the country in the league of the world's fastest-growing economies. However, in the last three decades, a substantial proportion of South-East Asia's impressive economic growth can be attributed to a "one-off fire-sale of natural resources," which means that it may be difficult to maintain such growth when the trees, fish, and soil are depleted. For the individual basin resident, there are more personal concerns: he or she remembers fishing in a river or drinking from a stream as a child and regrets what has been lost when contemplating today's poisonous waters (Mallet 1999).

Unfortunately, most of the damage to the environment and rural communities in the region has been inflicted by governments. In public-sector, top-down development projects, with no public input, the government often acted as an independent interest group and was unaccountable to people at the grass-roots level. Without consulting affected localities, such projects have typically allocated resources to one group of people (often urban) to the detriment of another (often rural), leaving the latter with insufficient resources to sustain livelihoods.

River-basin development has followed the prevailing trend toward top-down management. Public-sector construction of dams, reservoirs, weirs, and irrigation infrastructure and the expansion of protected areas into upper water catchments have been deemed necessary to maximize the resource value of the system. For four decades, Thailand's river basin plans have focused solely on public-sector creation of large, medium, and small water-storage areas, whether for flood control or for dry-season water use. This process of river basin development has been confined to a small group of technocrats, economists, and irrigation engineers, along with foreign experts brought in by international and regional development agencies.

This situation may be said to have originated from three key factors: (1) centralization of the social and economic planning framework; (2) overdependence on dominant "expert knowledge" of river basin management; and (3) export-oriented economic development efforts that tie production to the global economy. In this centralized context, government agencies develop water resources unilaterally. Likewise, in Thailand, the relatively new National Economic and Social Development Board (NESDB), the overall planning body, has been able to draw up development projects without any reference to, or involvement of, people living in areas affected by the planned projects. Even in the current era of openness ushered in by the "People's Constitution" of 1997, government approaches to river basin management often exclude popular participation or allow only "stage-managed" participation.

The (Me)Kong–Chi–Mun diversion scheme in the north-east, and plans to divert Mekong headwaters from the northern Kok and Ing tributaries into the Chaophraya (the country's main river) via the Nan river – two schemes that are likely to affect the Mekong main stream – illustrate the practice. Direction by experts has complemented centralized river basin planning; in this process, traditional water-management knowledge has been discounted and dismissed. Despite the numerous salinity, flooding, and water-storage problems that public-sector water-development projects have created for effective water management, there are limited opportunities to challenge the mainstream, technological thinking (Chantawong 2002; Wangvipula 2003).

The top-down approach is typical of river basin development elsewhere in the Mekong River Basin. For example, the Chinese Government has not given the affected public a role in decision-making regarding a cascade of dams under construction across the Chinese segment of the river, to the extent that local communities have been denied access to EIAs (Panwudhiyanont 2002). The Lower Mekong riparian nations also have been kept in the dark about upstream developments.

Man Wan, the first dam in the Chinese cascade to be completed, has had serious adverse impacts on areas immediately downstream of China. After the dam's construction, the river's hydrological pattern underwent a radical transformation: the water level paradoxically rises in the dry season but falls in the wet season, and there is a disturbing uncertainty about it at any given time, upstream release for hydropower generation being the determining factor (Panwudhiyanont 2002). Uncertainty in water levels has left the Upper Mekong fishery in ruins, because fishermen can no longer read the water level and select the right fishing gear, enabling fish easily to take evasive action. There has been a negative correlation between the Mekong's hydrological upheaval and the size of the catch. The reduced fishing opportunities also entail an irreparable loss of a culturopolitical heritage for Thai and Laotian fishing communities that have managed the river's shared natural resources for centuries.

Perhaps more disastrous in its downstream impact than Man Wan has been that of Yali Falls, built in Viet Nam on the Se San River about 70 km above its border with Cambodia, where the Se San flows into the Sre Kong River before the latter's confluence with the Mekong. With an installed capacity of 720MW (less than half the size of Man Wan), it is, nevertheless, the lower basin's largest dam built on one of its largest tributaries. With the commencement of power generation in 1998, the irregular releases of water radically altered the hydrological regime and the water quality of the Se San River downstream. The transboundary environmental and socio-economic effects have been diverse and significant – severe flooding, flash-flooding (even in the dry season), forced evacuation, low river levels, human and livestock illness associated with contaminated river water, loss of cropland and concomitant nutritional impacts, increased turbidity, and waning fish stocks (the primary source of animal protein for most living along the river) (Panwudhiyanont 2002). In fact, some fish species have disappeared from the river altogether.

Theory and practice of public participation

Modern public participation in theory

As a reaction to the élitist and externally oriented river basin development, popular scrutiny of river basin development projects has emerged over the past decade, particularly in Thailand. Such scrutiny has been about both the projects themselves and the process. On the substantive side, salient issues include efficiency of dams and irrigation structures, environmental and social assessment, economic efficiency, the Royal Irrigation Department's water-allocation principles, compensation mechanisms for those adversely affected by projects, and water-demand forecasting. At the same time, process concerns have included overly centralized state-centric decision-making systems, inability of people to gain access to (and be involved in) decision-making at all levels, and an absence of opportunities for community-based knowledge to be employed in river basin development (Chantawong 2002).

Although the critique of the top-down approach to river basin development may be valid, the point is frequently overlooked that people do not always sufficiently appreciate what is in their best interest. The "merit goods" approach may be paternalistic in that the government compels people to consume certain things considered to be meritorious; nevertheless, it helps to generate the necessary consumption of meritorious things. Of course, beside the demerit of paternalism, there is always the danger that special-interest groups may make use of the government's coercive power to further their own views (Stiglitz 2000). This point, however, may be less relevant with regard to the importance of water and the advisability of the means to deliver it, since people at the grass-roots level know full well the value of water to them and the potential impacts of the proposed means of delivery.

Modern participatory principles and practice arose out of the growing recognition of the inadequacies of a top-down approach, particularly in rural development. By the 1980s, academic literature had recognized that externally imposed and expert-oriented forms of development were not effective (Chambers 1983, 1994).

More generally, modern participatory principles in Thailand have their origins in the abrupt (albeit, practically bloodless) transformation in 1932 from an absolute form of government to a limited, constitutional monarchy. After tumultuous decades that bore witness to abuse of political power, in 1997 a new constitution (the sixteenth since 1932) was adopted in the hope that it would lead the country on a path of participatory government and sustainable development. This was the first constitution to emerge from a process of public consultation (Mallet 1999). Civil rights and civil liberties are augmented so that they may come to life with popular participation. They include, *inter alia*, access to information (spelt out in the Information Act of 1997) that is in the public domain and held by a government entity. In particular, the Constitution guarantees local participation in environmental protection so that indigenous communities have the right to take part in the management of natural resources and the environment and to demand information, clarification, and justification from a government entity before it proceeds to approve, license, or carry out a project affecting the environment or human health and hygiene. Any activity or project that can seriously affect the quality of the environment is prohibited unless an environmental study is undertaken with a view to its endorsement by independent agencies, including representatives from environmental NGOs and university academics.

Other parts of the Mekong River Basin similarly espouse principles of public participation. For example, the principle appears to be accepted in China: as local-level management challenges cannot always wait for national institutions to meet, local governments and people should be encouraged to manage their own environment (Ting 2001). The Laotian Constitution is the product of discussion by the people throughout the country, and there are four levels of public participation guaranteed in public-sector projects: (1) information gathering; (2) information dissemination; (3) consultation; and (4) participation (IRI 2001). In Cambodia, the government has indirectly delegated a certain amount of authority downwards to civil society (Hourn 2001). In Viet Nam, the requirements for public participation in environmental decision-making are fourfold – knowledge, participation, discussion, and control (Can, Phan, and An 2001). Thus, regardless of whether public participation is constitutionally guaranteed, most countries in the Mekong River Basin recognize the significance of public participation and its key elements.

A prerequisite of public participation is the government's dissemination of information. A government's duty to disseminate information derives partly from the principle of freedom of information, which is widely accepted in many democracies and embedded in international human-rights law. It also finds support in international instruments related to international common-pool resources (ICPRs) (Benvenisti 2001). Dissemination of information on the conditions of transboundary waters, measures taken or planned to address transboundary impacts, and the effectiveness of these measures, nurtures domestic debate within the countries participating in their use regarding the range of options available to their governments. This process increases the governments' ability to assess public support, at the same time constraining possible attempts to diverge from national interests.

Modern-day participation, however, requires much more than dissemination of information. The right to participate not only is the freedom of speech but also addresses specific issues that could affect the lives of those involved: it extends to the right to negotiate compensation, the right to negotiate a changing mode of life that may take place, the right to negotiate property rights, and the right to know the nature and degree of risks that people may incur (Turton 2000).

Public participation could be made more effective and less costly, particularly in small-scale institutions that are likely to be more sensitive to the concerns of those directly affected by the uses of such ICPRs (Benvenisti 2001), as are represented by the resources of the Mekong River Basin. The existence of a number of relatively small institutions, each responsible for a single sub-basin, could facilitate efficient intra- and inter-basin trade in shares of the resources, with the central institution in the form, for instance, of a national Mekong committee serving as a forum for negotiations and even as a clearing-house for transactions among sub-basin representatives.

Although river basin development planning can be nicely packaged as integrated river basin management (IRBM), total catchment management (TCM), or integrated water resources management (IWRM), or can be advanced through language of participation as participatory irrigation management, multi-stakeholder consultation, or civil-society involvement, they all subscribe to the merits of participation by people who inhabit the river basins in question and those whose livelihoods depend on their resources (Chantawong 2002; Editorial 2002). However, owing to the enhanced integration and interdependence of modern economies, the latter subgroup of the stakeholders can be very great indeed.

The implantation of civil society in the Mekong River Basin, however, may not be as simple as it may seem. Civil society generally is embedded in specific social, economic, and political contexts, and civil society in river basin development is no different. To transfer participatory principles from one context to another can be difficult (Editorial 2002). In the Mekong River Basin there is a solid groundwork of traditional participation, and the basin's inhabitants have an intimate familiarity with their part of their watershed. For this reason, efforts by academics and social activists to introduce modern public-participation practices without proper grass-roots orientation and appreciation could appear to be an imposition of alien institutions, ultimately backfiring.

The real substance and significance of public participation ultimately hinges precariously on *de facto* power relations among the range of societal institutions and groups (Chantawong 2002). There is no guarantee that the formal openness of the new Thai Constitution or other legal guarantees in other parts of the Mekong River Basin will filter through, as general pronouncements can be far removed from the operational level of river basin planning and management. A real participatory process for popular or civil-society involvement is thus at heart a process of challenging the existing structures of river basin planning authority (Chantawong 2002).

Whether these institutions will ensure effective public participation remains to be seen. Hopes are high that the decision-making of the Mekong River Commission (MRC) will take into account civil-society's interests. Since 2002, civil-society representatives have been invited to attend the MRC Joint Committee and Council meetings as observers. Actual power does not, however, rest with the MRC, nor has it been invested with supranational authority. If participation is not granted, for example as a result of the rigidity of existing power relations, it may perforce have to come from genuine, earnest, and persistent efforts of civil society (Chantawong 2002), the ultimate upshot being open confrontation, social unrest, and instability.

Modern public participation in practice

Even where public participation is guaranteed by a constitution, authorities have generally been reluctant to allow it to operate on a consistent basis, even in democracies in the basin. To the extent that it does take place, it does so on an ad hoc basis.

The practice of public participation is less common in the basin's countries that are not democracies. For example, as noted above, China did not ask the inhabitants of the Man Wan area to participate in the decision-making process. Similarly, the second dam on the Lancang after Man Wan (Dachaoshan) was built in 1997 and came on stream in 2001, without people in the locality being consulted. Although the *People's Daily* keeps people abreast of such developments, many of the basin's inhabitants are illiterate or have no ready access to this government newspaper (Wongruang 2002).

Transboundary cases are handled no better than purely domestic ones. Thus, although the Cambodian government had been informed by Viet Nam that Yali Falls would be constructed, Cambodia failed to warn its own people downstream of the impending ecological effects. In some exceptional cases, the affected inhabitants are consulted, particularly when donors require it; however, the impacts of the public hearings are unclear. Thus, as a condition of financial support to the Government of Laos for the Nam Theun 2 hydropower dam, the World Bank required that the project meet with concurrence from social and environmental groups (Ganjanakhundee 2002).

Joern Kristensen, the outgoing Chief Executive Officer (CEO) of the MRC, refers not to the difficulty in principle that most governments in the Mekong River Basin find in agreeing to public participation but to practical problems of adopting it (Kristensen 2002). He maintains that the problem of how to involve stakeholders effectively in environmental decisions and the planning process has confronted governments across the basin. The difficulties are confounded, he argues, by potential conflicts of interest between communities at different levels, which may be local, national, and international. Local communities may oppose projects planted in their midst that are in the national interest, or they may support projects that are not; they may thus put local interest above national ones or even serve as proxies for unidentified vested interests. Unfortunately, all too frequently, such communities do not speak with one voice, and it is up to the authorities to decide which segments to listen to and to take seriously. Again, concerned outsiders may go out of their way to support or oppose such projects, while the majority may remain silent.

Transboundary implications are even more intractable than domestic ones. Decisions taken within one country may well spill over into neighbouring ones. Although any large water resource development project in the upper Mekong could adversely affect millions of people in downstream countries, Kristensen argues that it is difficult to imagine how to involve the masses to be affected downstream in decision-making upstream. While such issues are addressed with difficulty in developed countries in Europe and North America, it is unlikely to be any easier to deal with them in South-East Asia. In the particular context of the Lower Mekong Basin, problems of public participation have been compounded by poverty and the presence of multiple countries with differing national interests. Kristensen notes that the poor have both limited access to the media and generally low literacy, and many lack the skills and confidence to participate readily in public debate. Accordingly, it is difficult to disseminate information effectively and equally difficult to secure responses to any proposed initiative, especially in countries such as Cambodia where civil society has been disrupted by warfare and where the basic infrastructure is being rebuilt.

Kristensen may have exaggerated the practical difficulty of public involvement in decision-making on a transboundary basis. However, in addition to the difficulties he raised, there are transboundary communication and transaction costs that limit the effectiveness of environmentalists' intervention (Benvenisti 2001). Of course, a question prior to this is that of principle: there is no clear obligation on the part of any Mekong Basin state to allow the public in another basin state to participate in its water resources development activities.

When *ex ante* public involvement has been skirted even in Thailand, where democratic institutions and civil society are better developed than in the rest of the Mekong Basin (Panwudhiyanont 2002), people affected by public-sector water resource projects have had to resort to *ex post* protests. Such protests have been effective, especially where environmental effects from public-sector projects are not entirely irreversible. Indeed, the utility of *ex post* protests may, in the short run, be ad hoc in nature but could, in the long run, tilt the power balance in favour of *ex ante* public participation. Despite the potential drawbacks of *ex post* public participation (including protests), two instances in the Mekong River Basin point to the apparent effectiveness of protests against *faits accomplis* in the form of public-sector projects, since they have been able to undo much that has been done.

In its pre-regulation and pristine state, the bed and wetlands of the Mun (a major Mekong tributary in Thailand) served as an ideal habitat for fish in the flood season when fish migrated upstream for spawning. Inhabitants in the area regularly trapped large fish, with each wet-season catch formerly so plentiful as to allow the fisherfolk to distribute it among relatives and to either sell the leftover or preserve it for subsequent bartering for rice. In the dry season, movement of "hibernating" fish permitted another large-scale fishing expedition. Claims of ancestral rights to fish-trapping areas and possessory rights to man-made structures were generally recognized and were bought and sold openly. In addition to fishing, rice farming was practised on the banks of the Mun, even in the dry season, because of the ubiquity of water. Equally, dry-season vegetable horticulture took place on both banks of the Mun. This ecological and cultural balance was dramatically altered by the construction of two dams, Rasi Salai and Pak Mun, which between them constitute a cascade.

Seven years of water impoundment behind Rasi Salai Dam, one of the most controversial public-sector projects, caused extensive environmental harm; irrigation water distribution, the chief benefit claimed for it, has not been effective (Chuskul 2001). Public protests led to the opening of its seven sluice gates in July 2000 to alleviate the environmental and social effects of impoundment and to allow a land-rights survey and stocktaking of the situation to begin. In the wet season of 2000, after the opening of the dam gates, people reported sightings of huge fish (of 70–80 kg each) and even the much larger giant catfish migrating upstream from the Mekong and "spectacular" catches after seven years of interruption. In one case, the person reporting the sighting could only stand idly by and watch, since he no longer had the right gear with which to catch the big fish. Witnessing the fish's homecoming, inhabitants of the area hold high hopes of the return of the "good old times" and the restoration of the natural ecological balance.

In a similar vein, in response to popular pressure, the government also authorized the opening, during the four months of the flood season, of the sluice gates of the Pak Mun Dam situated close to the Mun's confluence with the Mekong. The Pak Mun project has been even more controversial than Rasi Salai, which is located further upstream. It remains to be seen whether the ecological balance can similarly be restored. It has been claimed, for example, that opening the Pak Mun Dam gates substantially improved the catches in 2001, an improvement that was attributed to the removal of the physical constraint on the natural migration of fish. Fishermen's annual household income is estimated to have increased from US$80 in 2000 to US$240 in 2001, although this was still a fraction of the pre-impoundment income of US$442 (Panwudhiyanont 2002).

Emerging measures under the MRC

To support civil-society development in the basin, the MRC not only has invited civil-society representatives to attend sessions of its Joint Committee and Council as observers (discussed above) but also has incorporated development of public participation as a component of all its core programmes. The MRC places a special emphasis on promoting participation at the sub-basin and local levels, and it has provided assistance to agencies of member governments to develop their capacity to institute effective public-participation activities (Kristensen 2002).

Internationally, the MRC is being assisted by the Murray–Darling Basin Commission (MDBC) to develop a public-participation strategy for the Mekong River Basin through joint workshops, study tours, and training programmes. The MDBC model is being scrutinized, and its relevant approaches are being adapted. The MRC is also seeking to learn from some of the mistakes made in the Australian context (Kemp 2002).

Even during the tenure of the Mekong Committee (the immediate predecessor to the MRC), modest beginnings were made with public involvement in project identification and development. It was here that national Mekong committees played a strategic liaison role between the stakeholders in each member country and the Committee. This should continue to be a crucial role of the national Mekong committees operating under the MRC.

In recent years, the MRC has incorporated the principle of public participation as a principal objective of its overall work plan. Public involvement "is believed to be a prerequisite for the overall aim and vision of our Mekong Agreement, i.e. sustainable development of the Mekong River Basin" (MRC 2003).

The MRC is currently developing its "Draft Basin Development Plan," which seeks to institutionalize the planning process required for sustainable development in the Mekong River Basin. In particular, the draft plan aims to balance socio-economic development with environmental concerns. To this end, the plan is developing an effective means to incorporate a participatory approach that integrates technical knowledge as well as stakeholder and other political views. In this context, a "Study on Public Participation in the Context of the MRC" was initiated in late 1996.

The MRC broadly defines the stakeholders who may be affected by MRC decision-making: a stakeholder is any person, group, or institution that has an interest in an activity, project, or programme. This includes both intended beneficiaries and intermediaries, those positively affected, and those involved in and/or those who are generally excluded from the decision-making process (MRC 2003). Stakeholders may also include those who live outside MRC member countries, or the basin.

The MRC public-participation policy is to go through the four stages of (1) information gathering, (2) information dissemination, (3) consultation, and (4) participation (MRC 2003). However, this is not a check-list and, (depending on the particular project) different levels of public participation may be deemed appropriate and applied.

Guidelines for applying public participation in the MRC have also been developed, which take into account the specific needs of individual countries. Capacity is an important consideration in applying the guidelines.

In fact, the capacity of the MRC to make information publicly available requires careful consideration of MRC's planning process and capacity: "While the will may be there to do so, it requires planning and resources to make the right documents available at the right time" (MRC 2003). For this and other reasons, the guidelines are adaptable.

It is intended that these initiatives to improve public participation will facilitate cooperation among all stakeholders throughout the Lower Mekong River Basin. The approach to this work by the MRC focuses on four objectives, namely:

- to achieve basin-wide benefits while taking account of national interest
- to balance development opportunities with resource conservation
- to broaden public participation, and
- knowledge sharing and capacity building.

In addition to the MRC initiatives, there are a few efforts to promote public participation in the region more broadly. These include a proposed regional framework for ensuring transparency, public participation, and accountability (AEETC 2002; Hilden and Furman 2002), as well as efforts by an NGO coalition (Access Initiative 2003). These broader initiatives promote planning instruments such as environmental impact assessment (EIA), as well as transparency and public participation more generally, and stand to inform and reinforce the ongoing efforts of the MRC.

Ultimately, the crux of the matter is whether the government of a co-basin state voluntarily accepts the principle of public participation, particularly in a transboundary context. While the MRC has successfully brokered an agreement among the lower riparian countries on preliminary procedures for notification and prior consultation, it remains to be seen whether this will filter through to other arenas. As yet, the MRC does not include China, which (like Myanmar also in the Upper Mekong Basin) is no more than a "dialogue partner" to the MRC. Moreover, China is not relying on international donor agencies for financing its Lancang cascade programme. Circumstances being what they are, civil society will have to be particularly resourceful to obtain opportunities for public input from China.

Conclusions

Although inhabitants of the Mekong River Basin have been among the staunchest believers in, and practitioners of, traditional participatory principles, ensuring public participation in modern management of the river has been challenging. Modern participatory principles and practice are not susceptible of instantaneous implantation; to turn these into home-grown counterparts calls for long-term development of civil society, the present state of which gives cause for cautious optimism.

Civil-society participation in resource management in Thailand has increased substantially over the past decade, although the situation elsewhere in the Mekong River Basin has moved more slowly. Nevertheless, throughout the basin, international NGOs have started to play a role and local NGOs and specialized research institutes are emerging with their supportive roles. Recent developments within the MRC are also promising, as it incorporates participatory principles into its projects and fosters a dialogue on domestic and international avenues for promoting public participation within the watershed.

REFERENCES

Access Initiative. 2003. Internet: ⟨http://www.accessinitiative.org⟩ (visited 14 August 2003).

Asia–Europe Environmental Technology Centre (AEETC). 2002. Public Involvement in Environmental Issues in the ASEM – Background and Overview. Internet: ⟨http://www.vyh.fi/eng/intcoop/regional/asian/asem/asem.pdf⟩ (visited 15 August 2003).

Bangkok Post. 2000. 15 October.

Bangkok Post. 2001. "Fight the Power." 15 March. Outlook section, page 1.

Benvenisti, Eyal. 2001. "Domestic Politics and International Resources: What Role for International Law?" In Michael Byers (ed.). 2001. *The Role of Law in International Politics*. Oxford: Oxford University Press.

Can, Le Thac, Do Hong Phan, and Le Quy An. 2001. "Environmental Governance in Vietnam in a Regional Context." In *Resources Policy Research Initiative*, pp. 13–26. Chiang Mai, Thailand: REPSI.

Chambers, Robert. 1983. *Rural Development: Putting the Last First*. Harlow: Longmans.

Chambers, Robert. 1994. "Participatory Rural Appraisal: Challenges, Potentials and Paradigm Shift." *World Development* 22(10).

Chantawong, Montree. 2002. "Civil Society Participation in River Basin Management: A New Blueprint?" *Mekong Update and Dialogue* 5(2).

Chuskul, Sanan. 2001. "Opening of Rasi Salai Dam: Return of Communal Life, Man, Fish and Wetlands." *(Bangkok) Weekend Matichon* 21(1064). 8 January [in Thai].

Delannoy, Pierre. 2001. "Les Indiens Kogis, Nouveaux Sorciers du Management." *Paris Match*, 1 February.

Editorial. 2002. "Civil Society and River Basin Development." *Mekong Update and Dialogue* 5(2). April–June. Sydney.

Ekachai, Sanitsuda. 2003. "Desert's Hidden Wealth." *Bangkok Post*. Outlook Section. 26 March.

Ganjanakhundee, Supalak. 2002. "Laos Signs $2-bn Deal for Nam Theun II Dam." *(Bangkok) Nation*. 4 October.

Hardin, Garrett. 1968. "The Tragedy of the Commons." *Science* 162:1243–1248.

Hilden, Mikael, and Eeva Furman. 2002. "Towards Good Practices for Public Participation in the Asia–Europe Meeting Process." In Carl Bruch (ed.). 2002. *The New "Public": The Globalization of Public Participation*. Washington, DC: Environmental Law Institute.

Hourn, Kao Kim. 2001. "The Impact of Regional Integration on the Governance Processes in Cambodia: The Environmental Perspective." In *Resources Policy Research Initiative*, pp. 91–99. Chiang Mai, Thailand: REPSI.

IRI (Environmental Research Institute). 2001. "Public Participation in Development Projects in Lao PDR." In *Resources Policy Research Initiative*, pp. 52–58. Chiang Mai, Thailand: REPSI.

Kaosa-ard, Mingsarn, and Pornpen Wijukprasert (eds). 2000. *The State of Environment in Thailand: A Decade of Change*. Bangkok: TDRI.

Kemp, Susan. 2002. "Reflections on the Murray–Darling–Mekong Liaison." *Mekong Update and Dialogue* 5(2). April–June.

Kristensen, Joern. 2002. "Civil Society and River Basin Development." *Mekong Update and Dialogue* 5(2).

Kungsawanich, Ukrit. 2001. "Out of the Woods." *Bangkok Post*. Outlook Section. 10 February.

Mallet, Victor. 1999. *The Trouble With Tigers: The Rise and Fall of South-East Asia*. London: HarperCollins.

MRC (Mekong River Commission). 2003. "Public Participation in the Context of the MRC" Internet: ⟨http://www.mrcmekong.org/pdf/Public%20Participation%20the%20context%20of%20the%20Mrc.pdf⟩ (visited 12 November 2003).

Panwudhiyanont, Wiwat. 2002. "Impoundment of the Pakmun Dam has Brought Untold Misery to Mekong Inhabitants." *Sarakadee Magazine*. Bangkok, December [in Thai].

Resources Policy Research Initiative. 2001. *Mekong Regional Environmental Governance: Perspectives on Opportunities and Challenges*. Chiang Mai, Thailand: REPSI.

Ruangdit, Pradit, and Chatrudee Theparat. 2003. "Think-tank Says B 70 bn Gone down the Drain." *Bangkok Post*. 25 March.

Sluiter, Liesbeth. 1992. *The Mekong Currency*. Bangkok: Project for Ecological Recovery.

Stiglitz, Joseph E. 2000. *Economics of the Public Sector*, 3rd edn. New York: W.W. Norton and Company.

Ting, Zuo. 2001. "Cases of Local Transboundary Environmental Management in Border Areas of the Mekong Watershed in Yunnan, China." In *Resources Policy Research Initiative*, pp. 28–33. Chiang Mai, Thailand: REPSI.

Turton, Andrew. 2000. "Introduction to Civility and Savagery." In Andrew Turton (ed.). 2000. *Civility and Savagery: Social Identity in Tai States*. London: Curzon Press.

Wangvipula, Ranjana. 2003. "People to Have More Say in Water Use." *Bangkok Post*. 12 March.

Wongruang, Piyaporn. 2002. "Mekong Fisherman Left High and Dry." *(Bangkok) Nation*. 16 October.

8

Public participation in Southern African watercourses

Michael Kidd and Nevil W. Quinn

Introduction

There are 11 shared watercourse systems in Southern Africa, occupying about 70 per cent of the land area (Chenje and Johnson 1996) (see fig. 8.1). Cooperation among neighbouring states that share these watercourses is relatively well developed, both with respect to individual systems and regarding water management generally. Many of these initiatives provide for public participation in the management of the water systems, but these are often merely paper commitments that are not being reflected in practice. This chapter describes the shared watercourse systems in the region and their shared management structures [the Southern African Development Community (SADC) Protocol on Shared Watercourses], and provides some analysis of the extent to which the public is participating in management of the systems.

For the purposes of this chapter, Southern Africa is regarded as the SADC nations excluding the Democratic Republic of the Congo (DRC). SADC comprises 14 member states: Angola, Botswana, the DRC, Lesotho, Malawi, Mauritius, Mozambique, Namibia, Seychelles, South Africa, Swaziland, Tanzania, Zambia, and Zimbabwe. The DRC is excluded from this study, as the data relied upon for this chapter deal only with the other terrestrial Southern African states and their agreements. Mauritius and Seychelles do not have shared watercourses, other than the Indian Ocean. The 11 shared watercourses in the region are reflected in

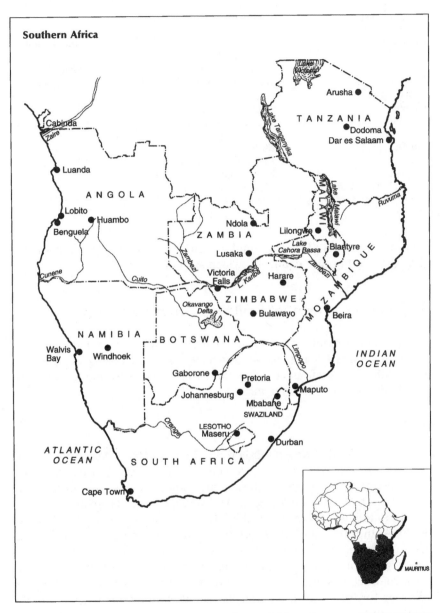

Figure 8.1 Shared watercourses in Southern Africa (Source: SADC/IUCN/ SARDC 1994)

Table 8.1 Basin countries of Southern African watercourse systems

Name of system	Countries sharing system
Zambezi	Angola, Botswana, Malawi, Mozambique, Namibia, Tanzania, Zambia, Zimbabwe
Orange	Botswana, Lesotho, Namibia, South Africa
Limpopo	Botswana, Mozambique, South Africa, Zimbabwe
Okavango	Angola, Botswana, Namibia
Cunene	Angola, Namibia
Rovuma	Mozambique, Tanzania
Inkomati	Mozambique, South Africa, Swaziland
Songwe	Malawi, Tanzania
Save	Mozambique, Zimbabwe
Pungwe	Mozambique, Zimbabwe
Lake Chilwa Basin	Malawi, Tanzania

table 8.1. [Angola, Tanzania, and Malawi have other watercourses that they share with other countries, which are not reflected here.]

Shared management of Southern African watercourses

Of the 11 watercourses set out in table 8.1, at least seven have arrangements for shared management. Each of these are briefly described in turn, with special reference being made to provisions for public participation in the relevant instruments.

Zambezi

Zambia and Zimbabwe established the Zambezi River Authority (ZRA) in 1987, which in practice is concerned mainly with the Kariba Dam. Another major dam on the Zambezi – the Cahora Bassa – is subject to an agreement between South Africa, Mozambique, and Portugal.

The founding agreement of the ZRA is incorporated in the Zambezi River Authority Act of 1987, promulgated simultaneously in both countries (Republic of Zambia 1987). According to Article 4 of the Agreement, the Council of the ZRA consists of two cabinet ministers from each contracting state, but there is no express provision in the agreement allowing for (or requiring) public participation in the management of the Authority.

In 1987, a cooperative agreement on the Zambezi Basin was endorsed by all states, referred to as the Zambezi River Basin System Action Plan (ZACPLAN) (Chenje 2000). One of the key elements of this plan was to develop an integrated water management plan as well as regional legislation and institutional structures (Leleka 1995). There have been subsequent efforts to establish the Zambezi Basin Commission (ZAMCOM); however, various factors have caused delays in establishing this institution (Chenje 2000). Nevertheless, the initial discussions on ZACPLAN served as the impetus for developing the SADC Protocol, discussed below, which seeks to provide a regional legal framework on which all river basin agreements for the region would be based (Ramoeli 2002).

Orange

There are several agreements between South Africa and her neighbours relating to the Orange River (Chenje and Johnson 1996), probably the most important being the Lesotho Highlands Water Project. The treaty establishing the project (between South Africa and Lesotho) does not provide for public involvement in management of the scheme (Treaty on Lesotho Highlands 1986). The project is implemented by two bodies – the Lesotho Highlands Development Authority (LHDA) on the Lesotho side and the Trans-Caledon Tunnel Authority (TCTA) in South Africa. Under Article 7(33) of the treaty, Members of the Board of the LHDA are appointed on the basis of their managerial, technical, and financial qualifications and experience, and there is a similar provision for the TCTA under Article 8(22).

Limpopo

The Limpopo Basin Permanent Technical Committee (LBPTC) was signed by all basin states in June 1986, but the organization was moribund for 10 years. From 1995 onwards, meetings have sought to resurrect the LBPTC; however, Mozambique has raised difficult, unresolved, issues, and Zimbabwe has been absent from these meetings, so little progress has been achieved. The fact that not even the basin states themselves can reach agreement for the Limpopo Basin (Heyns 1995; Turton 1999) suggests that the involvement of the public in the shared management of the Limpopo is not high on the current agenda.

Okavango

The main management instrument for the Okavango is the Okavango Commission (OKACOM) involving Angola, Botswana, and Namibia, which was formalized in September 1994. According to Article 1.2 of the agreement establishing OKACOM, the purpose of the commission is to provide technical advice on matters relating to the conservation, development, and utilization of water resources and to perform other relevant functions pertaining to the development and utilization of such resources. The OKACOM has been described as:

generally functioning satisfactorily despite the highly visible and hostile public exchange that occurred as a result of Namibian plans to build the Rundu–Eastern National Water Carrier [ENWC] pipeline that were announced during the drought that coincided with Namibian independence. Tempers seem to have cooled since then and relations have become more cordial over time. (Turton n.d.)

Efforts are currently under way in the Okavango Basin to ensure effective public involvement in decision-making processes for the basin, including the appointment of the Okavango Basin Steering Committee (OBSC) to manage a "transboundary diagnostic analysis in order to identify the key areas of concern and the gaps in the knowledge of the biophysical, social and economic environment in the Okavango Basin" (Pinheiro, Gabaake, and Heyns n.d.).

Cunene

Angola and Namibia entered into a shared management agreement in 1969, which has been supplemented by a 1990 agreement. The latter agreement requires the countries to adopt "the best joint-utilisation schemes during planning, execution and operation of projects for water resources development in the basins of common rivers" (Chenje and Johnson 1996).

Inkomati

The principal mechanism for addressing issues on the three rivers (Incomati, Limpopo, and Maputo) shared by South Africa, Mozambique, and Swaziland has been the Tripartite Permanent Technical Committee (TPTC), established in 1983 (Gleick 1998). According to Gleick and others, the TPTC has not functioned effectively, to the extent that, in the mid-1990s, Mozambique threatened to take South Africa to the International Court of Justice if its water needs were not addressed more adequately, and also refused to sign a proposed Memorandum of Understanding (Gleick 1998; Atkins 2001). More recently, improved relations between Mozambique and South Africa have resulted in the signing of a tripartite interim agreement (Tripartite Interim Agreement 2002). This interim agreement retains the TPTC as the joint body for cooperation between the parties [in Article 5(1)]. With respect to the defined responsibilities of the parties, the interim agreement does not have any particular requirement for participation beyond general statements referring to promoting partnership, exchange of information (Article 12), and cooperation with the SADC organs and other shared watercourse institutions (Article 4). To give effect to the provisions of Article 12 on information, the parties also signed a Resolution of the Tripartite Permanent Technical Committee on Exchange of Information and Water Quality in August 2002. Under Annex V of the interim agreement, timetables are established for signing a Comprehensive Agreement for the Incomati and Maputo Watercourses by July 2005 and August 2009, respectively. Although the renewed cooperation in this basin must be seen as a positive development, the fact that the TPTC remains the institutional structure is a matter of concern: membership of this structure comprises the respective water-affairs departments, with no other sectoral representation or any other public input.

Songwe

The Songwe River forms the boundary between Malawi and Tanzania, and ultimately flows into Lake Malawi, thereby forming part of the Zambezi Basin. Because of persistent flooding in the lower Songwe, the Songwe River Stabilization Agreement was signed between these two countries (Global Water Partnership–Southern Africa 2001). According to Simwanda, this agreement was negotiated under the existing Joint Permanent Commission, an intergovernmental/ministerial structure (Simwanda 1999).

General conclusions

From the brief examination of the various shared water basin management arrangements outlined above, it is evident that the basin-specific instruments currently operate with differing degrees of effectiveness. By and large, the instruments are not expressly concerned with public involvement in the management of the river basins, and this is possibly a contributing factor to the lack of success of those that are not currently operating well. Public participation in shared watercourse management is, however, an express aim of the SADC Water Protocol, to which our attention now turns.

The SADC Protocol on Shared Watercourses

The SADC Protocol was the first sectoral protocol developed by SADC, and it was adopted and signed by ten SADC members in Johannesburg,

South Africa in 1995 (Ramoeli 2002). It came into force in September 1998. Angola, DRC, and the Seychelles have not signed it, and Mauritius signed when it became a member of SADC in 1996. The basic purpose of the Protocol is to ensure equitable sharing and efficient conservation of water. A full discussion of the Protocol is beyond the scope of this chapter, but a reasonable summary of the Protocol is that ...

[I]t follows principles laid out in international rules and conventions and is premised on the effort to maintain a balance between development needs in the national interest of member states, the needs for conservation, as well as the needs for sustainable development. It aims to achieve and maintain close co-operation between member states. It also sets out the rights and obligations of member states with regard to shared watercourse systems in the region. (Ramoeli 2002)

Article 3 of the Protocol focuses on establishing river basin management institutions for shared watercourse systems. In addition to a monitoring unit (SADC Environment and Land Management Sector), member states undertook to establish river basin commissions between basin states in respect of each drainage region, as well as river authorities or boards for each drainage basin. As far as public participation is concerned, Article 5 of the Protocol sets out the functions of river basin management institutions, which includes: "Stimulating public awareness and participation in sound management and development of the environment and including human resources development."

Influenced by developments in international law (particularly the adoption of the 1997 United Nations Convention on the Law of Non-Navigational Uses of International Watercourses), and problems that some of the parties to the SADC Protocol had with the original document, a revised Protocol was adopted in 2000. As of February 2003, the revised Protocol has been signed by all member states and ratified by eight. Once ratified by two-thirds of the SADC members, it will replace the existing Protocol one year after it comes into force. The revised Protocol has improved provisions relating to the environment, including recognition of the environment as a legitimate user of water, as well as downstream and upstream rights, roles, and responsibilities. Article 5 sets out a revised institutional framework for implementation, including establishment of four SADC water-sector organs, as well as shared watercourse institutions. The four water-sector organs include the Committee of Water Ministers, the Committee of Water Senior Officials, the Water Sector Coordinating Unit, and the Water Resources Technical Committee and Sub-Committees. The previous reference to public participation has been omitted in the revised Protocol.

Perhaps one of the most encouraging developments with respect to regional participation in water resource management has been the publication of a draft of the SADC Regional Indicative Strategic Development Plan (RISDP) in March 2003. This document seeks to deepen regional integration in SADC. It provides SADC member states with a consistent and comprehensive programme of long-term economic and social policies. It also provides the Secretariat and other SADC institutions with a clear view of SADC's approved economic and social policies and priorities (SADC Secretariat 2003).

According to this document, the Protocol is being put into operation through a Regional Strategic Action Plan (RSAP) for Integrated Water Resources Management and Development. The seven key priority areas are as follows:
1. Improving the legal and regulatory framework.
2. Institutional strengthening and sustainable development policies.
3. Information acquisition.
4. Management and dissemination.
5. Awareness building, education, and training.
6. Public participation.
7. Infrastructure development (SADC Secretariat 2003).

The promotion of awareness and public participation in policy and programme formulation and implementation has been identified as a priority intervention area with an explicit target for "increased awareness, broad participation and gender mainstreamed in water resources development and management by 2005" (SADC Secretariat 2003). The institutional requirement of public participation in watercourse management is present, but much flesh still has to be added to the skeleton. The reality in Southern African watercourse management is that there is still much work to be done to involve the public in a meaningful way in management decisions. There are several obstacles to achieving successful public participation, as set out below.

Impediments to public participation

Several factors can be identified as presenting difficulties for improving public involvement in water management in the region These include language, war, historic distrust, the fact that public participation remains an emerging concept in the region, and selective public participation.

In most of Southern Africa, the lingua franca, if not the official language, is English. For the river basins covered in this study, the exceptions are Mozambique and Angola, where Portuguese is the primary language. However, uniformity of official language is more important for the purposes of intergovernmental or other institutional water basin management than it is for community involvement. Many members of the public in the region do not speak English (or Portuguese), and language differences undoubtedly contribute to problems in communication.

Some countries in the region have been embroiled in civil war or similar internal unrest for many years, and others are still at war. This means not only that water basin management is probably not a high priority with respect to other goals (ranging from rebuilding broken economies to clearing landmines) but also that there is a widespread distrust of others that has to be overcome.

Because of past tension in the region, there is still uneasiness between official bodies within neighbouring states that can lead to communication breakdowns. For example, in 2000, severe flooding in Mozambique was exacerbated by actions taken by South African water management authorities upstream that were not communicated to Mozambique. If the official bodies are not communicating properly, the task for the public is that much more difficult.

Public participation remains an emerging concept in much of the region. In South Africa, for example, the concept of public participation in managing water resources became a reality only with the promulgation of the National Water Act in 1998. Although institutional structures for public participation are provided for in the Act, their establishment requires a lengthy process of feasibility assessment; as of April 2003, no catchment management agency (CMA) has been formally proclaimed. With the concept still emerging in a domestic context, it is perhaps not surprising that transboundary public participation has yet to become well established.

Even in cases where public involvement is provided for, it may well be that those sectors of the public whose participation is facilitated are not a representative cross-section of the community. In the Inkomati region, for example, public participation is provided for in the South African National Water Act (representation on CMAs) and in the proposed Mozambican law, but the representation on these institutions tends to favour commercial agriculture and sectors such as electricity companies at the expense of the poorer water users such as subsistence farmers (Leestemaker 2000).

Public participation in watercourse management in South Africa

South Africa recently overhauled its water legislation, so that the National Water Act 36 of 1998 makes express provision for involving the public in water management in several regards. Chapter 7 of the Act provides for the establishment of CMAs, the purpose of which is to delegate water resource management to the regional or catchment level and to involve local communities within the framework of the national water resource strategy, as determined by Chapter 2 of the Act. Section 80 provides that the functions of a CMA are as follows:

- to investigate and advise interested persons on the protection, use, development, conservation, management, and control of the water resources in its water management area;
- to develop a catchment management strategy;
- to coordinate the related activities of water users and water management institutions within its water management area;
- to promote the coordination of its implementation with the implementation of any applicable development plan established pursuant to the Water Services Act (Act No. 108 of 1997); and
- to promote community participation in the protection, use, development, conservation, management, and control of the water resources in its water management area.

The governing board of the CMA, appointed by the Minister of Water Affairs and Forestry, is required by Section 81 to achieve a balance among the interests of water users, potential water users, local and provincial government, and environmental interest groups. According to the Department,

Public participation in decision-making concerning integrated water resource management is one of the basic principles of catchment management in South Africa. Public participation must actively engage the need for participation of all segments of society, including those that have historically been disadvantaged and marginalised, in accordance with the principles in the Policy and the Act. In particular, effort must be made to include women, rural communities and the poor in all decision-making structures and consultation processes. (Department of Water Affairs and Forestry 2003)

The Act also provides, in Chapter 8, for water user associations (WUAs), which are, in effect, cooperative associations of individual water users who wish to undertake water-related activities for their mutual benefit. The previous Water Act (Act No. 54 of 1956) provided for irrigation boards, and several of these have been transformed into WUAs. The process of establishing these bodies in South Africa is ongoing, and the Minister of Water Affairs and Forestry is currently reviewing the first of a number of proposals for the formal establishment of CMAs.

Conclusions

Hey has highlighted that Chapter 18 of Agenda 21 focuses on public participation and subsidiarity (management at the lowest appropriate level) as essential prerequisites for attaining sustainable use of water resources (Hey 1995). The importance of participation in contemporary water resource management was also recognized in one of the four 1992 Dublin Principles: "Water development and management should be based on a participatory approach, involving users, planners and policy-makers at all levels" (Dublin Principles 1992). This was reiterated in the 2000 Hague Declaration, in which Ministers recognized that "[i]ntegrated water resource management depends on collaboration and partnerships at all levels, from individual citizens to international organisations" and that "there is a need for coherent national and, where appropriate, regional and international policies to overcome fragmentation, and for transparent and accountable institutions at all levels" (Ministerial Declaration 2000). Yet, in their analysis of the discussion of the statements and pledges of nations at the Hague Ministerial Conference, Soussan and Harrison note that, although the need for improving public participation was widely recognized, the actual "number of commitments on this issue was perhaps surprisingly small. Those that were made were often vague in both intent and actions" (Soussan and Harrison 2000).

The most recent international commitment to water security, the 2002 Stockholm Statement, sets out as the first of four priority principles: "Water users must be involved in the governance of water resources" (Urgent Action Needed 2002). Since at least 1990, participation in water management has been seen as a critical need. Although some countries in Southern Africa have risen to this challenge, there is still some way to go in implementing effective public participation on a catchment basis within countries. Because of historical conflict, concern over diminishing water resources, and the paradigm of state sovereignty, the concept of participation in water resource management across international boundaries has even further to go.

REFERENCES

Atkins, Stephen. 2001. "Community Scale, Dynamics and the Reality of Participation: Towards Transboundary Water Management Agreement for the Maputo Basin." Presented to RGS-IBG Annual Conference, Plymouth.
Chenje, Munyaradzi. 2000. State of the Environment in the Zambezi Basin 2000. Maseru/Lusaka/Harare: SADC/IUCN/ZRA/SADRC.
Chenje, Munyaradzi, and Phyllis Johnson (eds). 1996. *Water in Southern Africa*. Harare: SADC/IUCN/SARDC.

Department of Water Affairs and Forestry. 2003. "Directorate: Catchment Management – Public Participation." Internet: ⟨http://www.dwaf.gov.za/cm/public%20particip.htm⟩ (visited 11 October 2003).

Dublin Principles. 1992. Internet: ⟨http://www.srh.ce.gov.br/dublin.htm⟩ (visited 11 October 2003).

Gleick, Peter. 1998. *The World's Water*. Washington, DC: Island Press.

Global Water Partnership – Southern Africa. 2001. "Progress Report for Phase 1 (1999–2001)." Internet: ⟨http://www.gwpsatac.org.zw/DFID%2099-2001%20Progress.pdf⟩ or ⟨http://www.gwpsatac.org.zw/progreports.html⟩.

Hey, Ellen. 1995. "Sustainable Use of Shared Water Resources: The Need for a Paradigmatic Shift in International Watercourse Law." In Gerald H. Blake, William J. Hildesley, Martin A. Pratt, Rebecca J. Ridley, and Clive H. Schofield (eds). *The Peaceful Management of Transboundary Resources*. London: Graham and Trotman.

Heyns, P. 1995. "SADC Agreements in Existence Pertaining to Shared Water Resources." In L. Ohlsson (ed.). *Water and Security in Southern Africa*. Publications on Water Resources No. 1. Stockholm: Swedish International Development Authority (SIDA).

Leestemaker, H.J. 2000. Gaps between the UN-Convention, the SADC Protocol and the National Legal Systems in South Africa, Swaziland and Mozambique. Internet: ⟨http://www.africanwater.org/leestemaker.htm⟩ (visited 11 October 2003).

Leleka, B. 1995. "The Zambezi River System Action Plan (ZACPLAN): A Brief Outline." In T. Matiza et al. (eds). *Water Resource Use in the Zambezi Basin*. Gland: IUCN. 99.

Ministerial Declaration of The Hague on Water Security in the 21st Century. 2000. Agreed to in The Hague on 22 March. Internet: ⟨http://www.thewaterpage.com/hague_declaration.htm⟩ (visited 11 October 2003).

Pinheiro, I., G. Gabaake, and P. Heyns. n.d. "Co-operation in the Okavango River Basin: The OKACOM Perspective." Internet: ⟨http://www.up.ac.za/academic/libarts/polsci/awiru/opp/papers/okacom.pdf⟩ (visited 11 October 2003).

Ramoeli, Phera. 2002. "The SADC Protocol on Shared Watercourses: History and Current Status." In Anthony Turton and Roland Henwood (eds). *Hydropolitics in the Developing World: A Southern African Perspective*. Pretoria: African Water Issues Research Unit.

Republic of Zambia. 1987. Zambezi River Authority Act, 1987. Act No. 17 of 1987. Internet: ⟨http://faolex.fao.org/docs/pdf/zam2611.pdf⟩ (visited 11 October 2003).

SADC/IUCN/SARDC. 1994. *State of the Environment in Southern Africa*. Maseru/Harare: SADC/IUCN/SARDC.

SADC Secretariat. 2003. Southern African Development Community (SADC) Draft Regional Indicative Strategic Development Plan (RISDP). March.

Simwanda, Lovemore. 1999. "Shared Scarce Watercourse Systems: Zambezi River – Possible Source of Conflict in the SADC Region." Internet: ⟨http://www.katu-network.fi/Artikkelit/kirja2/tekstit/Simwanda.htm⟩ (visited 11 October 2003).

Soussan, John, and Rachel Harrison. 2000. "An Analysis of Pledges and State-ments Made at the Ministerial Conference and World Water Forum, The Hague, March 2000." Internet: ⟨http://www.unesco.org/water/wwap/targets/ watersecurity.pdf⟩ (visited 22 May 2003).

Treaty on the Lesotho Highlands Water Project between the Government of the Kingdom of Lesotho and the Government of the Republic of South Africa. 1986. Signed at Maseru on 24 October.

Tripartite Interim Agreement between the Republic of Mozambique and the Re-public of South Africa and the Kingdom of Swaziland for Co-operation on the Protection and Sustainable Utilisation of the Water Resources of the Incomati and Maputo Watercourses. 2002. Signed in Johannesburg on August 29. Inter-net: ⟨http://www.dwaf.gov.za/Docs/Other/IncoMaputo/INCOMAPUTO%20 AGREEMENT%2029%20AUGUST%202002.doc⟩ (visited 11 October 2003).

Turton, A.R. 1999. "Impediments to Inter-state Co-operation in International River Basin Commissions within Arid Regions: Can Existing Theory Allow for   Predictability?"   Internet: ⟨http://www.up.ac.za/academic/libarts/polsci/ awiru/op28.html⟩ (visited 11 October 2003).

Turton, Anthony. n.d. "The Okavango River Basin." Internet: ⟨http://www.gci. ch/GreenCrossPrograms/waterres/pdf/WFP_Okavango.pdf⟩ (visited 11 Octo-ber 2003).

Urgent Action Needed for Water Security: 2002 Stockholm Statement. 2002. In-ternet: ⟨http://www.siwi.org/downloads/2002_Stockholm_Statement.pdf⟩ (vis-ited 11 October 2003).

9

Public involvement in water resource management within the Okavango River Basin

Peter Ashton and Marian Neal

Introduction

Throughout Southern Africa, escalating water scarcity is widely regarded as posing one of the greatest challenges to sustainable development in the region (Falkenmark 1989; Conley 1995; SARDC 1996; Shela 1996). Competing demands for water are especially acute in the more arid portions of the subcontinent, where water scarcity and associated increases in water pollution have also been linked to poverty, hunger, and disease (Pallett 1997; Gleick 1999; FAO 2000; Ashton 2003). The New Partnership for Africa's Development (NePAD) and the member states of the Southern African Development Community (SADC) have recognized the links between water shortage and poverty, and they have placed strong emphasis on the need to relieve regional water shortages (GWP 2000; NePAD 2001; SADC 2001). However, it is particularly difficult to meet the growing human needs for water in those situations where sufficient water is also needed to maintain the functioning of sensitive aquatic ecosystems and to protect the integrity of water resources (Falkenmark 1994, 1999; Ashton 2000a). Attempts to resolve the increasing competition for progressively scarcer water resources are often achieved in ways that damage or degrade the ecosystems concerned (Khroda 1996; Ashton and Neal 2003). The situation is further complicated by the fact that most of the larger river basins in Southern Africa are shared by several countries: the Zambezi, Okavango, Orange, and Limpopo rivers are all transboundary. The question of who should be allowed to use how much water and for what purpose becomes extremely sensitive and emotionally charged under these circumstances (Biswas 1993; Ashton 2000a; FAO 2000).

In Southern Africa, the water-rich Okavango Delta and its major inflow, the Okavango River, provide a classical example of a transboundary river system where human and ecosystem needs compete for scarce water supplies (Ellery and McCarthy 1994; McCarthy and Ellery 1998; McCarthy, Bloem, and Larkin 1998; McCarthy et al. 2000; Ashton and Neal 2003). The Okavango system spans three countries – Angola, Botswana, and Namibia – and, because of its perennial flows, the Okavango River and the world-renowned Okavango Delta function as a form of "linear oasis" in an otherwise arid area (Bethune 1991). The relative abundance of water in this system has inspired numerous plans and attempts to divert or abstract water from the system for domestic, agricultural, and industrial uses (UNDP/FAO 1976; JVC 1993; Heyns 1995a). Most of these attempts have not proceeded because of concerns that adverse social, economic, or environmental consequences could arise [International Union for the Conservation of Nature (IUCN) 1993]. To date, very small quantities of water are withdrawn from the system, and the Okavango River and Okavango Delta remain largely intact from an ecological viewpoint, whereas the need for water remains acute or is worsening in many surrounding areas (MGDP 1997; Ashton 2003).

The scenic beauty and extraordinarily rich biodiversity of the Okavango Delta and its component ecosystems have attracted widespread national and international concern about the future of this unique system (Ellery and McCarthy 1994; Ramberg 1997). In particular, local and international attention has emphasized the need to avoid forms of manipulation or management that could lead to adverse ecosystem changes; in short, the Okavango system can be considered to be an "internationalized" basin with a range of stakeholders that extends beyond that for most transboundary rivers in Africa (Pallett 1997; Ashton and Neal 2003; Turton, Ashton, and Cloete 2003).

However, despite the growing local and international interest in the Okavango Basin, recurring droughts and escalating regional water shortages in Botswana and Namibia continue to pose enormous challenges for water resource managers in these countries (Ashton 2000a,b, 2003). In addition, the recent cessation of civil war in Angola now means that Angolan authorities need to consider options for rehabilitating the country's economy: the development of agriculture, water supply, and hydropower facilities in the upper catchment of the Okavango River offer ideal opportunities to assist in achieving this goal (Turton, Ashton, and Cloete 2003). Taken together, these tensions, coupled with mounting local and international anxiety for the biological integrity of the Okavango Delta and its inflowing rivers, have accentuated the need to reach consensus on appropriate ways of managing the system (Ellery and McCarthy 1994). Clearly, both human and ecosystem perspectives must be taken into account if an equitable and sustainable solution is to be found (Ellery and McCarthy 1994; Ashton 2000b).

In recent years, integrated water resource management (IWRM) approaches have become widely accepted as offering the best way to achieve sustainable water resource management (Ohlsson 1995; Van der Zaag and Savenije 2000; Van der Zaag, Seyam, and Savenjie 2000). The IWRM approach promotes consideration of all components of the hydrological cycle and encourages wide public participation and transparency of decision-making. In addition, IWRM also advocates that responsibility for water resource management should be delegated to the lowest appropriate level, and it recommends the use of joint fact-finding and consensus-seeking approaches for the resolution of problems. Ultimately, sustainable management of the shared water resources and aquatic ecosystems within the Okavango Basin will require all stakeholders within each of the three basin states to participate in the development and implementation of a management plan for the system (Ashton and Neal 2003).

The sustainable management of a transboundary river system that is shared by more than one country depends on the collaborative efforts and collective goodwill of all the basin states involved (Wolf 1999; Lundqvist 2000; Ashton 2002). The activities of individual countries sharing a river basin are guided and directed by the provisions of national and international water law, as well as any international or regional watercourse-management treaties and protocols that may have been ratified by the basin states (Wouters 1999). Within this statutory and legal framework, however, it is the decisions, attitudes, and actions of national governments and individual stakeholders that usually play a decisive role.

The three basin states have signed and ratified several international accords and treaties, as well as the revised SADC protocol on shared watercourse systems. Although these instruments recognize the sovereignty of individual states, they also provide a framework for collaboration and impose specific obligations on the signatory states (Ashton and Neal 2003). In addition, the basin states have signed the OKACOM accord establishing the Permanent Okavango River Basin Commission – a formal institutional basis for the joint development of a management plan for the Okavango Basin (OKACOM 1994). A key part of the activities promoted and supported by OKACOM has been the initiation of extensive processes of public participation.

This chapter examines the degree to which stakeholders within the basin states are involved in the development of water resource management strategies for the Okavango Basin and highlights some of the challenges that have been encountered. Because of widespread confusion over prevailing circumstances in the basin, this chapter focuses attention on the prevailing geographical and political context within which local and regional initiatives seek to promote public participation in decision-making processes.

Geographical context

The Okavango system forms part of the Makgadikgadi Basin, which drains portions of four countries – Angola, Namibia, Botswana, and Zimbabwe (see fig. 9.1). The Makgadikgadi Basin is internally draining (endorheic), receiving inflows from one perennial river system in the

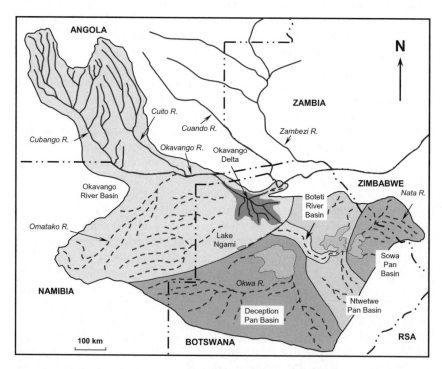

Figure 9.1 Sketch map showing the extent of the four sub-basins comprising the Makgadikgadi Basin, the Okavango Delta, and the various tributary rivers. Episodic and ephemeral rivers are shown as dashed lines; perennial rivers are shown as solid lines (Redrawn from UNDP/GEF 2001)

north-west (the Okavango River), as well as several smaller, ephemeral or episodic rivers in the drier southern portion of the basin. These smaller rivers contain surface-water flows only for short periods after heavy rainfall and have not contributed water to the Okavango Delta in living memory (Pallett 1997). Based on its topographic and hydraulic characteristics, the Makgadikgadi Basin can be divided into four distinct sub-basins or catchments that seldom have direct hydraulic contact with one another, and a small river basin (the Boteti River) that directs occasional outflows from the Okavango Delta towards the Makgadikgadi pans. Ntwetwe and Sowa pans comprise the Makgadikgadi Pan system in the east; the Deception Pan complex forms the southern portion of the basin; Sowa Pan, the easternmost sub-basin of the Makgadikgadi Basin, receives seasonal inflows from the Nata River system that rises in western Zimbabwe (Pallett 1997; Ashton 2000a).

The areas of the different sub-basins of the Makgadikgadi Basin are shown in table 9.1. The Makgadikgadi Basin covers an area of approximately 725,293 km^2, with Botswana providing the largest proportion (46.9 per cent), followed by Angola (27.6 per cent), Namibia (22.7 per cent), and Zimbabwe (2.8 per cent). The Okavango catchment or sub-basin covers an area of some 413,550 km^2 (in Angola, Botswana, and Namibia), with an additional 15,844 km^2 contributed by the wetland area of the Okavango Delta plus its islands in Botswana (Gumbricht et al. 2004). The combined area of the Okavango sub-basin and the Okavango Delta comprises approximately 59 per cent of the Makgadikgadi Basin (table 9.1). This chapter focuses on the Okavango sub-basin and does not deal with other components of the Makgadikgadi system.

The quantity and quality of water that enters the Okavango Delta depends on climatic factors (Wilson and Dincer 1976) and is influenced by any water-development activities that may take place in the upstream basin states (Ashton and Manley 1999; Ashton 2000a). Under international law (ILA 1966; Biswas 1993; ILC 1994; UNCLNUIW 1997), Angola and Namibia are technically entitled to withdraw water from, and to develop, water systems to which they are riparian; this right is entrenched and confirmed by the revised SADC Protocol on Shared Watercourse Systems (SADC 2001). As the lowermost riparian state, Botswana is theoretically in a vulnerable position and would clearly like to ensure that its interests are not unduly prejudiced by any developments that may take place in Namibia or Angola [Snowy Mountains Engineering Corporation (SMEC) 1987; IUCN 1993; Turton, Ashton, and Cloete 2003]. To date, very little water (less than 1 per cent of the mean annual run-off) is abstracted from the Okavango River, and the system remains in a near-pristine state (Ashton and Neal 2003).

Good interstate cooperation among Angola, Botswana, and Namibia

Table 9.1 Comparison of the area of each Makgadikgadi sub-basin[a] within the different countries comprising the basin, and their proportional contribution to the area of the Makgadikgadi Basin

| Sub-basin | Country contribution (km^2) | | | | Total area (km^2) | Proportion (%) |
	Angola	Botswana	Namibia	Zimbabwe		
Okavango River	200,192	59,575	153,783	0	413,550	57.02
Okavango Delta	0	15,844	0	0	15,844	2.18
Boteti River	0	10,920	0	0	10,920	1.51
Deception Pan	0	153,302	11,241	0	164,543	22.69
Ntwetwe Pan	0	74,028	0	0	74,028	10.21
Sowa Pan	0	26,389	0	20,019	46,408	6.39
Totals	200,192	340,058	165,024	20,019	725,293	100.00
Proportion (%)	27.60	46.89	22.75	2.76	100.00	

Source: Ashton and Neal 2003.
a. For the position of each sub-basin, see fig. 9.1.

jointly to resolve issues relating to the Okavango River not only is highly desirable but also is essential if sustainable solutions are to be achieved in the long term (OKACOM 1994; Heyns 1995a,b; FAO 2000; Ashton and Chonguiça 2003). However, there are perceptions in certain quarters that the relative costs and benefits of such cooperation may be unevenly distributed among the three countries (Ohlsson 1995; Ali 1996; Shela 1996; Ramberg 1997; Turton 1999). Nevertheless, although the three basin states may not have the same economic, technical, and personnel resources at their disposal, each country has pledged itself to cooperate with its neighbours on the matter of water resources (Republic of Botswana 1990; Heyns 1995a,b; Republic of Namibia 1995, 2000a,b; SARDC 1996; Pallett 1997; Ashton 2000b).

Before the formal ratification of the SADC protocol on shared river basins (SADC 1995) and its subsequent revision (SADC 2001), Botswana and Namibia had a relatively long history of amicable interstate cooperation on matters relating to their shared water resources (Taylor and Bethune 1999). The first, mostly informal, instances started in the early 1950s and were expanded over time to include joint flow-gauging exercises on the Okavango, Chobe, and Cuando rivers, as well as concerted efforts to control the invasive aquatic weed *Salvinia molesta* that infested rivers shared by the two countries (Taylor and Bethune 1999). A similar level of goodwill exists between Angola and Namibia concerning their joint interests in shared watercourses, most notably the Cunene, Okavango, and Cuando rivers.

Political context

The international context

The International Law Association (ILA) drafted the Helsinki Rules in 1966 in an attempt to bring greater uniformity to international water law by providing a comprehensive code for the use of transboundary drainage basins (Eckstein 2002). Since their introduction, these rules have formed the basis for negotiations among riparian states over the reasonable and equitable use of shared water resources (ILC 1994). Because of its involvement in the development of the Helsinki Rules, the General Assembly of the United Nations commissioned the International Law Commission (ILC) in 1970 to draft a set of articles to govern the non-navigational uses of transboundary water (Eckstein 2002). After some 25 years of debate among UN member states, the text of the UN Convention on the Law of Non-Navigational Uses of International Watercourses (UNCLNUIW 1997) was finally adopted on 21 May 1997.

Although the majority of member states adopted the text of the convention, the voting results showed that factors such as economic conditions and geographic position, as well as other national interests, played a major role in deciding the vote. It is interesting to note that all three basin states of the Okavango system voted for the Convention, even though Angola and Namibia are classified as upper and lower riparian states, and Botswana is classified as mostly lower (Eckstein 2002). However, five years after its adoption by the UN General Assembly, only 12 states have ratified the Convention although another 10 states have signed it. It is important to note that, even if this Convention never enters into force, it is the product of a democratic vote and it has had an obvious influence in the development of other water resource agreements. For example, the SADC Protocol on Shared Watercourse Systems in the Southern African Development Community Region is closely aligned with the Helsinki Rules and the principles of the then-near final Convention (SADC 1995).

Other international conventions that can potentially influence the management of the Okavango Basin include the Ramsar Convention (UNESCO 1971), the United Nations Convention on Biological Diversity (UNCED 1992a), the United Nations Convention to Combat Desertification (UNCCD 2001), and the United Nations Framework Convention on Climate Change (UNCED 1992b). All of these conventions contain provisions that specify obligations and responsibilities for the riparian countries to use their international water resources reasonably and equitably, while also promoting open and active participation and collaboration between riparian states. The fact that the three Okavango Basin states have signed and/or ratified these conventions, or are contemplating doing so, provides clear evidence of political goodwill and a shared spirit of cooperation and collaboration among the basin states (table 9.2).

The regional context

Southern African Development Community (SADC)

The Southern African Development Community (SADC) comprises 14 member states, including Namibia, Angola, and Botswana, and the objectives of the organization are outlined in the SADC Treaty (SADC 1992). In summary, SADC aims to promote sustainable development and economic growth, alleviate poverty, enhance the standard and quality of life of the people of southern Africa, and support the socially disadvantaged through regional economic integration. These objectives are not easy to attain, and SADC has recognized that the people and institutions within the region must be encouraged to take the initiative to develop bilateral and multilateral economic, social, and cultural ties across the region,

Table 9.2 Ratification dates of key international conventions by Angola, Botswana, and Namibia

| Country | Ratification date of convention[a] | | | | |
	Ramsar	UNCBD	UNCCD	UNCLNUIW	UNFCC
Angola	N/P[b]	1 Apr 1998	30 Jun 1997	N/P[b]	17 May 2000
Botswana	9 Dec 1996	12 Oct 1995	11 Sep 1996	N/P[b]	27 Jan 1994
Namibia	23 Aug 1995	16 May 1997	16 May 1997	29 Aug 2001	19 May 1995

Source: modified from Ashton and Neal 2003.

a. Ramsar, Ramsar Convention on Wetlands of International Importance; UNCBD, United Nations Convention on Biological Diversity; UNCCD, United Nations Convention to Combat Desertification; UNCLNUIW, United Nations Convention on the Law of the Non-Navigational Uses of International Watercourses; UNFCCC, United Nations Framework Convention on Climate Change.

b. N/P, not yet party to convention (based on available information).

while also participating fully in the implementation of SADC programmes and projects (SADC 1992).

The SADC Protocol on Shared Watercourse Systems (SADC 1995) was the first protocol to be developed by SADC and emphasizes the region's strong commitment to ensuring that member states collaborate with each other in the management of their shared watercourses. The rationale behind the development of this protocol was based on the recognition that a single legal instrument for river basin management would be more beneficial for the region than the development of individual management plans for each basin as and when the need arose. As a consequence of this decision, SADC member states initiated a process of negotiation to formulate the SADC Protocol, which was adopted and signed by 10 SADC member states in 1995. Because some SADC countries expressed concerns or reservations regarding certain specific articles within the Protocol, a process of amendment was initiated in 1997; this included a series of consultative workshops at the regional and national levels (Ramoeli 2002). In addition, the amendment process was influenced by concurrent developments in international water law – specifically, the adoption of the United Nations Convention on the Law of Non-Navigational Uses of International Watercourses (UNCLNUIW 1997) – and SADC member states felt that it was essential for the Protocol to be closely aligned with the Convention. A discussion paper was then circulated to all SADC member states for comment and the results were incorporated into the Revised SADC Protocol on Shared Watercourse Systems (SADC 2001). The Revised Protocol, which contains all the key elements of the United Nations Convention on Shared Watercourse Systems, has been signed by all 14 SADC member states and has already been ratified by 3 member states. The original Protocol, which entered into force in 1998, will remain in force until 12 months after the Revised Protocol has come into force (Ramoeli 2002).

The overall objective of the Revised SADC Protocol on Shared Watercourse Systems is to advance the sustainable, equitable, and reasonable utilization of shared watercourses and to promote coordinated and integrated environmentally sound development and management of shared watercourses. In order to do this, it is recognized that the processes of research and technology development, public participation, information exchange, capacity building, and the application of appropriate technologies in shared watercourses management need to be promoted (SADC 2001; Ramoeli 2002).

The Permanent Okavango River Basin Commission (OKACOM)

Both the original (SADC 1995) and revised (SADC 2001) versions of the SADC Protocol on Shared Watercourse Systems require the establishment of river basin institutions to manage shared water resources (Ramoeli 2002). Within this context, as well as in response to public opinion and perceived and actual threats to the Okavango Basin, Angola, Botswana, and Namibia signed an agreement in 1994 to form the Permanent Okavango River Basin Commission (OKACOM). The objectives of the OKACOM are to advise the respective governments on technical issues relating to the conservation, sustainable development, and utilization of the shared water resources of the Okavango Basin (OKACOM 1994). A key part of the importance of OKACOM is that it provides a regional example of a river basin commission that includes all riparian states and establishes a precedent that places a burden of responsibility on their commitment jointly to manage the basin in a participatory and sustainable manner (Ramoeli 2002; Ashton and Neal 2003). Articles 2 and 5 of the OKACOM Agreement state that the commission may liaise with advisors on particular issues that are relevant to the Okavango Basin to ensure sound decision-making (OKACOM 1994). Although the onus lies with OKACOM to ensure appropriate stakeholder involvement in discussions held by the commission, it is encouraging that non-contracting party delegates may be included in decision-making processes.

The national context

In the context of national policies and legislation, Angola, Botswana, and Namibia have clear policies and laws that govern the ownership and use of water resources. These are summarized in table 9.3.

Within Angola, the management and use of water are currently regulated as part of the Environmental Framework Law (Republic of Angola 1998). This law is based on the Angolan Constitution (Republic of Angola 1992) and falls within the ambit of the Department of Water Affairs in the Ministry of Fisheries and Environment (Russo, Rogue, and Krugmann 2002). The Department will administer Angola's new water law when it comes into effect (ANGOP 2002). Meanwhile, water use in Angola is administered by the Department of Agriculture, because agriculture is the largest water-use sector in the country (Ashton and Neal 2003). Water resource management is decentralized to provincial authorities wherever possible (ANGOP 2002) and is guided by two key documents – the National Environmental Management Programme (PGNA) and the National Environmental Strategy (ENA) – both of which include provisions for public consultation and participation processes (Russo, Rogue, and Krugmann 2002).

The Ministry of Mineral Resources and Water Affairs, through the Department of Water Affairs, is responsible for the conservation and protection of water resources in Botswana (Khupe 1994). Botswana's Constitution and its national policies stipulate that all activities that can

Table 9.3 National policies, legislation, and management plans pertaining to water resource management in Angola, Botswana, and Namibia

Country	Policies, legislation, and management plans
Angola	Constitution of the Republic of Angola (Republic of Angola 1992). Environmental Framework Law (Republic of Angola 1998). Angolan Water Law (ANGOP 2002). National Environmental Management Programme (PGNA) (Russo et al. 2002). National Environmental Strategy (ENA) (Russo et al. 2002).
Botswana	Constitution of the Republic of Botswana (Republic of Botswana 1990). Water Act of 1968. National Water Master Plan (SMEC/KPB/SGAB 1992).
Namibia	Constitution of the Republic of Namibia (Republic of Namibia 1989). Water Act No. 54 of 1956 (including amendments up to 1979). Water Amendment Act No. 22 of 1985. Namibia's Second National Development Plan (NDP2) (Republic of Namibia 2001).

Source: adapted from Ashton and Neal 2003.

affect the use of water resources must be coordinated through the Department of Water Affairs (Republic of Botswana 1990, 1991; Khupe 1994). In addition, all developments related to water are required to meet the provisions of the National Water Master Plan (SMEC/KPB/ SGAB 1992) and the objectives of Botswana's National Development Plans (Khupe 1994). These documents were drawn up after an intensive public consultation process.

In Namibia, the Department of Water Affairs is part of the Ministry of Agriculture, Water and Rural Development (MAWRD), and is responsible for water resource management (Heyns 1995b). The control, conservation, and use of water is currently regulated by the Water Act No. 54 of 1956 and the Water Amendment Act No. 22 of 1985 (Heyns et al. 1998), which were originally promulgated in South Africa prior to, and shortly after, Namibia's transition to independence (Ashton and Neal 2003). A new Water Act for Namibia is in the process of being finalized (Republic of Namibia 2000a) and is aligned with the Constitution of the Republic of Namibia, which expresses the need to preserve the environment and prevent natural resource degradation (Republic of Namibia 1989). The key role that water plays in Namibia's development plans and the need to align the activities of all government departments that have an influence on the country's water resources is made explicit in Namibia's Second National Development Plan (NDP2) (Republic of Namibia 2001). Once again, processes of public participation and consultation are promoted in each of these documents.

Public-participation processes

Within the Okavango Basin states, public participation in local and national decision-making processes related to natural resource management is promoted through specific provisions in the national constitution of each country (Republic of Namibia 1989; Republic of Botswana 1990; Republic of Angola 1992). Typically, responsibility for management and decision-making are devolved to the lowest appropriate level (usually a local authority), although responsibility and accountability for "strategic" decisions (i.e. those with international implications) are still taken at national level. Each country has specific provisions in its national water and environmental management policies regarding the ownership and management of natural resources, including water, and defines responsibilities for achieving sustainable development goals (Turton, Ashton, and Cloete 2003).

Within each country, local and regional levels of government promote public participation in decision-making processes, while non-governmental organizations (NGOs) also play an important role in informing and shaping public opinion. Until recently, the civil war in Angola hampered effective participation by government and civil society; cessation of hostilities has revealed that the Angolan Government must now deal with a number of key priorities to rehabilitate the country's economy and infrastructure. Widespread poverty and pervasive ill health, combined with economic and infrastructural damage and the presence of numerous displaced communities and ex-combatants, have hampered government activities in the Angolan portion of the Okavango Basin (Ashton and Neal 2003; Turton, Ashton, and Cloete 2003).

In contrast, Botswana and Namibia have very active and extensive processes of public participation within the Okavango Basin. These involve individuals and communities; traditional leaders; local, regional and national government officials; and NGOs. Specific public concerns around the need for effective management of the Okavango Basin and the Okavango Delta were sparked by earlier attempts to withdraw water from the Okavango Delta in Botswana and the Okavango River in Namibia, as well as steadily declining river inflows to the Okavango Delta over the past twenty years. In both countries, several active NGOs and community associations are involved in activities to promote public awareness, as well as in the development and expansion of projects and actions designed to enhance the socio-economic status of rural communities.

Perhaps the most well-known NGO project in the Okavango Basin is the "Every River Has Its People" Project, which is funded by the Swedish International Development Cooperation Agency (SIDA) and run jointly by the Kalahari Conservation Society (KCS) in Botswana and the Namibia Nature Foundation (NNF) in Namibia ("Every River Has Its People" Project 2003). Jointly, these two NGOs have succeeded in developing an excellent basis of knowledge and information sharing among water resource managers in OKACOM, government departments, local communities, and traditional leaders. In both countries, the participants are enthusiastic about the project and its objectives and see this as an ideal example of how to involve communities and all levels of government in appropriate types of decision-making processes. In the near future, project participants and facilitators will visit selected sites in the Angolan portion of the Okavango Basin to initiate similar processes of public participation. Here, a partnership arrangement will be initiated with a suitable Angolan NGO that can facilitate the public participation process in Angola (Dr C. Brown, CEO of NNF, personal communication, 12 May 2003).

To date, participants from Botswana and Namibia have visited sections of the catchment within both countries and held meetings with the relevant OKACOM commissioners; they now have a far better appreciation of the situation in each country and more fully understand the needs and aspirations of local residents. This process of reciprocal participatory visits will be repeated (and extended) when the Angolan partners have been brought into the association ("Every River Has Its People" Project 2003).

The "Every River Has Its People" Project aims to promote and facilitate the effective participation of Okavango Basin stakeholders in natural resource decision-making and management, with a particular (though not exclusive) emphasis on water resources. In order to achieve this aim, its two primary objectives are as follows:

1. To increase the capacity of communities and other local stakeholders to participate effectively in decision-making processes at local, national, and regional (basin-wide) levels;
2. To develop mechanisms to assist communities and other local stakeholders to participate in natural resource management and decision-making activities, particularly those related to water resources, at local, national, and basin-wide levels ("Every River Has Its People" Project 2003).

The KCS is the leading Botswana environmental NGO, with a particular emphasis on the sustainable utilization of natural resources to benefit local communities. Overall guidance of project implementation in Botswana is provided by the Botswana Steering Committee, which meets on a quarterly basis. The committee functions on a consensus basis and works to ensure that there is close communication and coordination among project partners within Botswana ("Every River Has Its People" Project 2003).

The NNF is a non-profit, environmental NGO, the mission of which is to promote sustainable development, the conservation of biological diversity and natural ecosystems, and the wise and ethical use of natural resources for the benefit of all Namibians ("Every River Has Its People" Project 2003). The Project Manager at NNF provides overall project management for activities within Namibia, with guidance by the Namibian Steering Committee. This committee is made up of representatives from NNF, Integrated Rural Development and Nature Conservation (IRDNC), the Desert Research Foundation of Namibia (DRFN), and the Rössing Foundation. The IRDNC is a Namibian NGO and Trust that links conservation and the sustainable use of wildlife and other natural resources to the social and economic development of rural communities in Namibia. The DRFN is an NGO dedicated to creating and furthering awareness and understanding of arid environments and developing the capacity, skills, and knowledge of people to manage arid environments appropriately. The Rössing Foundation aims to promote education in general, foster greater understanding among Namibians, and, through environmental education and networking, improve living standards of Namibians, the sustainable use of natural resources, community-based natural resource management, capacity building, and training.

The "Every River has its People" Project targets riparian communities along or near the Okavango River in Namibia, and those living within or around the periphery of the Okavango Delta in Botswana. It works on a project-by-project level. The project team undertakes extensive socio-ecological surveys in partnership with local communities, regional and local authorities, line ministries, schools, traditional leaders, and NGOs. In this process, members of the project team help to improve people's understanding of the Okavango system as a whole and of the management challenges faced by each country and the OKACOM commission. From these partnerships, a basin-wide community forum was established in 2001 to support and liaise with OKACOM ("Every River Has Its People" Project 2003).

This project has demonstrated clearly that the active incorporation of all stakeholders, from local community members to international funding agencies, is the key to ensuring the success of conservation and management initiatives in the Okavango Basin.

Challenges that have to be faced

The socio-economic conditions within each of the three basin states are quite different, and this poses different scales of challenges that must be overcome. In practical terms, this means that the priorities of each national government vary from those of its neighbours. In addition, government institutions tend to focus most of their attention on more strategic, national-scale issues while NGOs and community-based organizations (CBOs) provide a large measure of the technical and logistical support needed to promote local processes of public consultation and participation in decision-making. In contrast to the situation that prevails in Botswana and Namibia, the Angolan Government faces an enormous humanitarian crisis as it struggles to deal with the aftermath of a protracted civil war (Porto and Clover 2003; Turton, Ashton, and Cloete 2003).

Following the signing of the peace accord between the Government of Angola and the leaders of the UNITA [União Nacional para a Independência Total de Angola (National Union for the Total Independence of Angola)] movement on 4 April 2002, a clearer picture is gradually beginning to emerge of the extent and nature of the tasks needed to rehabilitate, sustain, and develop the country's economy (Porto and Clover 2003). The viciousness, severity, and duration of armed conflict in Angola left a legacy of over 1.5 million casualties, some 4 million internally displaced people (IDPs or *deslocados*, amounting to approximately one-third of the population), and close to 0.5 million refugees in neighbouring countries (Porto and Parsons 2003). In addition, it has been estimated that 8–10 million land-mines, both anti-tank and anti-personnel, have been laid in mostly unmarked minefields across some 50 per cent of Angola, making it one of the most heavily mined countries in the world (Porto and Clover 2003). Angola's challenges are therefore as great as they are varied. Importantly, most development priorities will have to wait until the Angolan authorities have addressed the critical issues of re-settling internally displaced people, extending and consolidating government administration processes to areas previously controlled by UNITA, and the socio-economic reintegration of ex-combatants from both sides.

A peaceful Angola is often considered as having all the necessary resources and conditions to become an economic powerhouse in Southern Africa (Porto and Clover 2003). However, socio-economic development will be severely constrained for many years by present conditions, including the fragmentation and destruction of much of the country's transport, communications, health, and administrative infrastructure. In fact, apart from the problems linked to the removal of land-mines and the rehabilitation of ex-combatants and displaced persons, Angola faces enormous social, economic, and humanitarian challenges. According to the United Nations common country assessment (UN 2002), some of the medium- to long-term challenges include the following:

- reduction of urban and rural poverty through policies that promote improved access of the poor to employment and other resources;

- adequate responses to high levels of urbanization and other demographic problems;
- economic diversification, away from excessive dependence on oil revenues, through policies that promote development of the non-oil sectors;
- rebuilding of social sectors, with a particular emphasis on basic social services;
- mounting of an effective national response to the HIV/Aids pandemic;
- development of political participation and democratic accountability; and
- strengthening of public administration, including systems for ensuring rigour and transparency in the management of public resources.

The Angolan portion of the Okavango Basin (see fig. 9.1) contains some of the most remote and sparsely populated portions of the country, which were referred to as "the lands at the end of the earth" during colonial times (Porto and Clover 2003). However, this region was a UNITA stronghold, and some of the most ferocious battles of the civil war were fought here. Numerous mines have been laid along all of the roads and encircling each urban centre, as well as along many parts of the border with Namibia and at all bridges and river-crossing points. As a result, road travel and access to the towns in the catchment (Menongue, Longa, Cuito Canavale, Mavinga, Savata, and Caiundo) is dangerous, and air transport to Menongue remains the most reliable means to access the catchment (Brown, personal communication, 2003).

Assessments and surveys conducted in the Angolan sector of the Okavango Basin have revealed high levels of malnutrition and pervasive poverty, as well as extremely poor health status due to the widespread incidence of malaria, diarrhoea, anaemia, and tuberculosis (UN 2002; Porto and Clover 2003). Population estimates vary widely and, since the cessation of hostilities, large numbers of people have migrated out of the area (Brown, personal communication, 2003).

In the process of restructuring and rehabilitating the national economy, the Angolan Government will probably seek to initiate hydropower and irrigation development projects on some of the catchment's river systems so that the local population can be fed and provided with basic services. Inevitably, this development scenario will require the construction of new water-management infrastructure (dams, pipelines, irrigation schemes, and water-treatment plants). Ironically, any concerted attempt to alleviate the humanitarian crisis in this region by improving people's access to food, shelter, energy, and wholesome water supplies in the basin could pose potential problems for the water resources of the lower Okavango Basin (Porto and Clover 2003; Turton, Ashton, and Cloete 2003).

The situation in the catchment is compounded by the difficulty in

Table 9.4 Comparison of some socio-economic characteristics for Angola, Botswana, and Namibia that illustrate potential challenges to attaining effective public participation

Characteristic	Angola	Botswana	Namibia
Population and language			
Population in catchment (no.)	850,000	135,000	150,000
No. of languages spoken in catchment (indigenous + official)	8 (+2)	5 (+2)	8 (+2)
Economic issues[a]			
Per capita GDP (PPP $ in 2000)	1,031	7,566	4,661
Population proportion below poverty line (PPP US$2/person/day)	75	50	56
Health issues			
Malaria prevalence per 1,000 people	288	49	265
Adults with HIV/Aids (%)	5.5	35.2	20.5
Access and communications			
Telephones and cellular phones per 1,000 people	6	240	85
Radios and TV per 1,000 people	60	175	140
Internet users per 1,000 people	0.5	14	5
Paved roads (km/1,000 people)	0.6	4.9	5.7

Source: data from CIA 2000, FAO 2000, UNAIDS 2002.
a. PPP, purchasing power parity.

accessing communities and the almost total lack of telecommunications infrastructure in the region. A brief summary of typical statistics that compare the relative ease of access, state of health, and availability of infrastructure in Angola, Botswana, and Namibia is shown in table 9.4. It is also important to understand that, although government authorities and NGOs in the catchment may have reasonable access to the communications infrastructure, few local residents enjoy such access. Against this backdrop, it is clear that it will be challenging to ensure that effective processes of consultation and public participation in decision-making are able to take place in the Angolan sector of the catchment.

The situation in the Namibian and Botswana sectors of the Okavango catchment is far more stable than that in Angola, and the communities in each country are able to actively participate in decisions that affect the ways in which they exploit the water and other natural resources available to them (Turton, Ashton, and Cloete 2003). Each country faces a range of challenges, although these are expressed as site-specific issues in different parts of the catchment. The Namibian and Botswana sectors of the Okavango catchment represent a relatively arid environment, and most communities tend to be located close to the available water resources. This concentration of human activities in close proximity to the

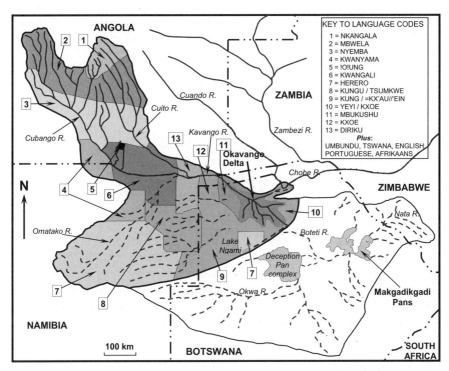

Figure 9.2 Sketch map of the Okavango Basin showing the spatial distribution of indigenous languages spoken by residents within the basin (Drawn from data taken from Summer Institute of Linguistics 2002)

water resources of the Okavango River and the Okavango Delta represents a growing dependency on these resources and could represent a potential threat to the ecological integrity of these systems if resource exploitation patterns are not carefully balanced by resource protection (Ashton and Neal 2003; Turton, Ashton, and Cloete 2003).

An additional layer of complexity is added when attempts are made to communicate effectively with local stakeholders living in the Okavango Basin. As illustrated in figure 9.2, catchment residents represent a wide variety of linguistic and cultural groupings, to the extent that there are 13 different indigenous languages in use as well as five "official" languages (Summer Institute of Linguistics 2002). This poses several practical problems to water resource managers in the three basin states, in terms of both communicating with individuals and respecting the various cultural norms and practices that prevail in different communities. To be fully effective, communication processes have to rely on the services of local translators, while electronic media such as radio programmes are used to inform communities living in more remote regions. This process appears to work well in Namibia and Botswana, but is less effective in Angola because of the scarcity of effective communications media (see table 9.4).

Externally, many international environmental NGOs and interest groups have focused their attention on the unique Okavango Delta ecosystems and have expressed their opposition to any form of water resource development in the upper catchment within Namibia and Angola (Greenpeace 1991). Although this interest and concern is understandable, it is not always helpful to the basin states that are trying their best to reach an agreement on equitable water sharing and joint management approaches in the basin. In addition, many of the press releases and articles written by these well-intentioned organizations and individuals contain factually incorrect information and personal perceptions that confuse local stakeholders, rather than clarifying the issues at stake. In particular, many of the articles create the impression that international interests can (and should) override the priorities and decisions of national governments in the basin states (Greenpeace 1991). At best, this situation can be misleading; at worst, it is counter-productive to effective public participation and decision-making processes.

The road ahead

The preceding discussion has highlighted a hierarchy of possible levels of participation in decision-making processes, ranging from the participation of state delegates in developing international conventions to local community members working with NGOs to address issues pertaining to resource use within the Okavango Basin. Although not all policy development and management decisions are open to all the interested and affected parties (owing, in part, to the strategic nature of some decisions), there are numerous mechanisms within the various institutional structures that encourage public participation in the Okavango Basin. While this hierarchy of decision-making exists, local community support is critically important since it is the community members on the ground, rather than state ministries or institutions, that ultimately determine whether water resource management principles, policies, and programmes are effective.

Some of the difficulties of attaining effective public participation within the Okavango Basin arise from the fact that Angola, Botswana, and Namibia are at different levels of social, political, and economic development, and that each country has different priorities and objectives in terms of their future needs for water. The challenges facing Angola's need for post-war rehabilitation of the state and its people differ completely from those of Namibia and Botswana. Although these differences are extreme in the case of the Okavango Basin, it is not uncommon for riparian states that share water resources to have different national needs and priorities. In order to reconcile these differences and develop a common understanding of water resource management, it is essential that the states concerned adopt an approach that encourages wide public participation and transparency of decision-making. Integrated water resource management (IWRM) approaches are based on these principles and have become widely accepted as offering the best way to achieve sustainable water resource management. In order to achieve these principles, the important roles played by NGOs should be recognized and encouraged, because they provide a vital link between individual and community water users, central government ministries and institutions, and multilateral institutions.

In essence, the Okavango Basin states have to reach consensus on how the water (and other) resources within the basin should be managed. This agreement should ensure that all the provisions and requirements of international treaties and accords, as well as regional (SADC) protocols, are complied with, within the sovereign limits of each state (Ashton and Neal 2003). Ideally, the basin states should work as partners to manage the resources of the Okavango Basin; this could best be achieved by the creation of a formal management structure such as a river basin organization (RBO). This RBO should function independently, though within the agreed mandate set by the three states concerned, and any external attempts to interfere, control, or direct the decisions and actions of the RBO should be resisted. Through concerted, joint decision-making, supported by effective processes of joint fact-finding and public participation, it will be possible to ensure that the Okavango system continues to meet the human needs for water while sustaining its unique array of ecosystems and the socio-economic activities that depend on these systems.

REFERENCES

Ali, S.H. 1996. "Water and Diamonds: The Lost Paradox of Resource Scarcity and Environmental Degradation in Southern Africa." Master of Environmental Studies Thesis. New Haven, CT: Yale University (unpublished).

ANGOP. 2002. "Law on Water Approved. News from Angola." Luanda: Angolan National Press Agency. 22 February. Internet: ⟨http://www.angola.org/news/newsdetail.cfm?NID=7615⟩ (cached version visited 8 November 2003).

Ashton, P.J. 2000a. "Potential Environmental Impacts Associated with the Proposed Abstraction of Water from the Okavango River in Namibia." *Southern African Journal of Aquatic Science* 25:175–182.

Ashton, P.J. 2000b. "Water Security for Multi-national River Basin States: The Special Case of the Okavango River." In M. Falkenmark and J. Lundqvist (eds). *2000 SIWI seminar – Water Security for Multi-National River Basin States – Opportunity for Development*. Stockholm: Swedish International Water Institute.

Ashton, P.J. 2002. "Avoiding Conflicts over Africa's Water Resources." *Ambio* 31(3):236–242.

Ashton, P.J. 2003. "The Search for an Equitable Basis for Water Sharing in the Okavango River Basin." In M. Nakayama (ed.). *International Waters in Southern Africa*. Tokyo: United Nations University Press, pp. 164–188.

Ashton, P.J., and E. Chonguiça. 2003. "Issues and Trends in the Regional Harmonization of EIA Processes in Southern Africa, with Special Reference to Trans-boundary and Cumulative Impacts." In E. Chonguiça (ed.). *The Need for Regional Approaches to Environmental Impact Assessment in Southern Africa*. Maputo: IUCN – The World Conservation Union.

Ashton, P.J., and R.E. Manley. 1999. "Potential Hydrological Implications of Water Abstraction from the Okavango River in Namibia." In *Proceedings of the Ninth South African Hydrological Symposium*. Stellenbosch: University of Stellenbosch.

Ashton, P.J. and M. Neal. 2003. "An Overview of Key Strategic Issues in the Okavango Basin." In A.R. Turton, P.J. Ashton, and T.E. Cloete (eds). *Transboundary Rivers, Sovereignty and Development: Hydropolitical Drivers in the Okavango River Basin*. Geneva: Green Cross International, pp. 31–63.

Bethune, S. 1991. "Kavango River Wetlands." In *Madoqua* 17(2), (Special Wetlands Edition: *Status and Conservation of Wetlands in Namibia*) pp. 77–112.

Biswas, A.K. 1993. *Management of International Water: Problems and Perspective*. Paris: UNESCO.

CIA. 2000. *The World Factbook: Country Listing*. Washington, DC: Central Intelligence Agency of the USA. Internet: ⟨http://www.cia.gov/cia/publications/factbook/geos/⟩ (visited 8 November 2003).

Conley, A.H. 1995. "A Synoptic View of Water Resources in Southern Africa." In *Proceedings of the Conference of the Southern Africa Foundation for Economic Research on Integrated Development of Regional Water Resources*, held at Nyanga, Zimbabwe, 13–17 November, 1995. Nyanga: Southern African Foundation for Economic Research. 32 pages.

Eckstein, G. 2002. "Development of International Water Law and the UN Watercourse Convention." In A.R. Turton and R. Henwood (eds). *Hydropolitics in the Developing World: A Southern African Perspective*. Pretoria: African Water Issues Research Unit (AWIRU), Centre for International Political Studies (CIPS).

Ellery, W.N., and T.S. McCarthy. 1994. "Principles for the Sustainable Utilization of the Okavango Delta Ecosystem, Botswana." *Biological Conservation* 70:159–168.

"Every River has its People" Project. 2003. "Objectives." Internet: ⟨http://www.everyriver.net/objectives.htm⟩ (visited 7 November 2003).

Falkenmark, M. 1989. "The Massive Water Scarcity Now Threatening Africa: Why Isn't It Being Addressed?" *Ambio* 18(2):112–118.

Falkenmark, M. 1994. "The Dangerous Spiral: Near-Future Risks for Water-Related Eco-Conflicts." In *Proceedings of the ICRC Symposium Water and War: Symposium on Water in Armed Conflicts.* Montreux: International Committee of the Red Cross, 21–23 November.

Falkenmark, M. 1999. "Competing Freshwater and Ecological Services in the River Basin Perspective – An Expanded Conceptual Framework." In *Proceedings of the SIWI/IWRA Seminar "Towards Upstream/Downstream Hydrosolidarity."* Stockholm: Stockholm International Water Institute.

FAO. 2000. *Aquastat Information Service on Water in Agriculture and Rural Development.* Rome: United Nations Food and Agriculture Organization. Internet: ⟨http://www.fao.org/ag/agl/aglw/aquastat/main/index.stm⟩ (visited 8 November 2003).

Gleick, P.H. 1999. "The Human Right to Water." *Water Policy* 1(5):487–503.

Greenpeace. 1991. *Okavango – Delta or Desert? A Question of Water.* Amsterdam: Greenpeace International.

Gumbricht, T., P. Wolski, P. Frost, and T.S. McCarthy. 2004. "Forecasting the Spatial Extent of the Annual Flood in the Okavango Delta, Botswana." *Journal of Hydrology* 290:178–191.

GWP. 2000. *A Vision of Water for Food and Rural Development.* The Hague: Global Water Partnership.

Heyns, P.S.v.H. 1995a. "Existing and Planned Development Projects on International Rivers within the SADC Region." In *Proceedings of the Conference of SADC Ministers Responsible for Water Resources Management,* held in Pretoria, South Africa, 23–24 November, 1995. Pretoria: Southern African Development Community. 14 pages.

Heyns, P.S.v.H. 1995b. "The Namibian Perspective on Regional Collaboration in the Joint Development of International Water Resources." *International Journal of Water Resources Development* 11(4):483–491.

Heyns, P.S.v.H., J. Montgomery, J. Pallett, and M. Seeley. 1998. *Namibia's Water: A Decision-Maker's Guide.* Windhoek: Desert Research Foundation of Namibia and Department of Water Affairs.

Hirji, R., P. Johnson, P. Maro, and T.M. Chiuta. 2002. *Defining and Mainstreaming Environmental Sustainability in Water Resources Management in Southern Africa.* Maseru/Harare/Washington, DC: SADC/IUCN/SARDC/The World Bank.

ILA. 1966. *Helsinki Rules on the Uses of Waters of International Rivers.* The Hague: International Law Association.

ILC. 1994. *Draft Articles on the Law of Non-Navigational Uses of International Watercourses.* The Hague: International Law Commission.

IUCN. 1993. *The IUCN Review of the Southern Okavango Integrated Water Development Project.* Gland: IUCN.

JVC. 1993. Central Area Water Master Plan: Phase 1, Volume 3: Water Demand. Report DIR/1/93/3 by Joint Venture Consultants, CES, LCE and WCE, for the Department of Water Affairs. Windhoek, Namibia.

Khroda, G. 1996. "Strain, Social and Environmental Consequences, and Water Management in the Most Stressed Water Systems in Africa." Internet: ⟨http://

web.idrc.ca/ev.php?ID=31133_201&ID2=DO_TOPIC⟩ (visited 8 November 2003).

Khupe, B.B. 1994. "Integrated Water Resource Management in Botswana." In A. Gieske and J. Gould (eds). *Proceedings of the Workshop on Integrated Water Resources Management*. Gaborone: Department of Water Affairs.

Lundqvist, J. 2000. "Rules and Roles in Water Policy and Management – Need for Clarification of Rights and Obligations." *Water International* 25(2):194–201.

McCarthy, T.S., and W.N. Ellery. 1998. "The Okavango Delta." *Transactions of the Royal Society of South Africa* 53(2):157–182.

McCarthy, T.S., A. Bloem, and P.A. Larkin. 1998. "Observations on the Hydrology and Geohydrology of the Okavango Delta." *South African Journal of Geology* 101(1):117.

McCarthy, T.S., G.R.J. Cooper, P.D. Tyson, and W.N. Ellery. 2000. "Seasonal Flooding in the Okavango Delta, Botswana: Recent History and Future Prospects." *South African Journal of Science* 96(1):25–33.

MGDP (Maun Groundwater Development Project). 1997. *Maun Groundwater Project Phase 1: Exploration and Resource Assessment – Executive Summary*. Gaborone: Department of Mineral, Energy and Water Affairs.

New Partnership for Africa's Development (NePAD). 2001. *Document of Accord*. 3 July. Internet: ⟨http://www.nepad.org⟩ (visited 8 November 2003).

Ohlsson, L. 1995. *Water and Security in Southern Africa. Publications on Water Resources, No. 1*. Stockholm: Swedish International Development Agency (SIDA), Department for Natural Resources and the Environment.

OKACOM. 1994. Agreement Between the Governments of the Republic of Angola, the Republic of Botswana and the Republic of Namibia on the Establishment of a Permanent Okavango River Basin Water Commission (OKACOM). Windhoek, Namibia, 15 September.

Pallett, J. (ed.). 1997. *Sharing Water in Southern Africa*. Windhoek: Desert Research Foundation of Namibia.

Porto, J.G., and J. Clover. 2003. "The Peace Dividend in Angola: Strategic implications for the Okavango Basin Cooperation." In A.R. Turton, P.J. Ashton, and T.E. Cloete (eds). *Transboundary Rivers, Sovereignty and Development: Hydropolitical Drivers in the Okavango River Basin*. Geneva: Green Cross International.

Porto, J.G., and I. Parsons. 2003. *Sustaining the Peace in Angola: An Overview of Current Demobilisation, Disarmament and Reintegration*. Pretoria/Bonn: Institute for Security Studies/Bonn International Centre for Conversion.

Ramberg, L. 1997. "A Pipeline from the Okavango River?" *Ambio* 26(2):129.

Ramoeli, P. 2002. "The SADC Protocol on Shared Watercourses: Its Origins and Current Status." In A.R. Turton and R. Henwood (eds). *Hydropolitics in the Developing World: A Southern African Perspective*. Pretoria, South Africa: African Water Issues Research Unit (AWIRU), Centre for International Political Studies (CIPS).

Republic of Angola. 1992. *Constitutional Law of the Republic of Angola*. Luanda: Government Printer. Internet: ⟨http://www.angola.org/referenc/constitution/constit.htm⟩ (visited 8 November 2003).

Republic of Angola. 1998. *Environmental Framework Law No. 5/98 of 19 June 1998*. Luanda: Government Printer.

Republic of Botswana. 1990. *Botswana National Conservation Strategy: National Policy on Natural Resources Conservation and Development*. Government Paper 1 of 1990. Gaborone: Government Printer.

Republic of Botswana. 1991. *National Development Plan 7 (1991–1997)*. Gaborone: Ministry of Finance and Development Planning.

Republic of Namibia. 1989. *The Constitution of the Republic of Namibia*. Windhoek: Government of the Republic of Namibia.

Republic of Namibia. 1995. *Namibia's Environmental Assessment Policy for Sustainable Development and Environmental Conservation*. Windhoek: Ministry of Environment and Tourism.

Republic of Namibia. 2000a. *National Water Policy White Paper*. Windhoek: Ministry of Agriculture, Water and Rural Development.

Republic of Namibia. 2000b. "Securing Water for the Future." Namibian Representative's Statement to the Second World Water Forum. The Hague, Netherlands, 16–22 March. Windhoek: Ministry of Agriculture, Water and Rural Development.

Republic of Namibia. 2001. *Second National Development Plan (NDP2): 2001/02–2005/06*. Windhoek: National Planning Commission (NPC).

Russo, V., P. Rogue, and H. Krugmann. 2002. "Environmental Assessment in Angola." In P.W. Tarr (ed.). *Environmental Assessment in Southern Africa*. Windhoek: Southern African Institute for Environmental Assessment (SAIEA).

Southern African Development Community (SADC). 1992. *Declaration and Treaty of the Southern African Development Community (SADC) Region*. Windhoek: Southern African Development Community. Internet: ⟨http://www.sadc.int⟩ (visited 8 November 2003).

Southern African Development Community (SADC). 1995. *Protocol on Shared Watercourse Systems in the Southern African Development Community (SADC) Region*. Gaborone: SADC Council of Ministers. Internet: ⟨http://www.sadc.int⟩ (visited 8 November 2003).

Southern African Development Community (SADC). 2001. *Revised Protocol on Shared Watercourse Systems in the Southern African Development Community (SADC) Region*. Windhoek: Southern African Development Community. Internet: ⟨http://www.sadc.int⟩ (visited 8 November 2003).

SARDC. 1996. *Water in Southern Africa*. Harare: Southern African Research and Documentation Centre.

Shela, O.N. 1996. "Water Resource Management and Sustainable Development in Southern Africa: Issues for Consideration in Implementing the Dublin Declaration and Agenda 21 in Southern Africa." In *Proceedings of the Global Water Partnership Workshop*. Windhoek, Namibia, 6–7 November.

SMEC (Snowy Mountains Engineering Corporation), Australia. 1987. *Southern Okavango Integrated Water Development Project Phase 1: Final Report, Technical Study*, five volumes. Report to the Department of Water Affairs, Gaborone, Botswana.

SMEC/KPB/SGAB. 1992. *Botswana National Water Master Plan Phase II: Final Report.* Gaborone: Ministry of Mineral Resources and Water Affairs.

Summer Institute of Linguistics. 2002. *Ethnologue: Languages of the World,* 14th edn. Dallas: Summer Institute of Linguistics (SIL). Internet: ⟨http://www.ethnologue.com⟩ (visited 8 November 2003).

Taylor, E.D., and S. Bethune. 1999. "Management, Conservation and Research of Internationally Shared Watercourses in Southern Africa: Namibian Experience with the Okavango River and the Rivers of the Eastern Caprivi." *Southern African Journal of Aquatic Sciences* 24(1, 2):36–46.

Turton, A.R. 1999. "Water and Social Stability: The Southern African Dilemma." Paper presented at the 49th Pugwash Conference on Science and World Affairs: "Confronting the Challenges of the 21st Century." Rustenburg, South Africa, 7–13 September.

Turton, A.R., P.J. Ashton, and T.E. Cloete. 2003. "An Introduction to the Hydropolitical Drivers in the Okavango River Basin." In A.R. Turton, P.J. Ashton, and T.E. Cloete (eds). *Transboundary Rivers, Sovereignty and Development: Hydropolitical Drivers in the Okavango River Basin.* Geneva: Green Cross International, pp. 6–30.

UN. 2002. *Angola: The Post-War Challenges: Common Country Assessment.* Luanda: United Nations County Office.

UNAIDS. 2002. Epidemiological Fact Sheets on HIV/AIDS and Country Profiles. Geneva: UNAIDS/WHO. Internet: ⟨http://www.unaids.org/⟩ (visited 8 November 2003).

UNCCD (United Nations Secretariat for the Convention to Combat Desertification). 2001. "United Nations Convention to Combat Desertification in Countries Experiencing Serious Drought and/or Desertification, Particularly in Africa." Bonn. Internet: ⟨http://www.unccd.org⟩ (visited 7 November 2003).

UNCED (United Nations Conference on Environment and Development). 1992a. Convention on Biological Diversity. Rio de Janeiro. Internet: ⟨http://www.biodiv.org⟩ (visited 7 November 2003).

UNCED (United Nations Conference on Environment and Development). 1992b. United Nations Framework Convention on Climate Change. Rio de Janeiro, Brazil. Internet: ⟨http://unfccc.int/resource/convkp.html⟩ (visited 8 November 2003).

UNCLNUIW. 1997. *Convention on the Law of the Non-Navigational Uses of International Watercourses. United Nations General Assembly Document A/51/869.* April 11. New York: United Nations Publications.

UNDP/FAO. 1976. *Investigation of the Okavango as a Primary Water Resource for Botswana.* Rome: Technical Report to United Nations Development Programme and United Nations Food and Agricultural Organization, UNDP/FAO, BOT/71/506, three volumes.

UNDP/GEF. 2001. *1:2,000,000 Scale Map of the Okavango River Basin.* Prepared by the Cartographic Section of the Department of Public Information of the United Nations, for the Permanent Okavango Basin Commission (OKACOM), as part of the UNDP/GEF Regional Project RAS/96/G42 on Integrated Management of the Okavango River Basin. Washington, DC: Global Environment

Fund. Internet: ⟨http://www.un.org/Depts/Cartographic/map/other/Okavango.
pdf⟩.

United Nations Educational, Scientific and Cultural Organization (UNESCO).
1971. Ramsar Convention on Wetlands of International Importance Especially
as Waterfowl Habitat. Internet: ⟨http://www.ramsar.org⟩ (visited 8 November
2003).

Van der Zaag, P., and H.H.G. Savenije. 2000. "Towards Improved Management
of Shared River Basins: Lessons from the Maseru Conference." *Water Policy*
2:47–63.

Van der Zaag, P., I.M. Seyam, and H.H.G. Savenije. 2000. "Towards Objective
Criteria for the Equitable Sharing of International Water Resources." In *Proceedings of the Fourth Biennial Congress of the African Division of the International Association of Hydraulic Research*. Windhoek: International Association
of Hydraulic Research.

Wilson, B.H., and T. Dincer. 1976. "An Introduction to the Hydrology and Hydrography of the Okavango Delta." In *Proceedings of the Symposium on the
Okavango Delta and its Future Utilization*. Gaborone: Botswana Society.

Wolf, A.T. 1999. "Criteria for Equitable Allocations: The Heart of International
Water Conflict." *Natural Resources Forum* 23:3–30.

Wouters, P. 1999. "The Legal Response to International Water Scarcity and Water Conflicts: The UN Watercourses Convention and Beyond." In *The German
Year Book: International Law 293*. Dundee: International Water Law Research
Institute.

Part III

International institutions

10

Access to information, public participation, and conflict resolution at the World Bank

Charles E. Di Leva[1]

Introduction

At the conclusion of the Second World War, new structures were created to address the global economic order. The International Monetary Fund (IMF) and the International Bank for Reconstruction and Development (IBRD) were established at the Bretton Woods Conference. The IMF would serve to help achieve global currency stability and correct balance-of-payment disturbances (Shihata 1991). The IBRD (now commonly referred to as the "World Bank") was initially conceived of largely by the United States and established in 1946. [The World Bank Group consists of the IBRD, International Development Association (IDA), the International Finance Corporation (IFC), Multilateral Investment Guarantee Agency (MIGA), and the International Center for Settlement of Investment Disputes (ICSID). For purpose of this chapter, though, the World Bank means IBRD and IDA.]

The original aim of the IBRD was to reconstruct countries damaged by the war. Since that time, the Bank's Articles of Agreement have been left largely unchanged. They state that the Bank's purpose, in part, is

[t]o assist in the reconstruction and development of territories of members by facilitating the investment of capital for productive purposes ... and the encouragement of the development of productive facilities and resources in less developed countries. (IBRD 1949, IDA 1960)

The Bank's Articles also state that the Bank shall not interfere in the political affairs of any of its members.

The commitments of governments to the Bank's Articles means that the Bank was created under the principles and rules of public international law (Shihata 1991). In addition, the Bank carries out its operations to be in compliance with applicable international law principles and rules.

Today, the World Bank stands as the world's largest multilateral development bank, with more than 180 countries as participating members. Consistent with the objectives set forth in its Articles, in recent years, the Bank's mission has been restated so that its goal is to fight global poverty using its financial and technical resources in collaboration with a wide range of partners to support sustainable development in developing countries around the world.

As with other multilateral development banks, the World Bank provides assistance to developing countries for their infrastructure and other needs, and such assistance has long included hydropower development. A number of these hydropower projects have been situated within transboundary watersheds. To seek to ensure that these projects are prepared, developed, and implemented in a way that comports with its Articles and the objectives of sustainable development, the World Bank applies environmental, social, financial, and technical policies, procedures, and guidelines. The World Bank website (http://www.worldbank.org) sets out the full text of its operational policies and procedures as well as supporting documents (Di Leva 1998). These policies and guidelines have continually evolved, and are considered by many to represent international best practice in the field of development finance.

The World Bank has sought to ensure harmony in the application of its environmental and social guidelines with its sister organizations of the World Bank Group. This harmonious arrangement is apparent in the similarity of the environmental and social policies and guidelines of the IFC, the World Bank's private-sector counterpart. The general recognition of these policies and guidelines as international best practice was most recently evident in June 2003, when ten major commercial banks announced that they would voluntarily adhere to the "Equator Principles," a set of guidelines that are based on the environmental and social policies and guidelines of the World Bank and the IFC. The Equator Principles will apply to all project finance above US$50 million (World Bank 2003b). The reference to "guidelines" implies that the Principles also include reference to the industrial-sector guidelines set forth in the World Bank Group's *Pollution Prevention and Abatement Handbook* (World Bank 1998). This handbook includes, *inter alia*, a series of industrial-sector guidelines that are "normally applicable" to Bank- and IFC-financed investments (World Bank 1999).

Even in those cases in which the Bank is not the only financier of a project, the Bank still seeks to ensure that project development be conducted in a manner that complies with a set of policies that includes 10 environmental and social "safeguard policies." In this manner, the Bank has sought to ensure access to relevant information, sound public participation, and meaningful consultation; that the impacts of the projects be addressed and managed; and that, if the Bank has failed to comply with these policies and people are adversely affected as a result, they have recourse to an independent inspection mechanism to address their claims.

The Bank's approach to these issues was not always as all-encompassing and geared toward ensuring meaningful public participation in the design and implementation of large-scale hydropower and other infrastructure projects. It arrived at this stage after a long and sometimes arduous process during which the Bank was often the target of extensive criticism by a range of sectors of civil society. The Bank's experience mirrors some fundamental changes in development finance. This chapter provides a brief background to these activities, followed by an overview of how, in the setting of large-scale hydroelectric and other projects on international waters, the three interrelated objectives of access to information, public participation, and conflict resolution are currently addressed in the World Bank.

Historical background

Beginning with the Bank's involvement in 1950s in the Indus River Basin and the riparian dispute in that area, the Bank has long been involved in major infrastructure projects, including those on international, or transboundary, watersheds (Krishna 1998; Salman and Boisson de Chazournes 1998). Throughout its early years, these projects were largely prepared pursuant to economic models that recognized the urgency of accelerating development as soon as possible, and in which concerns about environmental and social impact were not always at the forefront. As a result, some commentators have noted that these projects were often dominated by the views of central state planners, who focused almost solely on the economic costs and benefits of state-centric activities. It is viewed by some that this was certainly the case in projects that would dominate the early years of the IBRD (Bradlow 2001).

As the understanding about the impacts of large-scale infrastructure grew, the policies and procedures to address them also evolved. The World Bank was at the forefront of the international financial organizations in designing the new policy instruments. None the less, the Bank was often criticized for not fully complying with these policies and for not recognizing the environmental and social impact of large infrastructure projects.

This issue was highlighted in the mid-1980s in connection with the Bank's financial investments in the Sardar Sarovar Dam and irrigation projects in the Narmada Valley in India. Because of the difficulties concerning these projects, the World Bank undertook a precedent-setting measure to set up an independent review of the projects. The Morse–Berger Commission undertook a lengthy review in the Narmada Valley and in the surrounding states, producing perhaps the most extensive report of its kind (Morse and Berger 1992). The Commission noted, in its covering letter to the World Bank President Lewis T. Preston, that the Bank deserved credit for striving to implement policies that would properly resettle and care for people that had to be involuntarily relocated owing to the project and to properly address environmental impacts. At the same time, despite these efforts, the report also found "fundamental failures in the implementation of the ... Projects." The report noted that:

We think the Sardar Sarovar Projects as they stand are flawed, that resettlement and rehabilitation of all those displaced by the Projects is not possible under the prevailing circumstances, and that the environmental impacts of the Projects have not been properly considered or adequately addressed. Moreover, we believe the Bank shares responsibility with the Borrower for the situation that has developed.

Although this form of criticism of Bank activity had been voiced in other projects, it was the first time that the Bank had voluntarily put itself on the firing line and that the critics were in such a prominent position to shed light on the difficulties of large-scale development finance.

The difficulties related to this project provided ammunition for the view that the Bank needed to strengthen its policies and be held to greater accountability. In particular, non-governmental organizations (NGOs) pressed for greater access to information, public participation, and development of an accountability mechanism. Each of these areas has since been extensively addressed by the Bank. Since the Morse–Berger Commission, the Bank has revised many of its policies and procedures, opened up many avenues of communication and public participation, and set up measures of accountability (Bernasconi-Osterwalder and Hunter 2002). In each of these areas, other multilateral banks as well as many private-sector institutions have since followed suit. As an especially pertinent recent example, the Bank had a large part to play in the establishment of the World Commission on Dams (WCD) (Salman 2001). The WCD was established by the international community to review the history and impact of large dams and to propose "a new framework for decision-making" for the siting of hydroelectric facilities. The Bank and IUCN (The World Conservation Union) served as secretariat agencies for the WCD, which was born out of a workshop that they initiated in Gland, Switzerland in 1997. Although the Bank could not completely endorse all aspects of the WCD final report, the Bank's important role with WCD and other activities related to access to information and public participation were recently noted by the US Department of the Treasury in its 2001 Annual Report to the US Congress on the Environment and Multilateral Development Banks (Annual Report 2003).

Perhaps even more significant has been the Bank's continuing work on river basin initiatives. Over the past decade, the Bank has worked on many of the major international waters in Africa, most recently helping to complete the first Transboundary Environmental Analysis for the Nile River Basin (Nile Basin Initiative 2001). Through this extensive effort, stakeholders along the length of the Nile participated in the process of looking at transboundary issues.

Against this background, one can deduce that the issues of access to information, public participation, and accountability mechanisms are increasingly recognized as key to the sustainability of development finance. It is well established that, when the public is meaningfully involved during the project preparation and implementation stages, the success of such projects increases. Further detail on each of these three areas is discussed below.

The Bank's approach to sustainable development has kept the Rio Principles of Environment and Development at the forefront of its efforts. Many, if not all, of the Rio Principles are reflected in the World Bank's environmental and social policies – the previously mentioned "safeguard policies" of the Bank. Rigorous implementation of these policies is intended to ensure that Bank projects "do no harm" to people and the environment and that they try to deliver positive environmental and social benefits to those affected by the project.

Indeed, public participation and access to information is addressed through a series of ten "safeguard" policies and procedures as well as a policy on information disclosure. These safeguard policies include policies on environmental assessment, international waterways, natural habitats, indigenous people, forests, safety of dams, and involuntary resettlement. In this regard, the Bank's safeguard framework supports implementation of Rio Principle 10, which states, in pertinent part:

Environmental issues are best handled with the participation of all concerned citizens, at the relevant level. At the national level, each individual shall have appropriate access to information concerning the environment that is held by public authorities, including information on hazardous materials and activities in their communities, and the opportunity to participate in decision-making processes by making information widely available. (UNCED 1992)

Most recently, the Bank took further steps to support Rio Principle 10 during the 2002 World Summit on Sustainable Development in Johannesburg, South Africa. At the Summit, the Bank stated that it would become a partner in the Partnership for Principle 10, a partnership officially listed by the Commission on Sustainable Development as part of the Type II outcomes (Partnership for Principle 10 2003). As a partner, the Bank commits itself to join in efforts to help improve the policies and practices of international organizations as they relate to access to information, participation, and justice.

Access to information

Taken as a whole, the Bank's safeguard policies express the need to ensure public participation in Bank-financed projects. They contain numerous requirements for access to be provided to the public at a stage at which they can have meaningful input. For example, the Bank's policies on Environmental Assessment and Disclosure of Information require that a draft environmental assessment be made available to the affected communities in a language that they understand while it is still in draft form (World Bank 1999). In addition, the Bank recently revised its disclosure policy to increase access to information. The revised Policy on Disclosure of Information became effective on 1 January 2002. Under this new policy, the Bank continues its presumption in favour of disclosure, and certain types of information that were previously not disclosed will now be made available to the public. At the same time, the revised policy continues to recognize certain "constraints," such as that the Bank will keep confidential certain materials, including those "provided to the Bank on the explicit or implied understanding that they will not be disclosed outside the Bank ..." (World Bank 2002). Such constraints recognize the continuing tension between the need to protect certain proprietary information and the benefits of public access to information.

Increasingly, the public is of the view that economic information previously considered proprietary should be disclosed. These concerns were highlighted by recent events of corporate wilful withholding of information, such as those associated with the Enron débâcle. In addition, disclosure advocates contend that economic information, such as rates of return on investment, have a direct link with the environmental and social sustainability of a project. Thus, particularly when projects receive public finance, these advocates argue that the public should have access to the information upon which financial planners rely. On the other hand, profit-seeking enterprises contend that they need to withhold this information, because to divulge it would provide an unfair advantage to their competitors.

In general, all World Bank projects must disclose to the public project-related information as soon as the project concept has been developed. This information must be made available at the location of the relevant country office; at Bank headquarters in Washington, DC; and on the Bank's website. It must also be made available in the local language, so that potentially affected people can understand it.

Public participation

Ensuring that the public has access to information is not the same as ensuring that the public can participate in a meaningful manner. The Bank recognized this important distinction during its preparation of the *World Bank Participation Sourcebook*. This manual was prepared based upon the experience of over 200 Bank staff and consultants (World Bank 1996). The manual provides a variety of measures to help ensure the engagement of affected people in the various plans that the Bank uses to implement environmental and social mitigation measures. These measures include ensuring that affected people can participate in the environmental impact assessment process that leads to the environmental management plan, as well as plans related to involuntary resettlement, indigenous people, and the safety of dams.

Recently, the issue of public involvement was highlighted in the work of the WCD. Following discussion among various members of the Bank, the Bank set out on its website its "Position on the Report of the World Commission on Dams" (World Bank 2001c). This statement noted that the Bank decided not to endorse all of the 26 specific guidelines of the WCD, but that it "shares the WCD enumerated five core values and concurs with the need to promote the seven strategic priorities." The importance of public participation is addressed in these values and priorities: the core values consist of equity, efficiency, participation, sustainability, and accountability; and the strategic priorities include "gaining public acceptance."

In relation to the 26 guidelines, the Bank noted that they were not considered to be binding, but should be viewed as "guidance" (World Bank 2001c). In addition, the Bank statement referred to certain WCD recommendations that caused specific concerns, particularly regarding the issue of state sovereignty. In response to these guidelines, the Bank noted that "the State retains the right to make decisions that it regards as being in the best interest of the community as a whole...." In addition, the Bank stated that, although the Bank's resettlement policy is built on the principle of informed participation, the Bank policy "does not require the negotiation of development and mitigation plans." The third WCD recommendation from which the Bank distanced itself concerned indigenous peoples. The WCD Report recommended that indigenous and tribal peoples be given the right of free and prior informed consent. World Bank policies make clear that there must be prior, meaningful participation of indigenous peoples ahead of detailed project preparation and that, where the project goes forward, plans must be made to ensure mitigation of any harm and the provision of benefits. These plans are incorporated in the project's legal agreements between the borrower and the Bank. At the same time, however, the Bank would not seek to infringe "the right of the State to make decisions which it judges to be the best solution for the community as a whole." Thus, although state sovereignty is respected, meaningful and prior participation remains a basic principle for all Bank projects.

As a concrete example of this principle put into practice, the Bank helped facilitate the negotiation of the recently concluded Water Charter of the Senegal River, described in more detail by Fall and Cassar in chapter 11 of this volume. This Charter is to be ratified by the Republic of Mali, the Islamic Republic of Mauritania, and the Republic of Senegal, all members of the Organisation Pour La Mise en Valeur du Fleuve Sénégal (OMVS) through Resolution No. 005 of the Conference of the Heads of the State and Government. The Charter expressly leaves open room for accession by the Government of Guinea, the most upstream riparian. Three of the four major riparian states within this transboundary river basin agreed to this historic precedent-setting agreement, which includes a number of advances, including providing for observer status for non-governmental representatives in the official body that determines water allocation. In addition, the Charter states that "the riparian States will ensure public access to all information pertaining to the condition of the Senegal River's waters, the measures planned or taken to ensure regular river flow, as well as water quality."

Nevertheless, critics of the Bank point to claims filed with the Inspection Panel in large-scale hydroelectric projects as examples that the World Bank failed to comply with its policies concerning public participation and meaningful consultation. It is, of course, difficult to capture objectively when consultation with the public has been "meaningful." To address this difficulty, many development organizations such as the Bank have taken steps to become more systematic in their approach to public meetings and consultations. In fact, the US Treasury Multilateral Development Bank (MDB) Report noted that "[t]he involvement of civil society in MDB activities has increased greatly over the past ten years." In some hydroelectric projects, this increased engagement has included the use of local universities to carry out public surveys to record the communities' views on a broad scale. Following claims of lack of adequate consultation in areas adjacent to the Yacyreta Hydro-Electric Facility located in both Argentina and Paraguay, the World Bank supported surveys conducted by the locally based Catholic University. In addition, the Bank and other development banks have increased the use of NGOs during the preparation and monitoring of projects.

The ability of NGOs to help ensure meaningful participation has led to their increased engagement in World Bank projects and is expressly recognized in Bank practice (Shihata 1994; World Bank 2000). In addition, the Bank has established an NGO and Civil Society Unit at Bank headquarters, and it has adopted guidelines for public consultations on Bank projects. The Bank's website carries and updates information about all active and formally proposed Bank projects, and the Bank's Infoshop is the depositary responsible for receiving and posting all information.

In response to the WCD Guidelines that set forth the basis upon which the WCD believed the participation values and priorities should be implemented, the Bank issued a position paper noting that the Bank believes that its operational policies and procedures provide the basis to ensure that key stakeholders are "systematically identified and involved in project planning and implementation: upstream meaningful consultations are held with affected groups to guide project decision making and their views and preferences are reflected in the plans developed as an integral part of the project." (World Bank 2001c). At the same time, the Bank recognizes the right of eminent domain, as well as the right of the State "to make decisions that it regards as being in the best interest of the community as a whole, and to determine the use of natural resources based on national priorities."

Similar to the Bank's Environmental Assessment policy, the Bank's policies pertaining to involuntary resettlement and indigenous peoples also require that affected people are given an opportunity to participate in planning and implementation (World Bank 2001b). Before becoming final, plans for resettlement must be shared with the affected community, which also must be given an opportunity to understand the plans in their local language and afforded a reasonable time to comment on the proposals. Implementation of agreed mitigation and development plans are negotiated and incorporated into a project's legal agreements. Further, despite a state's rights of eminent domain, the Bank participates in the monitoring of the various plans, and independent teams increasingly have been set up to review the implementation of the plans. Before a project is considered "complete," the agreed plans must be fully implemented and surveys must be completed to determine if the benefits of the plans have been delivered.

Similarly, where plans would affect indigenous people who live in or near the international watershed activity, Paragraph 8 of the Bank's policy on indigenous peoples (a policy that is currently being revised) states that the preparation of such plans should be prepared only after "affected hosts and resettlers have been systematically informed and consulted during preparation of the resettlement plan about their options and rights." (World Bank 1990; Davis 2001).

Although not directly related to civil-society access to information and public participation, two other important policies pertinent to international watercourses address projects on international waterways and the safety of dams. The policies in these two areas were consolidated and strengthened in the 1990s. Projects on international waterways frequently require extensive environmental assessment, as many of them involve engineering activity, such as irrigation and hydroelectric schemes. Thus, projects financed, in whole or in part, by the Bank must comply with the Bank's environmental assessment policy.

These projects must also address long-standing principles about riparian use of shared waterbodies, including those concerning the process by which states inform each other of their proposed use. The Bank has sought to support these principles through its "Projects on International Waterways" (World Bank 2001a). [For the purposes of this policy, with one exception, the Bank defines international waterways to be consistent with the "international watercourses" addressed by the UN Convention on the Law of the Non-Navigational Uses of International Watercourses.] In Paragraph 4 of the policy, the Bank encourages early notice from the riparian party undertaking a project to other riparian users. This notice may be required under Paragraph 8(c), even when the riparian party undertaking the project is not in a position to cause harm to other riparian parties, but to ensure that the project "will not be appreciably harmed by the other riparians' possible water use." Presumably, this requirement also protects the financial investment in the project by ensuring that the importance of maintaining adequate water flows has been recognized by upstream riparian nations. This provision is not explicit in the Convention. Thus, Paragraph 8 provides that notice may be required even if the project would not cause appreciable harm.

Before presenting the loan to the Bank's Board, Paragraph 8 requires the Bank staff to ensure that the project is either covered by an agreement between the relevant riparian states that the project received a positive consent or no objection from the other riparians, or that "the project will not cause appreciable harm to the other riparians." In those instances where states may object to riparian projects by other riparians, the Bank has traditionally offered its services to facilitate and assist countries or regional organizations on those issues to be negotiated.

Conflict resolution

In 1993, shortly after the Morse–Berger Commission had completed its independent review of the Sardar Sarovar projects, the Bank established an Inspection Panel to address complaints from claimants who can establish that they were adversely affected because the Bank violated its own policies and procedures during the design, appraisal, or implementation of a Bank-financed project (World Bank 1993). This was the first step by any multilateral financial institution to set up such an accountability mechanism.

The Inspection Panel has authority, under Paragraph 12 of the Resolution establishing it, to "receive requests for inspection presented to it by an affected party in the territory of the borrower ..." (World Bank 1993). In addition, under Paragraph 12, the "affected party must demonstrate that its rights or interests have been or are likely to be directly affected by ... the Bank as a result of a failure of the Bank to follow its operational policies or procedures...." Bank Management is authorized to respond to the Panel concerning the eligibility of the request. Within 21 days of the Management response, the Panel can recommend to the Bank's Board of Executive Directors that the Panel proceed with an investigation. If the decision is to proceed, the public is provided with copies of the request along with the Panel's recommendation to proceed and the Director's decision. After an investigation is concluded, the Panel's findings are submitted to the Board and Management, and Management is given an opportunity to respond to the Board concerning the findings. After the Board has been able to consider these documents, it informs the requestors of the results of the investigation and any action to be taken. Once the requestors have been informed, the Bank makes available to the public the Panel's Report, the Management's recommendations, and the Board's decision. It is a specified mandate of the Panel to "seek to enhance public awareness of the results of investigations through all available information sources."

In carrying out its mandate, the Panel has the right to access all relevant project information, as well as to interview any person connected with the project. To assist in its work, the Panel can hire consultants to carry out research and field studies and can travel to inspect the project sites. It has an independent budget allocated to carry out its work.

Since its inception, the Panel has handled 28 claims. At least six of these have focused principally on large-scale hydroprojects on or potentially affecting international watercourses (World Bank 2003c). These are (Nepal) Arun III Proposed Hydroelectric Project; (Chile) Pangue/Ralco Complex of Hydroelectric Dams; (Argentina/Paraguay) Yacyreta Hydroelectric Project (two claims); (Brazil) Itaparica Resettlement and Irrigation Project; (Lesotho) Highlands Water Project; and (Uganda) Bujagali Hydroelectric Project. In almost all of these claims, local citizens have alleged that they were not adequately consulted on certain aspects of the project. International watershed projects do not stand out as unique in terms of the policies that are implicated; however, they are somewhat unique in terms of their scale and the number of safeguard policies that must be applied to the project preparation and implementation. This is especially true of those hydroelectric projects with large reservoirs that generate resettlement and major social and environmental impacts.

Indeed, conflict over the siting and viability of projects involving international waters, especially hydroelectric projects, is almost inevitable given that (in many cases) one part of civil society adamantly opposes hydropower and another believes that it is the best form of power available, especially for developing countries with extensive, untapped resources. Yet, as the World Bank has noted, the future of energy supply for large parts of the developing world, especially in Africa, may well be through hydroelectric sources (World Bank 2003a). Hydropower may be a way to try to avoid or mitigate greenhouse-gas production typically associated with conventional energy sources. In this light, the Bank has also noted that "[w]hile about 70 per cent of hydropower potential in Europe and North America is already tapped, in Asia, Latin American and Africa, only 20 per cent, 15 per cent and 5 per cent, respectively, has been developed" (World Bank 2003a).

The model of the Inspection Panel has since been followed by many other financial institutions, albeit with varying structures (Bernasconi-Osterwalder and Hunter 2002). The IFC, the private-sector arm of the World Bank Group, set up the Office of Compliance Advisor and Ombudsman (CAO). The CAO's first project engagement concerned the Pangue Hydroelectric project in Chile. After the Inter-American Development Bank (IDB) had set up a conflict-resolution mechanism, the first claim that it received also concerned a hydroelectric facility for which they provided some financing on an international waterway. That project, the Yacyreta facility, was also financed by the World Bank, and the World Bank is currently addressing the second Panel claim filed concerning that facility (the first was filed in 1997 and the second in 2002).

The fact that the first and almost immediate engagement of many of these mechanisms concerns large-scale hydro facilities on international waterways highlights the importance of improving public involvement in projects affecting international watercourses. Indeed, as the International Court of Justice (ICJ) has noted in its decision on the Gabčíkovo–Nagymoros facility situated on the Danube, there is a new way of looking at the environment (International Court of Justice 1997). Norms pertaining to protection of the environment have evolved, and this is reflected in the modern concept of sustainable development. The old methods of addressing major infrastructure are changing. As the ICJ concluded in Paragraph 140:

> Throughout the ages, mankind has, for economic and other reasons, constantly interfered with nature. In the past, this was often done without consideration of the effects upon the environment. Owing to new scientific insights and to a growing awareness of the risks for mankind – for present and future generations – of pursuit of such interventions at an unconsidered and unabated pace, new norms and standards have been developed, set forth in a great number of instruments during the last two decades. Such new norms have to be taken into consideration, and such new standards given proper weight, not only when States contemplate new activities but also when continuing with activities begun in the past. This need to reconcile economic development with protection of the environment is aptly expressed in the concept of sustainable development.

Against this background, one of the more recent, widely reported, struggles for public participation seems to be reflected once again in the environmental concerns about a hydroelectric facility – in a project that is not receiving World Bank financing. It is reported that the Three Gorges Dam measures 7,600 feet ($\approx 2,280$ m) across and 600 feet (≈ 180 m) high, making it the world's largest hydroelectric dam. It is also reported that the reservoir behind it will be so large that more than 1.3 million people will have to be resettled by the time construction is finished in 2009 (Pomfret 2003). According to news reports, there are some Chinese who wish to enhance the role of the public in the decisions taking place regarding this project. One activist who has launched a campaign concerning the dam noted that "[t]his is a long-term struggle ...." However, "[o]ur organization's deepest purpose is to try to build a civil society and promote democracy. I am confident it will happen. I am sure we will survive" (Pomfret 2003).

Conclusions

The growing body of international and domestic norms, institutional practice, and scholarly literature highlight the great variety of environmental and natural resource issues that confront projects in international watersheds across all regions of the globe. Nevertheless, a unifying theme is that the sheer scope, importance, and size of these watersheds requires that we ensure that affected people have access to information about the human impacts on these watersheds, that people can meaningfully participate and (it is hoped) benefit in decisions affecting their lives and, where differences about decisions emerge, that there is a means to resolve these differences in a peaceful and just way. Painful lessons have been learned when these aspirations have been ignored.

Fortunately, the international community has sought to improve the development paradigm, especially for projects with large social and environmental impacts. As a general matter, because of these impacts, the Bank has tried to improve its operational policies and has expanded its policy on disclosure of information, sought to fully integrate environmental and social issues in its lending programmes, and decentralized many of its operations. For projects in international watersheds, the World Bank has benefited from participating in the work of the WCD and related activities.

In addition to policy and operational improvements, the Bank has also tried to learn from some of the most difficult projects. One such lesson may arise in the revenue-sharing context. The Bank and many others are closely tracking the developments in local community benefit and revenue-sharing schemes in projects such as the World Bank co-financed Chad–Cameroon oil pipeline, to help local people to realize that they can share in the anticipated gains. For many, this revenue-sharing development is a key change in modern project finance. Under this scheme, a portion of the revenue from the pipeline will be dedicated to a fund that is not under the control of the government, but that has members of civil society empowered to make decisions on how these resources are directed. Although the Chad–Cameroon project involves the flow of oil, its lessons appear relevant in the context of revenue that could be generated by large-scale hydroelectric development.

In view of these lessons learned, a concern of this author is that the Bank's lessons and efforts at improvements be matched by other sources of development finance that, unlike the Bank, do not always receive the same degree of public scrutiny. It is a positive development that multilateral development banks are jointly working to harmonize in the area of environmental and social policies (having also done so in the areas of procurement and financial management). It is also positive that many major commercial banks are now pledging through the Equator Principles to follow the World Bank safeguard policies to help ensure the sustainability of large-scale project finance. More entities should fully embrace these environmental and social policies.

Even if all project finance entities fully adopt modern environmental and social policies, more needs to be done to ensure that projects comply with these and other relevant environmental and social policies. The adoption of good policies is laudable, but the real challenge is to ensure that they are put into practice and monitored to stay that way. To date, there has not always been consistency of implementation by both the international public and private finance sector communities. Governments also have sometimes been inconsistent in ensuring that projects are implemented according to design. Thus, another lesson to learn is to find ways to maximize incentives for compliance. One such incentive is provided through modern tools to maximize information flow. The environment and local people can benefit from the globalization of information. It is clear, therefore, that, as we go forward, there is an obligation to make constant the flow of all relevant information and to use these data to try to avoid adverse effects and to enhance project benefits to all people.

In conclusion, throughout history, projects located in international watersheds have posed special legal and operational challenges. Although some of these challenges will always remain, there is evidence that a better method for addressing them has evolved, and continues to do so. For the results of this method of evolution to enable these projects to be successful, it is important to remain vigilant in identifying and implementing these methods, and to foster commitments for transparency and accountability.

Note

1. This chapter does not represent the views of the World Bank, but instead solely represents the views of the author.

REFERENCES

Annual Report to Congress on the Environment and the Multilateral Development Banks, 2001. 2003. Report Submitted in Compliance with Section 539(3) of Title V of Public Law 99–591 and Section 533(b) of Public Law 101–167. March.

Bernasconi-Osterwalder, N., and D. Hunter. 2002. "Democratizing Multilateral Development Banks." In Carl Bruch (ed.). *The New "Public": The Globalization of Public Participation.* Washington, DC: Environmental Law Institute.

Bradlow, D. 2001. "The World Commission on Dams and Its Contribution to the Broader Debate on Development Decision-Making." *American University International Law Review* 16:1531.

Davis, S.H. 2001. "Indigenous Peoples, Poverty, and Participatory Development." *Entrecaminos.* Spring. Internet: ⟨http://www.georgetown.edu/sfs/programs/clas/Pubs/entre2003/indigenous.html⟩ (visited 4 November 2003).

Di Leva, C. 1998. "International Environmental Law and Development." *Georgetown International Environmental Law Review* 10:501.

IBRD (International Bank for Reconstruction and Development). 1949. Articles of Agreement. Dec. 27. 2 U.N.T.S. 134, amended by T.I.A.S. No. 5929 (16 December 1965).

ICJ (International Court of Justice). 1997. Case Concerning the Gabcikovo–Nagymoros Project (Hungary/Slovakia). 1997 I.C.J. No. 92. Internet: ⟨http://www.icj-cij.org/icjwww/idocket/ihs/ihsjudgement/ihs_ijudgment_970925.html⟩ (visited 4 November 2003).

IDA (International Development Association). 1960. Articles of Agreement. 26 January. 439 U.N.T.S. 249.

Krishna, R. 1998. "The Evolution and Context of the Bank Policy for Projects on International Waterways." In S. Salman and L. Boisson de Chazournes (eds). *International Watercourses: Enhancing Cooperation and Resolving Conflict*. Washington, DC: World Bank, pp. 31–43.

Morse, B., and T.R. Berger. 1992. *Sardar Sarovar: The Report of the Independent Review*. Ottawa: Resource Futures International Inc.

Nile Basin Initiative. 2001. *Transboundary Environmental Analysis*. Washington, DC: Nile Basin Initiative, Global Environment Facility, United Nations Development Programme, and the World Bank. May.

"Partnership for Principle 10." 2003. Internet: ⟨http://www.un.org/esa/sustdev/partnerships/law/principle10reg.pdf⟩, ⟨http://www.pp10.org⟩ (visited 4 November 2003).

Pomfret, J. 2003. "Environmentalists Keep up Fight over Chinese Dam." *Washington Post*, June 22, at A15.

Salman, S. 2001. "Dams, International Rivers, and Riparian States: An Analysis of the Recommendations of the World Commission on Dams." *American University International Law Review* 16:1477.

Salman, S., and L. Boisson de Chazournes (eds). 1998. *International Watercourses: Enhancing Cooperation and Resolving Conflict*. Washington, DC: World Bank.

Shihata, I.F.I. 1991. *The World Bank in a Changing World: Selected Essays*. The Hague: M. Nijhoff.

Shihata, I.F.I. 1994. "The World Bank and Non-Governmental Organizations." *Cornell International Law Journal* 25:623.

UNCED. 1992. *Rio Declaration on Environment and Development*. Report of the United Nations Conference on Environment and Development, Rio de Janeiro, 3–14 June. 47th Sess., UN Doc. A/CONF.151/5/REV. 1, reprinted in *International Legal Materials* 31:874.

World Bank. 1990. *Indigenous Peoples*. Operational Directive 4.30. June.

World Bank. 1993. *Resolution Establishing the World Bank Inspection Panel*. International Bank for Reconstruction and Development (IBRD) Resolution No. 93-10, International Development Association (IDA) Resolution No. 93-6. September 22.

World Bank. 1996. *World Bank Participation Sourcebook*. Washington, DC: World Bank.

World Bank. 1998. *Pollution Prevention and Abatement Handbook*. Internet: ⟨http://lnweb18.worldbank.org/ESSD/envext.nsf/51ByDocName/PollutionPrev entionandAbatementHandbook⟩ (visited 4 November 2003).

World Bank. 1999. *Environmental Assessment*. Operational Policy 4.01. January.

World Bank. 2000. "Involving Nongovernmental Organizations in Bank-Supported Activities." *Good Practices*. World Bank Operational Manual GP 14.70. February.

World Bank. 2001a. *Projects on International Waterways*. Operational Policy 7.50. June.

World Bank. 2001b. *Involuntary Resettlement*. Operational Policy 4.12. December.

World Bank. 2001c. "The World Bank Position on the Report of the World Commission on Dams." December 18. Internet: ⟨http://lnweb18.worldbank.org/ESSD/ardext.nsf/18ByDocName/OfficialWorldBankResponsetotheWCDReport/$FILE/TheWBPositionontheReportoftheWCD.pdf⟩ (visited 4 November 2003).

World Bank. 2002. *Policy on Disclosure of Information*. Internet: ⟨http://www1.worldbank.org/operations/disclosure/policy.html⟩ (visited 4 November 2003).

World Bank. 2003a. *Water Resources Sector Strategy: Strategic Directions for World Bank Engagement*. February. Washington, DC: The World Bank.

World Bank. 2003b. *Leading Banks Adopt Equator Principles*. 4 June. Internet: ⟨http://www.ifc.org/ifcext/equatorprinciples.nsf/Content/corepoints⟩ (visited 4 November 2003).

World Bank. 2003c. *The Inspection Panel: Panel Register*. Internet: ⟨http://wbln0018.worldbank.org/ipn/ipnweb.nsf/WRegister?openview&count=500000⟩ (visited 4 November 2003).

11

Improving governance and public participation in international watercourse management: Experience of the African Development Bank in the Senegal River Basin

Aboubacar Fall and Angela Cassar

Introduction

African water resources are both variable and diverse. It is difficult to generalize overall trends for this continent of contrast. Of the total surface water in Africa, 50 per cent is contained within a single, water-plentiful river basin, the Congo River Basin, and 75 per cent of total water resources is confined to eight major river basins – those of the Congo, Niger, Ogooue (Gabon), Zambezi, Nile, Sanga, Chari-Logone, and Volta (Donkor and Wolde 1998).

Nevertheless, on average, the African continent is one of the driest in the world (Kabbaj 2000; UNECA 2000), with only Antarctica and Australia being drier. However, increasing populations in Africa present a challenge not faced in the other, drier continents that are sparsely populated. Africa experienced the largest regional population rise for the period 1990–2000; over the next 25 years, population projections indicate an expected increase of a further 65 per cent. Africa also has the lowest total water-supply coverage of any region, with only 62 per cent of the population having access to improved water supplies since 1990 (WHO/UNICEF 2000). This poses a challenge to future water supply, and at least nine basins, with a projected population of more than 10 million people, are expected to experience further water scarcity by 2025. In recent years, African countries have experienced both water stress and scarcity, whereas others have been subjected to frequent floods and inundation with devastating consequences: examples are the recent drought in the Horn of Africa and tragic floods in Mozambique (Kabbaj 2000).

Although the total available water resources in Africa is a significant factor, how these resources have been utilized is, perhaps, more relevant to this discussion. It is significant that, at present, less than 3.8 per cent of the potentially available water resources are developed and utilized (FAO 2002). It should be stressed that most of the freshwater resources in Africa are contained within transboundary river basins, requiring an examination of governance practice within and between countries, of their cooperative mechanisms, and of the role of public participation.

The African Development Bank (AFDB) has identified the untapped potential of transboundary water development in many contexts. These include transboundary hydroelectric power generation, multinational irrigation schemes, intra-state navigation, joint inland fisheries development, joint water-supply sources utilization, environmental protection and wildlife conservation, and recreation and ecotourism development.

African experience has shown that, when transboundary water resources are developed in an integrated manner and on the basis of a win–win principle, not only do they contribute significantly to the socioeconomic development of the riparian countries sharing these rivers and lakes but also they can enhance subregional and regional cooperation. For example, soon after independence, Mali, Mauritania, and Senegal (which share the Senegal River Basin) recognized the imperative for inter-State cooperation and proceeded to establish one of the first river basin organizations (RBOs) in Africa in 1972 – the Organisation pour la Mise en Valeur du Fleuve Sénégal (OMVS), often referred to as the Senegal River Development Organization.

More recently, the AFDB – among other financing bodies including the World Bank, the US Agency for International Development (USAID), and the UN Development Programme (UNDP) – has played a decisive role in assisting the OMVS to implement transboundary development programmes and projects. One of these projects, the construction of the Manantali Dam (which has been operational since September 2001), features prominently in this chapter owing to efforts to engage the public in its development (AFDB 1998).

International watercourse management is a vital aspect of sustainable development and an important component of the overall development goals and poverty-alleviation policies of the AFDB. In providing funding for watercourse development, the AFDB seeks to promote these policies. How this plays out in practice is a matter for further consideration, and the OMVS provides a good example of ways that governance can be improved through an innovative body that promotes public participation

and the involvement of community groups – the Programme d'Atténuation et de Suivi des Impacts Environnementaux (PASIE).

This chapter examines specifically the role that improved governance
and increased public participation has played as a practical matter in
Africa, tracing the recent experience of the AFDB in this realm. The
next section briefly examines the important aspects of AFDB policy
that relate to improving governance and public participation in international watercourse management in Africa, specifically policies on public
participation, economic cooperation, involuntary resettlement, and integrated water resource management (IWRM). The stated objectives of
the AFDB having been outlined, the subsequent section (pp. 222–227)
considers in more detail the management of the Senegal River Basin, focusing on the implementation of the OMVS Development Programme
and the scope to improve public participation. The section that starts on
page 227 then considers how public participation can be improved – first,
through the OMVS Development Programme and, second, by means of
one of the most innovative developments to emerge from these management initiatives through PASIE – the formation of local coordination
committees (or Comités Locaux de Coordination; CLCs), which are local
coordination and monitoring units in which stakeholder groups are represented to improve public participation. The potential role of such units in
future public participation and improved governance initiatives in international watercourse management is also discussed.

Background on AFDB policy

The AFDB adopted its Vision Statement in 1999, with a central objective
to reduce poverty in Africa (AFDB 1999). Within this vision, the good
governance is indispensable. The vision defines good governance as "respect for the rule of law, and human rights, enhanced accountability and
transparency in the management or public resources as well as a credible
legal and regulatory system" (AFDB 1999). This section briefly outlines
the AFDB policies on public participation, regional integration, involuntary resettlement, and IWRM as important means to achieve the AFDB's
vision. These general policies provide an important backdrop to the next
section (pp. 222–227), which focuses on the Senegal River Basin and the
public-participation structures and organizations that have developed in
that context.

Public participation

The AFDB has recognized that public participation not only is a fundamental component of achieving good governance – a central tenet of the
AFDB Vision – but also is essential to reduce poverty and realize sus-

tainable development (AFDB 1999; Fall 2002). In Africa, water resource management and development has shifted over time from being concerned primarily with mobilizing financial resources to being an important component to achieve sustainable development, of which necessary elements include greater emphasis on considerations of public participation and poverty reduction. The management of watercourses, and particularly of international watercourses, therefore constitutes a key context in which to consider implementation of the AFDB's policy on good governance (AFDB 2000b).

Participatory approaches have been shown to enhance project quality, ownership, and sustainability; to empower beneficiaries; and to contribute to long-term capacity building and self-sufficiency (Fall 2002). For these reasons, since the adoption of the AFDB Vision in 1999, principles of public participation have been applied to the different phases of the Bank's project cycle, including project identification, preparation, appraisal, implementation and management, supervision, monitoring and evaluation, completion, and portfolio review (Fall 2002). By including public participation at various phases throughout the cycle, the Bank seeks to involve all stakeholders more effectively in AFDB processes. The section on pages 227–235 of this chapter examines how this has played out in practice.

Having stated the Bank's objective, it should also be noted that the AFDB's policy on good governance in general, and particularly its policies relating to public participation, refer to "consultation" as well as "participation." These terms connote slightly different concepts, although they have been used interchangeably, causing some confusion when it comes to implementing public participation on the ground. "Consultation," as defined by the AFDB, involves a process of information sharing, listening and learning, and joint assessment, whereas "participation" is a more publicly accessible process that involves shared decision-making, collaboration, and empowerment (AFDB 2001; Fall 2002). If the effectiveness of public participation is considered and its potential to be implemented through AFDB initiatives, it is also necessary to consider access to justice and whether laws and institutions effectively give a voice to all – including the poor, women, children, and indigenous groups. There is not yet a mechanism to guarantee access to justice through AFDB policies; however, preliminary studies are under way to establish an inspection panel to address grievances of people and groups affected by AFDB projects.

Economic cooperation and regional integration

The AFDB has sought to improve and facilitate regional integration in Africa by financing regional studies and projects, as well as financial assistance to regional institutions. Regional integration is an important aspect of AFDB policy for international watercourse management, in which management of international river basins as a whole optimally requires the cooperation of all riparian states.

To facilitate regional integration and cooperation, the AFDB has financed several studies and projects relating to international water resource development and management. For example, the AFDB provided a loan to finance dam construction projects under the auspices of the Liptako Gourma Community intersecting Burkina Faso, Mali, and Niger. This project involved an irrigation scheme to promote regional agriculture, mining, energy, and health (AFDB 2000a).

Regional integration has also been facilitated by the AFDB through their financial assistance to various regional institutions, including regional RBOs such as the OMVS, which is examined in depth on pages 222–227. Other regional institutions that have benefited from the AFDB's policy of economic cooperation and regional integration include the Niger Basin Authority – which dates back to 1964 and has a membership of nine riparian states including Benin, Burkina Faso, Cameroon, Chad, Côte d'Ivoire, Guinea, Mali, Niger, and Nigeria. The Comité Inter-Etats pour la Lutte contre la Sécheresse dans le Sahel (CILSS) – a committee established to combat the devastating effects of drought in the Sahel (a huge region between the Sahara and the savannahs from Senegal in the west to Ethiopia in the east) – has similarly benefited. The CILSS was formed in 1973 to coordinate activities to fight drought in Senegal, Mali, Burkina Faso, Mauritania, the Gambia, Chad, and Cape Verde.

The international nature of these projects and institutions has provided an important forum to develop integrated principles utilized by the AFDB. In particular, these regional-integration projects and institutions are based on integrated principles, including IWRM, and good-governance principles. Accordingly, the public-participation element ideally involves the participation of all interested stakeholders in all riparian states in all phases of the AFDB's project cycle (AFDB 2000a).

Involuntary resettlement policy

The AFDB Group's Involuntary Resettlement Policy has been developed to cover involuntary displacement and resettlement of people caused by AFDB-financed projects (AFDB 2003). It applies when the implementation of an AFDB project results in relocation or loss of shelter by persons residing in the project area, as well as addressing lost assets and affected livelihoods. The policy is set within the framework of the Bank's Vision, in which poverty reduction represents the overarching goal.

A first version of the AFDB Guidelines on Involuntary Displacement and Resettlement was adopted in 1995, but later proved to lack clarity in various aspects. A revised draft was completed in October 2002, and the AFDB Board of Directors adopted it in January 2003. This policy contains some guiding principles that are relevant to the discussion at hand (particularly regarding public participation), as follows:

- conducting meaningful consultation with host communities early in the planning process and ensuring their effective participation in the resettlement programme;
- establishing a legal framework to compensate displaced persons for their losses at "full replacement" cost before taking land or related assets; and
- setting up a conflict-resolution mechanism designed to address disputes between host communities and resettlers (AFDB 2003).

The AFDB Group's experience on involuntary resettlement is limited; such experience as there is mostly relates to projects involving agriculture and rural development, such as irrigation schemes and medium-size infrastructure-development projects including dams for water storage and power generation. One such project is the Manantli Dam project of the OMVS, which is discussed in more detail in the section on "Joint works" (pp. 225–226).

Integrated water resources management

In 2000, the AFDB adopted its Policy for Integrated Water Resources Management (IWRM) (AFDB 2000c). The IWRM Policy functions as an important instrument to fulfil the AFDB Vision. It seeks to facilitate the sustainable development of water resources primarily for poverty reduction. By delineating the critical importance of shared waters in Africa and providing a suitable institutional framework for their management, the IWRM Policy has the potential to improve prospects for mutually beneficial regional cooperation and integration through the management of transboundary water resources. In the near future, the AFDB and the African Water Ministers' Council expect to conclude an agreement for the establishment of the African Water Facility Trust Fund that will be devoted, among other things, to the effective management of the shared water basins. The African Development Fund (ADF) has set aside significant resources for multinational projects including transboundary water-course management. These resources have yet to be utilized, but are intended to address issues that arise during the course of AFDB-funded projects, including mitigation measures in the event of an involuntary resettlement.

Management of transboundary river basins: The Senegal River Development Organization (OMVS)

In the early days of their independence, most African riparian states recognized the importance of basin-wide development and management of transboundary water resources. In the Senegal River Basin, three of the four riparian states – Mali, Mauritania, and Senegal – formalized the management of the shared waters of the Senegal River, and the OMVS formally came into existence in 1972.

This section traces the development of the OMVS and the importance of the Senegal River Basin to the livelihoods of people who live there. An outline of the institutional framework and its evolution is also provided. The discussion emphasizes the status, financing, and management of joint works such as dams, hydroelectric power stations, high-voltage lines, and navigation facilities in the Senegal River Basin. The most recent developments include the Senegal River Water Charter signed in May 2002, which establishes its legal and regulatory framework and clearly states that the river's water must be allocated to various sectors.

Description of the Senegal River Basin

The Senegal River Basin has a surface area of about 300,000 km^2 (OMVS 2002). There are three main sections of the basin (including the upper basin, the valley, and the delta), which must be considered as a whole to promote a fully integrated management of the basin. The average annual discharge from the Senegal River is low, estimated to be 24 BCM (OMVS 2002). However, in years of drought, human populations are almost totally deprived of this important water source. This critical environmental, social, and economic situation is made more vulnerable by desertification and an increasing imbalance between the resources available and population growth. Thus, the OMVS member states place the highest priority on the common development of the basin, focusing on large-scale improvement, protection, and restoration of the overall ecosystem for the long-term benefit (OMVS 2002). The need to establish a formal association for managing the Senegal River was recognized as particularly important owing to overall regional water scarcity.

The population within the Senegal Basin accounts for about 16 per cent of the total population of the three member states of OMVS. Overall, population density is low, although it does vary throughout the basin (OMVS 2002). The population in the basin has been declining over the past two decades, as the Senegal River Basin has witnessed large-scale migration. This has been attributed to drought and desertification in the region, making livelihoods difficult. Typically, the region has low agricultural productivity, and the traditional method of cropping utilizes rainfed and post-flooding methods, and these are to be found around the basin and in the hinterland (including those dependent on hill-based dams).

Sustaining the population of the river basin is a challenge. Currently, the needs of rural families are barely met from one year to the next. In the face of increasing population pressures, member states through the OMVS set up institutional and organizational frameworks to utilize the water resources of the Senegal River more effectively, particularly for agricultural development. To provide a reliable source of water for agricultural development and hydroelectric power generation, the OMVS states agreed to build two large dams – the Diama and Manantali dams – which started operating in 1986 and 1988, respectively.

Institutional framework and background: Development of the OMVS

Historical development

The present structure of the OMVS results from the changing cooperation among the four Senegal River Basin states, dating from colonialism in the 1960s to their independence. Prior to their independence, two of the four basin states – Senegal and Mauritania – had recognized the need to cooperate within the basin. The first bodies to deal with the development of the Senegal River valley date from colonial days: they are the Organisation Autonome de la Vallée (OAV), the Valley's Autonomous Organisation, and the Mission d'Aménagement du Bassin du Fleuve Sénégal (MAS), the Basin Development Mission. These organizations are of interest historically for the economic and technical data they provided; however, in legal and institutional terms, these arrangements have contributed little.

One of the most inclusive historical initiatives, which included all four basin states of the Senegal River, was the establishment of the Comité Inter-Etats (Inter-state Committee for the Development of the Senegal River). It was established following the Conakry Recommendation of 11 July 1962, which led Guinea, Mali, Mauritania, and Senegal to sign the Bamako Convention of 25 July 1963 relating to the overall development of the basin and establishing the Comité Inter-Etats. Further specifying the legal status of the Senegal River, the Bamako Convention was complemented by the Dakar Convention of 7 February 1964.

The Comité Inter-Etats laid the groundwork for subregional cooperation in development of the Senegal River Basin. The process that this institutional body established consisted of two main foci for future development of the Senegal River Basin: it established the international status of the Senegal River, as well as the need to stabilize the river's flow following dam developments built upstream.

The mandate of subregional cooperation was expanded by the Labé Convention of 26 May 1968, which established the Organization of Riparian States (OERS) to replace the Comité Inter-Etats. The establishment of the OERS broadened the subregional focus from development alone, and it was intended that this organization would be a cooperative body to promote the economic and political integration of the four member states.

Development of the OMVS and legal status of the Senegal River Basin

Following the signing of the convention to create the OMVS and an adjunct convention relating to the status of the river, the OMVS formally came into existence on 11 March 1972. Guinea, the fourth basin state, was not a party to these conventions, although these two conventions provide for the admission of any riparian state that so wishes. The conventions terminated previous agreements relating to the Senegal River Basin – namely, the Convention of 26 July 1963 relating to the general development of the Senegal River Basin and the Convention of 7 February 1964 relating to the status of the Senegal River. [The Convention relating to the status of the Senegal River was amended by Resolution 5/75 of 16 December 1975.]

Through the Convention of 11 March 1972 relating to the status of the Senegal River, the Conference of Heads of State and Government established the current status of the river. According to the first article of the Convention, on the Status of the Senegal River Basin, the Senegal River is an "international river, together with its tributaries" (OMVS 2002) on the territories of the three member countries. The international status of the river rests on two fundamental principles – freedom of navigation and the equal treatment of the users. The convention details the application of these principles, particularly relating to agricultural and industrial development.

A decade later, the original conventions were complemented by two additional conventions that further detailed the mandate of the OMVS in developing and managing the Senegal River Basin. The first convention, signed on 12 May 1982, relates to the conditions of funding for joint works; the second, signed on 21 December 1982, relates to the status of joint works. In this context, joint works are projects and management initiatives that are owned collectively by the member states. In addition to these conventions, a Charter relating to the waters of the Senegal River was signed by the heads of state of each member state and ratified by all member parties.

In summary, a comprehensive legal framework for cooperation among the OMVS member states has been negotiated and established over the past three decades, despite the continued non-membership of Guinea for various political reasons. Nevertheless, through the OMVS framework, Guinea has steadily increased its participation.

Joint works

The 1972 Convention on the Status of the Senegal River Basin affirmed a spirit of cooperation for areas of common interest. For example, Clause 5 provides for special conventions to be signed among member states relating to the implementation and use of works of "common interest." This provision has been important in providing a legal basis for "common" or "joint" works in the states. This provision enabled approval of the legal instruments relating to the Diama Dam in 1979 and the Manantali Dam in 1982 by the Conference of Heads of State and Government. Other joint works, which share common ownership, include the sea–river port of Saint-Louis (Senegal), the river port of Kayes (Mali), the ports and related works for developing a navigable channel, and related works and accessories.

Expanding on the status of the joint works, the notion of common property was defined by the Convention of 21 December 1978 between the three OMVS member states. Under this convention, common property is "[a]n approach to property right whereby each of the co-owners is entitled to a share of the same property, and all are entitled to the same property as a whole." The two additional OMVS conventions of 1982 set forth practical modalities for joint projects.

Generally speaking, joint works include dams, hydroelectric power stations, high-voltage lines, and navigation facilities built around the Senegal River Basin, all of which belong jointly to the member states. The rights and obligations of the co-owner states are based on the principles of equality and equity among the co-owners according to the ultimate profits of the states. However, there are provisions for adjustment, depending on the overall results of the operation of the regional infrastructure system as a whole.

Management of joint works is through the establishment of specific agencies that are accountable to the OMVS. For example, for the Manantali Dam, the Société de Gestion de l'Energie de Manantali (SOGEM) is granted the right to utilize the dam and its facilities to produce and transport electrical power. SOGEM, in turn, grants the Manantali Development Company (SEM) the right to oversee the operation of the Manantali hydroelectric facilities once these have been completed. SOGEM has signed a contract with the South African Company ESKOM to assist SEM in its development mission.

A similar arrangement exists for the Diama Dam, which has an Agency for the Management and Development of the Diama Dam (SOGED) that is responsible for the development, maintenance, and restoration of the dam, the intake works at Aftout-es-Sahel, the dykes along the Senegal River from Diama to Rosso, and related works.

The OMVS Water Charter

In addition to the legal and institutional framework that has developed over the past three decades, the recent Water Charter (Charte de l'Eau) has been one more important instrument. All three OMVS member states ratified it, and it entered into force in April 2003. The Water Charter supplements the legal provisions already mentioned, and supports the principles of sustainable development that have evolved from the 1992 Rio Summit to the 2002 World Summit on Sustainable Development (WSSD) in Johannesburg.

The Water Charter establishes the principles, methods, and mechanisms of managing the Senegal River. It establishes the framework for optimal and equitable allocation of the waters of the Senegal River among the various uses – drinking and sanitation, irrigation, energy, and artificial flooding. The Water Charter adds a specific dimension to the sustainable development of the Senegal River Basin in line with the principle of subsidiarity, establishes the optimal strategy for the provisional allocation of the water among its users, and sets forth two specific guidelines for managing the Manantali and Diama dams.

OMVS governance structures

The institutional and legal structures that influence the operation of the OMVS have been set out above in some detail. How this body operates in practice also requires brief consideration. The governance structure of the OMVS comprises essentially six entities:

1. *The Conference of Heads of State or Government* is the supreme authority responsible for working out the cooperation policy and for making all decisions related to general economic development.
2. *The Council of Ministers* is the decision-making body responsible for designing the general policy for developing the Senegal River and the resources of the basin, as well as for cooperation among the member states.
3. *The High Commission* is the implementing organ that carries out the Council of Ministers' decisions and submits reports thereon.
4. *The Permanent Committee on Water* establishes the principles and methods for allocating the Senegal River waters among the member states and the various sectors. This committee issues advisory notes to the Council of Ministers.
5. *The Regional Planning Committee* offers advice on the investment and pre-investment programme related to the development of the basin's resources. It is also responsible for monitoring the programme.

6. *The Advisory Committee* includes delegates from the member countries, funding institutions (such as the AFDB), and the OMVS. It assists the High Commission in mobilizing financial and human resources. It is also responsible for promoting information exchange on the rules and procedures for raising and using funds, on implementing projects, and on the prospects for cooperation between the OMVS and its development partners.

As regards practical matters, these bodies cooperate well with one another, and there appears to be minimal administrative overlap. For example, the Council of Ministers as the decision-making body takes an overarching, guiding role, while the High Commission's role is to implement these decisions.

Improving public participation

The OMVS development programme and PASIE

The OMVS has developed a preferred approach to development in the Senegal River Basin. After several studies on various forms of integrated development, the OMVS opted for a programme of simultaneous development of the three main water-based sectors – irrigation, hydroelectric production, and navigation. This programme rests on a basic infrastructure made up of an anti-salt dam at Diama near the delta (23 km from the river's mouth) and a second, regulating, multi-purpose dam at Manantali.

Although the mandate of the OMVS focuses mainly on the development of irrigation, hydroelectric production, and navigation, its activities have evolved to include a stronger emphasis on social and environmental issues. The primary reason for this been to address challenges that have emerged as a result of the dams' operation. The OMVS has been particularly innovative in this regard, initiating a relatively new programme – Programme d'Atténuation et de Suivi des Impacts Environnementaux (PASIE), the Programme for the Mitigation and Monitoring of Environmental Impacts. PASIE has taken a more participatory approach to the management of the Senegal River Basin, including a greater emphasis on environmental impact assessment (EIA).

Programme components and implementation: Diama and Manantali dams

The combined functions of both the Diama and Manantali dams fulfil a number of roles. Together, they irrigate 375,000 ha of land, while ensuring that there is an adequate draught for safe navigation between Kayes and Saint-Louis year-round. In nine out of ten years, the dams have yielded an annual hydroelectric power output of 800 GWh. Agriculturally, the dams maintain, through a transitional period, hydraulic conditions required for flooding the valley and for traditional crops when there is a drop in the water level. In addition to mitigating floods at Manantali, the dams halt intrusion of sea water into the river. In fact, a primary justification for initially building the Diama Dam was to mitigate saline intrusion at the mouth of the Senegal River, providing a reliable source of water to the riparian villagers for agricultural and domestic use. The dams also improve the filling of the Guiers and Rikiz lakes, as well as of hollows such as Aftout-es-Sahel. Finally, the dams seek to maintain acceptable ecological conditions around the river basin.

Prior to the construction of the Diama and Manantali dams in 1977 and 1978, respectively, the OMVS undertook several EIAs to assess the impact of the dams. More recently, following the creation of the Centre for the Study of Lakes and Marshes at Manantali in 1989, further EIAs were completed between 1993 and 1996. The most recent EIAs identified negative impacts from the development of irrigated agriculture as a result of the dams, and the EIAs recommended corrective measures. In response to this finding, the OMVS initiated PASIE to mitigate and monitor environmental impacts.

Over the past decade, the Centre for the Study of Lakes and Marshes at Manantali has been responsible for an ongoing follow-up and evaluation of the Manantali's ecology. The role of this body further evolved in 1996 with the development of PASIE. This groundbreaking programme aims to measure, remedy, maximize, and monitor the water of the Senegal River. More specifically, it comprises six components:

- reduction of the impacts of the hydroelectric power-generation project, including the impacts of power lines, and adoption of measures to minimize environmental impacts during any works;
- acquisition of areas for public use;
- optimization of the management of storage basins by development of a charter for water use and a *Handbook for Basin Management*;
- design and implementation of pilot projects for the control of water-borne diseases, particularly bilharzia, and the setting-up of a regional health action plan;
- facilitation and promotion of rural electrification, development of phase II hydroelectric sites, and the creation of small-scale projects to generate income and to reduce poverty; and
- promotion of monitoring and follow-up activities through the creation of a "watchdog" environmental committee, a General Action Plan for the Environment, and an Environmental Code.

PASIE receives funding primarily from the AFDB and the World Bank as part of a hydropower project, and most activities are already under way. PASIE has some exciting potential in the Senegal River Basin. The OMVS High Commission has been preparing a programme for the management of water resources and of the environment of the Senegal River Basin as a whole. This programme is being developed with partnership organizations including the Worldwide Fund for Nature (formerly the World Wildlife Fund; WWF), the World Bank, and the United Nations Development Programme (UNDP). There is also a real opportunity to significantly enhance the management of the Senegal River Basin: PASIE could facilitate the entry of Guinea into the OMVS, which would bring about complete basin-wide membership.

The new challenges

Public participation in the Senegal River Basin has, until recently, been quite limited. The OMVS has examined past practices and determined that a shift toward more sustainable practices is necessary. It is not enough to increase agricultural output through the construction of large-scale infrastructure programmes, such as dams and hydro-agricultural equipment; instead, it is to develop a more integrated and participatory development programme that maintains social economic and environmental benefits over the medium and long terms.

Working towards a participatory regime in the Senegal River Basin has been challenging. The approaches are not yet resolved, although new developments in management approaches taken by PASIE are going some way to address this. In the past, the participatory management scheme has been limited to irrigation agriculture and not inclusive of all basin-wide considerations. This specific scheme aimed to ensure a continuous and long-term supply of irrigation water to villages. This was advanced through policies that are intended to (1) grant farming communities the power to participate in the decision-making process and (2) ensure farming-community involvement in both the development works associated with the dam and the management of irrigated areas.

In practice, however, public consultation has taken place only after the initial planning has already occurred. Planning and programming is regularly carried out upstream by state and financial institutions before downstream populations are consulted. This has meant that state consultation with farmers has tended to occur after state consultation with financial institutions. Sometimes, this also has meant that consultation with farmers occurs after the decision to carry out a development has already been made and after the funding has been secured. Whether the farmers' submissions are taken into account has depended on the particular project, its profitability, and particular localized political situations.

The state plays the primary role in formulating choices and planning development. More generally, it is noteworthy that the difficulties in ensuring public participation in practice have prevailed largely due to a top-down management and regulatory culture in the region. This culture means that the central government performs the initial diagnosis, which then sets the framework within which the rural sector can operate. In this process, the government establishes the objectives to be achieved and the means of implementation. Thus, the key challenges to date in establishing a more participatory regime have been as follows:

- participation in the management of the Senegal River Basin's resources has, so far, been limited to the participation of farmers;
- participation has concerned only the promotion of irrigated agriculture through the establishment of villages in which irrigated areas are devoted to cultivation or farming;
- studies carried out so far have focused on public participation only in the context of existing village-development associations.

The following subsection examines how recent initiatives of PASIE have had (and have the potential to continue to have) positive effects on the breadth and nature of public participation in the Senegal River Basin.

PASIE activities: Local coordination committees

To implement PASIE effectively, the OMVS set up a coordination and monitoring programme based on a participatory approach and emphasizing social development. The programme has led to a system of public participation aimed at putting in place local coordination and monitoring units – local coordination committees (Comités Locaux de Coordination; CLCs), in which all socio-professional interests are represented. The following sections evaluate PASIE and the functioning of these coordination committees as mechanisms to promote public participation in the management of the Manantali Dam.

Conceptual framework, role, and composition of CLCs

There are, at present, five committees, which were established following a 1999 administrative decision by the Council of Ministers. They have the following objectives:

- to inform and raise awareness in all sectors of PASIE to facilitate its enforcement and the involvement of all local civil-society members;
- to follow up on-the-ground implementation of all aspects of PASIE;
- to encourage respect for decisions and recommendations made at all levels, including those made by PASIE's piloting committee;
- to act as a link between local populations, national authorities, and the OMVS; and
- to be involved in the execution of all aspects of PASIE, including the

formulation of general objectives of the PASIE programme (OMVS 2002).

The committees operate under the supervision of the national coordination committee (CNC). The CNC promotes national consultations to advance the objectives of PASIE. These consultations help to improve the acceptance of the final result by the parties involved in the process, thus helping to reduce risks of delay in executing the Manantali energy project. The CNC also links all relevant stakeholders at both the national and regional levels. Ultimately, the CNC seeks to ensure the viability of the OMVS programme (OMVS 2002).

Generally, the committees have the same composition. Their members include the prefect (some committees have an assistant prefect), presidents of rural councils, farmers' representatives, fishermen's representatives, craftsmen's representatives, the local president of women's promotion associations, relevant NGO representatives, and a representative of the local press. Other members may include a medical doctor and other people involved in natural resource management (OMVS 2002).

The degree of success that the CLCs enjoy appears to be related to the groups' sense of ownership of the operations and benefits that the committees can provide. In essence, the committees' degree of involvement; their sense of collective enterprise; and a desire to make the operation their own, according to their interests and benefits, seem to govern the level of success enjoyed by committees.

Evaluation of the functioning of the CLCs

For better examination of the functioning of the CLCs, the CNCs conducted a survey to enable those with particular concerns to express themselves. The objectives of the questionnaire were twofold. The first objective was to have local communities identify the limitations of the OMVS and the CLCs. This was the first participatory management experience of the OMVS. Second, the questionnaire sought to determine the best approach to ensure the participation of all stakeholders. The functioning of the committees was evaluated through three aspects – their legal status, their objectives, and their limitations and adaptive capacities. The results of this evaluation are discussed below.

The legal status of the CLCs

The CLCs are formal associations that are "legally" constituted by decree; however, they do not have formalized legal status. The committees cannot borrow from financial institutions, nor can they be contractors to execute a project in the name of the community to which they belong. They are, however, "legally" recognized by relevant authorities. Legal authority and discretion concerning the committees' operation lies with the prefect. The committees were created by a unilateral prefectorial decision and legal administrative order taken by a competent authority (the prefect) in the lawful exercise of his or her functions. The committees' authority can be withdrawn or repealed by the prefect or by a higher administrative authority at any time if the latter finds that it is inconsistent with the constitution, the statutes, or decrees of the country.

This limited legal recognition could be a handicap for the role that they are expected to play in managing the water resources and the environment of the basin. Decree No. 97-347, dated 2 April 1997, could help to resolve this matter by providing more autonomy to the committees to enable them to mobilize funds for their activities. By bringing more legal recognition to the committees, this decree is expected to assist them in independently implementing water resources management projects.

On the other hand, the advantage of the local coordination committees is that they have flexibility from the legal recognition procedure and they have the potential to bring development opportunities. In particular, these committees can identify potential partners for development and facilitate the creation of partnerships with local communities.

The name "local coordination committee" may be somewhat misleading, and potentially limiting. The scope of CLCs often extends beyond local areas. For example, "local" in the Senegalese sense of territory refers to the district, which has a higher legal authority than designated local communities. It would, perhaps, be more accurate to term them "department coordination committees," to be in conformity with the provisions of the existing administrative code.

Objectives
In the execution of PASIE, CLCs play a major role in managing water resources, environmental health problems, land acquisition, accompanying measures, micro-projects for poverty reduction, and so on. Because of their objectives, the CLCs are development-supporting associations, as well as development associations. They are development-supporting associations in that they support dynamics of development initiated by the OMVS. However, to distinguish their role from that of NGOs (which support these actions with expertise, and not for profit), CLCs do so because they profit thereby.

It is this role that gives CLCs the status of a development association. The committees seek to promote the economic and social interest of their constituents by supporting and helping the OMVS in its development programme, and particularly through the PASIE programme.

Limitations and adaptive capacities of CLCs
The major limitations of the CLCs relate to the limits of socio-professional participation. The two main socio-professional categories

are local unions that are grouped by the state through its supervising structures and local unions created through private initiatives.

The primary issue is the challenge of equitable representation of socio-professional and farmers' organizations. The number and type of socio-professional organizations that should participate is not clearly specified, and criteria to choose their representatives remain arbitrary. For example, a representative can be chosen during a meeting if that person is an expert; however, such recognition does not necessarily mean that the person is a popular representative. Moreover, if there was not much notice of the meeting, a representative can be chosen at random or according to proximity, as long as the representative belongs to the relevant socio-professional category. Finally, if the local president of a socio-professional group does not live in a given village, the departmental services, for practical reasons, are likely to select someone who does live in that village.

No decree to date has set the number of associations according to their socio-professional categories to participate as members; rather, the decree speaks only of "farmers' and cattle raisers' representatives." The committee needs to take diversity into account in order to reap the full benefit from the contribution of the farmers' organizations. The Committee President of Podor suggested an equitable approach: he recommended that, for good representative participation to take diverse interests into account, the first step should be to determine the number of people on the committee, and then to designate those who should serve on the committee.

Once the committee president has been informed of a meeting with the CNC located in Dakar, he alerts his councillors, who constitute the decentralized technical services of the state (rural development, communal development, departmental service for rural expansion, woods and forest service, and animal husbandry). In their turn, the councillors inform the socio-professional organizations corresponding to their respective sectors.

The failure to identify relevant participatory associations can contribute to coordination problems for PASIE activities, because there are inconsistencies in who attends the meetings. However, although there are some weaknesses in the selection of representatives, it should be borne in mind that the overall steps to involve members of the public (including the participation of socio-professional representatives) constitutes a significant improvement over previous practice.

Rural development

The operation of PASIE through CLCs is complex, and there are numerous stakeholders. However, rural development requires particular consideration as it is the sector with the most complex structural organization. This sector experiences particular challenges in coordinating its operations in the committee structure, especially relating to access and flow of information. This is due in part to the numerous socio-professional categories, some of which are detailed below.

Dissemination of information in both socio-professional and rural communities is a challenge. The socio-professional categories may have difficulty in accessing information, in part because the region's centralized political tradition has ensured that information has tended to remain guarded, often at the presidential level. The participation of an organization's representatives at the departmental or communal level does not guarantee good transmission of the information, because the departmental representative does not necessarily report to the districts.

To ensure effective transmission of information in socio-professional circles, it is necessary to convene all localities at the meetings – that is, a representative at the local district, communal, and departmental levels. In this way, each will be responsible for disseminating the information in the respective localities.

Equality of information is an issue in rural communities also. Political considerations and partisan divisions within rural councils affect the sharing of information, according to political conviction. Councillors and rural council members who share the same political views as the president will generally stand a better chance of obtaining superior information in a timely manner.

Further, the concentration of meetings at the regional, basin-wide level does not allow effective public participation at the local level. Meetings held at the departmental level have never been decentralized to the local level. Better dissemination of information could be promoted through local meetings at the district level that are convened by the local development committees.

Fishing organizations

Associations, federations, and cooperatives of fishermen are set according to the fishing sector. Within the delta regions of the Senegal River Basin, there are three fishing organizations – the Toroney and Guiers lakes, the Senegal River, and Makadiama. Each sector has a fishing council. At the federal level, the three councils form the Dagana Fishing Council. Information is easily distributed within the fishing areas.

Pastoral organizations (cattle-breeding)

Pastoral organizations are essentially cooperatives at the local level. These organizations allow farmers to benefit from subsidies granted by the state. In this field, along with cattle-breeders, there are professionals (such as butchers and traders) and private breeders who do not use traditional methods.

The Youth Departmental Council

In this sector, there are youth communal councils and youth homes of rural communities. The latter are mainly composed of sporting and cultural associations and other youth associations.

There is a strong tendency towards federations of groups in the region with similar interests. This may be seen as a solution to improving populations' representation through large groupings.

Conclusions

The legal status of the committees needs to be clarified and strengthened to promote the implementation of development projects. To achieve this end, perhaps the committees need to put in place devices to enable them to continue to exist and operate after PASIE concludes.

One of the actions that may engender change is reorganization of the farming and other components of the committees into larger associations to include service provisions in their fields of action. This could enable them to mobilize resources, because the committees have resources and logistical support provided by the OMVS. The microprojects-related loans designed to reduce poverty could also be invested in savings and credit associations to make them last in perpetuity.

The OMVS has resolutely adopted a new approach to establish synergies among all the agents involved in managing the resources of the Senegal River. However, some changes should be made to the committees' procedures. One of the major changes could be to identify all the actors to ensure equitable participation: for example, in preparing the Senegal River Water Charter, seminars and workshops were held early in the process to provide opportunities for a large number and range of people to participate.

Acknowledgements

The authors would like to acknowledge the critical assistance that they received from Carl Bruch (ELI), Mohamed Salem Ould Merzoug (High Commissioner of OMVS), Djibril Sall (Programme Director, OMVS), Abdoul Samboly Ba (Legal Counsel, OMVS), and Ndèye Dior MBacké and Ababacar Ndao (Experts at the Cellule Nationale OMVS du Sénégal).

REFERENCES

AFDB (African Development Bank). 1998. *Review of the Bank's Experience in the Financing of Dam Projects*. Operations and Evaluation Department.

AFDB (African Development Bank). 1999. *African Development Banks Vision*. Internet: ⟨http://www.afdb.org/knowledge/documents/The_Banks_Vision.htm⟩ (visited 30 June 2003).

AFDB (African Development Bank). 2000a. *Economic Cooperation and Regional Integration Policy Paper*. February.

AFDB (African Development Bank). 2000b. *Good Governance Policy*. Internet: ⟨http://www.afdb.org/projects/polices/pdf/governance.pdf⟩ (visited 27 June 2003).

AFDB (African Development Bank). 2000c. *Policy for Integrated Water Resources Management* (IWRM). April.

AFDB (African Development Bank). 2001. *Handbook on Stakeholder Consultation and Participation in African Development Bank Operations*.

AFDB (African Development Bank). 2003. *Guidelines on Involuntary Displacement and Resettlement*.

Donkor, S.M.K., and Y.E. Wolde. 1998. *Africa: Some Problems and Constraints in Integrated Water Resources Management in Africa*. Addis Ababa, Ethiopia: United Nations Economic Commission for Africa.

Fall, Aboubacar. 2002. "Implementing Public Participation in African Development Bank Operations." In Carl Bruch (ed.). *The New "Public": The Globalization of Public Participation*. Washington, DC: Environmental Law Institute.

FAO (Food and Agriculture Organization of the United Nations) AQUASTAT. 2002. *Water Resources, Development and Management Service: General Summary Africa*. Internet: ⟨http://www.fao.org/ag/agl/aglw/aquastat/regions/africa/index3.stm⟩ (visited 30 June 2003).

Kabbaj, Omar. 2000. "Opening Statements. Safeguarding Life and Development in Africa: A Vision for Water Resources Management in the 21st Century." Africa Caucus Presentation at the Second World Water Forum. The Hague, The Netherlands, 18 March.

OMVS (Cellule Nationale OMVS du Sénégal). 2002. *Comités Locaux de Coordination du PASIE: De Nouveaux Mécanismes d'Organisation de l'OMVS pour une Participation des Populations à une Gestion Intégrée du Bassin du Fleuve Sénégal*. August.

UNECA (United Nations Economic Commission for Africa). 2000. *Transboundary River/Lake Basin Water Development in Africa: Prospects, Problems and Achievements*. Addis Ababa, Ethiopia: United Nations Economic Commission for Africa. December. ECA/RCA/RCID/052/00.

WHO/UNICEF. 2000. *Global Water Supply and Sanitation Assessment 2000 Report*. New York: UNICEF.

12

A North American toolbox for public involvement in international watershed issues

Geoffrey Garver[1]

Introduction

In the early 1990s, the countries of North America took advantage of the opportunity that the debate over the North American Free Trade Agreement (NAFTA) presented to create an institutional structure for examining environmental issues on a continental scale. North America's Commission for Environmental Cooperation (CEC) was established in 1994 under the North American Agreement on Environmental Cooperation (NAAEC) at the frontier of the emerging and increasingly visible trade and environment debate. The CEC was graced at its inception with a unique set of tools for protecting, conserving, and enhancing the environment in a new era of liberalized trade among Canada, Mexico, and the United States. The CEC reflects a formal recognition of the reality that ecosystems and watersheds in North America, as elsewhere, often do not respect national boundaries.

Among the more innovative tools in the CEC toolbox are those giving effect to the NAFTA parties' emphasis on public participation. For example, the preamble of NAAEC emphasizes "the importance of public participation in conserving, protecting and enhancing the environment" (NAAEC 1994). Of the CEC's public-participation mechanisms, none has drawn more attention than the citizen submission process under Articles 14 and 15 of the NAAEC. Emphasizing the citizen submission process, this chapter examines the CEC's public-participation mechanisms, with a view to their application to international watershed management.

This chapter first describes the goals and structure of the CEC, with an emphasis on its strong built-in commitment to public participation. Key features of the CEC reflecting that commitment are the citizen submission process and the 15-member Joint Public Advisory Committee (JPAC), which advises the CEC Council on a broad range of issues. Second, this chapter describes how the CEC's unique toolbox (and, in particular, the citizen submission process) has been used to promote public involvement in issues facing the environment in North America. Past and ongoing initiatives involving international watersheds are emphasized. Third, this chapter examines the future potential for the CEC to enhance public involvement in international watershed issues.

The CEC's emphasis on public participation

Building on the emphasis in its preamble on the importance of public participation in environmental protection, the NAAEC weaves public participation into the fabric of the CEC. The agreement's explicit objective in Article 1(h) is to "promote transparency and public participation in the development of environmental laws, regulations and policies" (NAAEC 1994). In support of this objective, the NAAEC commits the parties to providing for public participation in various ways, establishes a unique public advisory committee to advise the CEC Council, and allows persons and non-governmental organizations (NGOs) in North America to bring directly to the CEC their concerns regarding enforcement of environmental laws in the three NAFTA countries. In practice, providing opportunities for public involvement in all aspects of the CEC's work has become a hallmark of the organization.

In terms of party obligations, NAAEC Article 4 commits the signatory parties, to the extent possible, to publishing, and giving the public a reasonable opportunity to comment on, proposed environmental measures (NAAEC 1994). In addition, Articles 6 and 7 commit the parties to ensuring that interested persons have private access to remedies for violations of environmental laws and regulations and that the parties' proceedings for seeking redress contain certain procedural guarantees.

Structurally, the CEC's emphasis on public participation is most evident in the establishment of the JPAC as one of the CEC's three primary bodies, along with the CEC Council and the CEC Secretariat. The JPAC is a 15-member committee, with five members appointed from each country. Article 16(4) of the NAAEC empowers the JPAC to "provide advice to the Council on any matter within the scope of [the NAAEC] ...". The JPAC typically holds three or four public meetings a year, rotating meeting locations among the three countries. In addition to providing advice to the Council on specific matters, the JPAC annually reviews, and provides advice regarding, the annual programme and budget of the CEC. Representing a cross-section of the North American public, the JPAC provides an important lens for bringing the public's concerns to the attention of the CEC's cabinet-level Council members – namely, the Minister of Environment of Canada, the Secretary of the Environment of Mexico, and the Administrator of the United States Environmental Protection Agency (USEPA).

The CEC's citizen submission process gives individual members of the public their most direct means for focusing the CEC's attention on a particular concern – as long as the concern is related to environmental enforcement in one of the three NAFTA countries and other basic requirements of the process are met. Article 14 of the NAAEC provides that the CEC Secretariat may consider a submission from any person or NGO asserting that Canada, Mexico, or the United States is failing to effectively enforce an environmental law. First, the Secretariat will seek to determine that NAAEC Article 14(1) requirements for submissions are met. That is, the submission:

- is in writing in a language designated by that Party in a notification to the Secretariat;
- clearly identifies the person or organization making the submission;
- provides sufficient information to allow the Secretariat to review the submission, including any documentary evidence on which the submission may be based;
- appears to be aimed at promoting enforcement rather than at harassing industry;
- indicates that the matter has been communicated in writing to the relevant authorities of the Party and indicates the Party's response, if any; and
- is filed by a person or organization residing or established in the territory of a Party.

Where these requirements are met, the Secretariat may then request a response from the government party concerned, taking into account the factors enumerated in Article 14(2):

- the submission alleges harm to the person or organization making the submission;
- the submission, alone or in combination with other submissions, raises matters whose further study in this process would advance the goals of [the NAAEC];
- private remedies available under the Party's law have been pursued; and
- the submission is drawn exclusively from mass media reports.

Where the Secretariat makes such a request for a response from a Party, it shall forward to the Party a copy of the submission and any supporting information provided with the submission. Based on the submission and the response, the Secretariat can recommend to the Council under Article 15(1) that a so-called "factual record" be prepared.

If a majority of the Council authorizes preparation of a factual record, the Secretariat, in accordance with Articles 15(4) and 21(1)(a) of the NAAEC, undertakes an in-depth investigation, gathering facts from the governments and other sources or developing information itself, often with the assistance of technical or legal experts. Ultimately, the Secretariat produces a factual record and, if a majority of the Council agrees, publishes it in accordance with Article 15(7) of the NAAEC. Factual records do not reach a conclusion as to whether the Party is failing to enforce its environmental law effectively; instead, they provide information (regarding asserted failures to effectively enforce environmental law in North America) that may assist submitters, the NAAEC parties, and other interested members of the public to reach their own conclusions and to take any action they deem appropriate in regard to the matters addressed.

The citizen submission experience to date

As of 31 July 2003, the Secretariat has received 4,038 citizen submissions since the CEC's creation, including 13 concerning Canada, 197 concerning Mexico, and 8 concerning the United States. The CEC has published five factual records, as follows:

- *the Cozumel factual record*, involving enforcement of Mexico's EIA legislation in connection with a pier terminal in Cozumel;
- *the BC Hydro factual record*, involving Canada's enforcement of the Canadian Fisheries Act in connection with hydroelectric facilities in British Columbia;
- *the Metales y Derivados factual record*, involving Mexico's enforcement of its hazardous-waste laws in connection with an abandoned lead smelter near the United States–Mexico border in Tijuana;
- *the Migratory Birds factual record*, involving the United States' enforcement of its migratory bird law in connection with logging operations; and
- *the Aquanova factual record*, involving Mexico's enforcement of EIA and other law in connection with a shrimp farm in Nayarit.

Seven final factual records await a Council decision on publication, and three additional factual records are in preparation as of 31 July 2003. Twice, the Council has voted against a factual record that the Secretariat recommended. A registry of the submissions, as well as factual records and the Secretariat's determinations and notifications to Council at various stages in the process, are available on the CEC's website (CEC 2003).

Sixteen of the submissions filed to date have primarily involved water-related enforcement issues. Seven submissions addressing enforcement in Canada have asserted that Canada is failing to effectively enforce the fish-habitat or pollution-prevention provisions of the Canadian Fisheries Act (Fisheries Act 1985). These include: SEM-96-003 (Oldman River I), SEM-97-001 (BC Hydro), SEM-97-006 (Oldman River II), SEM-98-004 (BC Mining), SEM-00-004 (BC Logging), SEM-02-003 (Pulp and Paper), and SEM-03-001 (Ontario Power Generation).

Submissions involving Mexico have addressed surface or groundwater pollution or management in the Lake Chapala Basin; the Magdalena River Basin, Guadalajara; and along the coast in Nayarit. Those involving the United States have raised issues concerning management of the San Pedro River Basin in Arizona, groundwater pollution due to leaking underground storage tanks in California, and deposition of airborne toxic pollutants from waste incinerators into the Great Lakes. Two of these submissions involving the United States – the Great Lakes and Fort Huachaca submissions – touched upon management of, or impacts to, transboundary watersheds and are elaborated here.

The steady stream of submissions that the CEC has received to date indicates that the community of potential submitters is aware of the process involved in Articles 14 and 15 and has begun to test its effectiveness in solving problems of concern to members of the North American public. A significant number of submissions have involved water resource problems, indicating that the citizen submission process has shown some early potential to be a valuable new tool for addressing problems of water management and pollution. As one commentator suggests, the value of the citizen submission process may go beyond other available tools in that, among other things, it operates on an international stage and therefore "may attract an audience that other mechanisms will not reach and have a different impact as a result." (Markell 2001).

The Great Lakes submission

The Great Lakes submission, filed initially in May 1998, asserted that the United States was failing to effectively enforce laws regarding the deposition into the Great Lakes of airborne emissions of dioxin and mercury from numerous upwind solid- and medical-waste incinerators. The submitters contended that the United States was taking insufficient action to inspect and monitor incinerators emitting dioxin and mercury, to notify certain states that they must reduce such dioxin and mercury emissions because of adverse impacts of the emissions in Canada, and to implement measures that would lead to the virtual elimination of all such dioxin and mercury emissions. The United States responded to the submission and, at the Secretariat's request, provided additional information, particularly regarding the implementation of a new, stricter, regulatory regime for controlling dioxin and mercury emissions from municipal and medical-waste incinerators in the United States. In light of this information, the Secretariat decided against recommending a factual record. However, reflecting the detailed information that the United States provided, the Secretariat's determination dismissing the submission provides extensive information regarding the United States' enforcement of, and compliance with, the Clean Air Act provisions cited in the submission.

A factual record would have added little to this extensive information. The Secretariat concluded that the information the United States had provided indicated that the submitters' principal claims regarding deficiencies in monitoring regulatory compliance – for example, the claim that emissions sampling of incinerators takes place only once, at start-up, and is designed to show only ideal incinerator operations – were unsubstantiated. Further, the Secretariat found no indication of serious, widespread non-compliance and no unaddressed compliance problems. Regarding the claim that the United States was failing to require states to reduce dioxin and mercury emissions because of adverse impacts in Canada, the Secretariat concluded that (a) the EPA's broad discretion in deciding whether to impose that requirement on states, (b) the complexity of source–receptor relationships involving deposition of airborne dioxin and mercury into the Great Lakes, and (c) continuing improvements in the control of those emissions as described in the information the United States provided, all weighed persuasively against preparation of a factual record. Although it raised important questions, the Great Lakes submission turned out to be one in which the need to shed light on enforcement concerns through a factual record was not compelling.

The Fort Huachaca submission and the Secretariat's Article 13 report on the San Pedro River

The submitters of the Fort Huachaca submission, filed in November 1996, asserted that the United States was failing to effectively enforce Sections 4321–4370 of the National Environmental Policy Act (NEPA 1969) with respect to the US Army's operation of Fort Huachuca, Arizona. The submitters' central concern was that the United States had insufficiently considered the impact of groundwater pumping associated with expansion of

Fort Huachaca on water flow in the San Pedro River (CEC 1996). Specifically, the submitters claimed:

The Army has not analyzed the cumulative impacts of the base expansion on the San Pedro River, the San Pedro Riparian National Conservation Area, the riparian ecosystem, the wildlife that lives in that ecosystem, the federally listed threatened and endangered species in the San Pedro corridor, or the San Pedro aquifer.

The San Pedro River has its source in Mexico and flows north into the United States. Apart from being a transboundary watershed, the San Pedro basin is significant on a continental level for its important migratory-bird habitat. The Fort Huachaca submitters described the ecological significance of the watershed as follows:

The San Pedro is the largest and best example of riparian woodland remaining in the southwestern United States. As such, it contains a unique assemblage of avian species. The San Pedro also forms a corridor between Mexico and the United States and helps funnel millions of neotropical migratory birds north to their breeding grounds in the U.S. and Canada. Specifically, it contains the densest remaining breeding populations of the western race of the yellow-billed cuckoo, a subspecies declining throughout its range. The San Pedro also harbors 40 percent of the breeding habitat for the gray hawk in the United States.

Along with their submission under Article 14, the Fort Huachaca Submitters asked the Secretariat to prepare a report under NAAEC Article 13. Although the citizen submission process is triggered exclusively by submissions, the Secretariat has discretion regarding if and when to initiate an Article 13 report. Article 13 authorizes the Secretariat to prepare an independent report, without Council approval, on any matter within the scope of the annual work programme or (subject to a majority-vote veto by the Council) on any other matter related to the cooperative functions of the NAAEC, as long as the matter is not related to whether a Party has failed to effectively enforce its environmental law. After the Secretariat announced that it was going to proceed with a report under Article 13, the submitters withdrew their Article 14 submission.

The Secretariat's Article 13 report on the San Pedro River – Ribbon of Life (CEC 1999) – was the ultimate result of the efforts of the Fort Huachaca submitters to focus the CEC's attention on pressures facing this important transboundary watershed. Unlike a factual record (which is restricted to a presentation of factual information), an Article 13 report can present the Council with concrete policy recommendations for addressing the environmental matter at issue. As well, the Article 13 report on the San Pedro River could examine (more easily than could a factual record) water-management practices on the Mexican side of the border; a factual record probably would have looked only at the United States' application of NEPA to the expansion of Fort Huachaca.

Although an Article 13 report is not formally an alternative to a factual record, an Article 13 report on management of the San Pedro River Basin was particularly appropriate because the basin is transboundary and is important on a North American continental level in providing a habitat for migratory birds. In this way, the CEC's focus on issues of continental significance distinguishes it from bilateral organizations in North America, such as the International Joint Commission (IJC), the International Boundary and Water Commission (IBWC), and the Border Environment Cooperation Commission (BECC), whose work focuses primarily on bilateral issues along the US–Mexico and US–Canada borders.

The process for producing the "Ribbon of Life" report, the Secretariat's third Article 13 report, was typical of the approach the Secretariat has taken in generating these reports. It used a three-phase initiative, the first phase of which was the commissioning of a technical report by an interdisciplinary expert study team on "the operative ecological, bio-hydrologic, socio-economic, and legal/institutional circumstances that characterize the availability of base water flows needed to sustain and enhance the riparian area along the upper San Pedro riverine ecosystem" (CEC 1999). The second phase involved public review of, and comment on, the draft expert report. Over 650 people participated in focus groups and workshops in the region, in both Mexico and the United States, and the Secretariat received more than 300 written comments on the report. In the final phase, the CEC convened a 13-member Upper San Pedro Advisory Panel to consider the issues raised in the expert report and to formulate policy recommendations for meeting goals identified by the public and expert team. The trinational Advisory Panel had American and Mexican co-chairs and included members from academia, the environmental community, the ranching sector, and local communities, as well as former government officials. The members are listed in the report (CEC 1999).

The Secretariat's final report contained several conclusions and policy recommendations. Drawing upon both the technical report of the expert study team and the Advisory Panel recommendations, the Secretariat's conclusions emphasized the need for coordinated, binational, resource management to protect habitats in the San Pedro Basin, additional research, support for regional stakeholders attempting to protect the riverine ecosystem, and outreach to the broader public regarding protection of valuable transboundary resources (CEC 1999). The Secretariat suggested that the Council could recommend that the governments take, or could direct the Secretariat to take, the following actions:

- Designate an interagency working group to develop an implementation strategy for selected panel recommendations, including a mechanism for binational consultation and cooperation;
- Provide direct support for local efforts, such as the Upper San Pedro Partnership and other emerging proposals, as part of the CEC's North American Bird Conservation Initiative';
- Direct the Secretariat to work with the parties and others to identify potential funding mechanisms to support the implementation of selected advisory panel recommendations;
- Organize a workshop on lessons learned in transboundary water management, with a particular emphasis on regional, basin-specific management frameworks for transboundary groundwater resources. Workshop attendees would include representatives from relevant local, state and federal government and others, including institutions involved in transboundary resources along the US–Mexico border, such as the IBWC, BECC, and North American Development Bank. In addition, the workshop should include certain key institutions from outside the US–Mexico border area that have acquired considerable experience in addressing similar issues, such as the International Joint Commission (IJC); and
- Initiate a pilot project to apply the principles and approaches developed in the CEC's work on Sustainable Tourism in Natural Areas. The upper San Pedro valley already attracts roughly US$6 million in tourism revenue, much of which is directly related to bird watching. However, the benefits from virtually all of the valley's tourism are currently incurred within the US portion of the basin. Although ecotourism will not, in itself, provide the ultimate solution for preserving the upper San Pedro River ecosystem, it does provide an important opportunity for economic betterment in both countries. (CEC 1999)

The Article 13 report created renewed interest (on both sides of the border) in improving management of the entire San Pedro River watershed. A May 2002 update from the Udall Center for Studies in Public Policy at the University of Arizona, which assisted the CEC with the Article 13 report, noted that, since the report had been published, "numerous efforts have sought to involve stakeholders in discussions aimed at bridging disagreements between advocates of riparian protection and economic development" (Udall Center 2003).

For example, in 1999, through its North American Fund for Environmental Cooperation (NAFEC), the CEC awarded a US$65,000 grant to the Nature Conservancy and two Mexican partners – the Institute of Environment and Sustainable Development in Sonora (IMADES) and the reserve staff of the Ajos–Bavispe National Forest and Wildlife Refuge, part of the National Institute of Ecology (INE) – for a project to build the foundation for creation of a protected area along the Mexican reach of the San Pedro River. The project sought to:

- compile information to understand the ecological processes, biological values, and social context of the San Pedro's Mexico sub-watershed;
- use that information to produce a conservation plan for the new protected area, with active participation from local communities, that will provide the foundation for a management plan;
- analyze potential private lands conservation mechanisms; and
- provide technical assistance and facilitate training and collaboration between protected area site managers from both sides of the border, including IMADES, INE, The Nature Conservancy, and the Upper San Pedro Partnership. (CEC 1999)

In 1999, the Udall Center initiated a project called "Dialogue San Pedro" to continue discussions of issues surrounding protection of the San Pedro River Basin (Udall Center 2003). Dialogue San Pedro eventually merged with the Upper San Pedro Partnership, which was formed in 1998 during the CEC's Article 13 study. The partnership continues to meet regularly to promote conservation efforts (Varady, Browning-Aiken, and Moote 2001; Nature Conservancy 2003). The CEC's study also helped to coalesce action in Mexico, including efforts such as the NAFEC grant to designate a portion of the San Pedro Basin as part of a national protected area (Varady, Browning-Aiken, and Moote 2001). Improvements in water conservation around Fort Huachaca have been mentioned recently in the Arizona press (*Arizona Republic* 2003). Although the precise role of the CEC's Article 13 report in these developments is difficult to discern, it clearly enhanced ongoing binational efforts to protect the San Pedro watershed.

The future of the CEC's public-participation toolbox

The citizen submission process under Articles 14 and 15 of the NAAEC is a bold innovation by the three NAFTA countries: to give the public an international tool that addresses the governments' practices and holds the countries accountable took courage. However, in addition to being an accountability mechanism, the process also has the potential to dislodge thorny environmental issues that have been difficult to resolve domestically and to invigorate responsive action by the public, government, and other stakeholders. That potential in part motivated the NAFTA countries to create the citizen submission mechanism.

Although use of the citizen submission process to address issues related to transboundary water management has been limited so far, the Great Lakes and Fort Huachaca submissions demonstrate that the process has the potential to be applied in this context. The suitability of the process to a particular situation depends on several factors, including a consideration of the inherent features of the process.

First, it is essential that any submission raising issues of transboundary watershed management must involve an assertion that one of the NAFTA countries is failing to effectively enforce an environmental law. For the purposes of Article 14, "environmental law" according to Article 45(2)(a) means:

[A]ny statute or regulation of a Party, or provision thereof, the primary purpose of which is the protection of the environment, or the prevention of a danger to human life or health, through (i) the prevention, abatement or control of the release, discharge, or emission of pollutants or environmental contaminants, (ii) the control of environmentally hazardous or toxic chemicals, substances, materials and wastes, and the dissemination of information related thereto, or (iii) the protection of wild flora or fauna, including endangered species, their habitat, and specially protected natural areas in the Party's territory, but does not include any statute or regulation, or provision thereof, directly related to worker safety or health.

However, Article 45(2)(b) excludes from the definition of "environmental law" any law for which "the primary purpose ... is managing the commercial harvest or exploitation, or subsistence or aboriginal harvesting, of natural resources." This definition might limit the range of transboundary water-management issues that could be addressed in the citizen submission process. The CEC has not yet had occasion to address that issue.

Another feature of the citizen submission process is that it is not well suited for emergencies or other situations in which a relatively quick response is desired. Although the first factual record, for the Cozumel submission, was published 21 months after the submission was filed, on average the factual records published to date have taken about three years from the date the submission was received to finalize and publish. Indeed, although voting within two months on some occasions, the Council has taken up to 15 months to take action on pending factual record recommendations. Although the creation of a separate unit to process submissions and the hiring of additional staff has improved the speed of processing, experience to date suggests that normally it will be difficult to produce a final factual record in less than approximately two years.

The citizen submission process touches upon two particularly sensitive areas for national governments – sovereignty and enforcement discretion. Although the governments consistently have expressed their support for the process, Articles 14 and 15 confront them with an inherent tension between their roles as both creators and overseers of the process and as

potential targets of it. That the process allows an international organization to present information regarding a country's enforcement of its own
laws probably increases this inherent tension. NAFTA Chapter 11 is ar-
guably a broader relinquishment of sovereignty in that it allows a private
investor whose investment is nationalized or expropriated in violation of
NAFTA's provisions to seek compensation from a NAFTA government
through binding arbitration (NAFTA 1992). Nevertheless, the citizen
submission process is at the frontier of North American accountability
mechanisms that give an international organization a degree of independence in reviewing the actions of one of the three NAFTA countries.
An attempt to shed light on a country's enforcement actions can be
hampered by the potential reluctance of governments to provide details
regarding enforcement strategies and the exercise of enforcement discretion. Further, the possibility, however remote, that a citizen submission
could trigger a dispute resolution proceeding under Articles 22–36 of
the NAAEC, which allow one NAFTA country to claim that another
has a "persistent pattern of failure by that other Party to effectively enforce its environmental law and seek sanctions or loss of trade benefits," potentially affects how the countries handle the citizen submission
process.

The critical juncture at which the Parties' potential concerns over sovereignty and their sensitivity regarding enforcement matters are most
likely to be reflected is when the Council votes on factual record recommendations. Although the Council has authorized preparation of factual
records for eleven submissions, in four of those cases – BC Logging, BC
Mining, Migratory Birds, and Oldman River II (CEC 2003) – it instructed
the Secretariat to prepare factual records that differed in scope from
what the submitters sought and the Secretariat recommended. The possibility that a factual record might not address the enforcement issues that
a potential submitter sought to raise might deter use of the process to
some extent, particularly when it is only at the stage where the Council
votes on a factual record recommendation that the scope of a factual
record is determined. Uncertainty in this regard might dissuade a potential submitter from investing resources into gathering the information
necessary to support its assertions.

A potential submitter must also take into account the likelihood that a
submission will not proceed through the process if the submitters have
not pursued private remedies available under the laws of the Party whose
environmental enforcement is questioned. Article 14(2) guides the Secretariat to consider whether private remedies have been pursued in deciding whether to request a response from the Party. Although there is
no explicit requirement that private remedies be pursued, let alone

exhausted, the NAAEC strongly suggests that a submitter should seek domestic relief before filing a submission with the CEC.

A final major feature of the process that could affect its suitability in a particular situation is that a factual record cannot impose sanctions or force a Party to do anything in regard to the matters addressed. Indeed, as noted above, a factual record does not even decide whether the Party is failing to effectively enforce its environmental law. As a result, reflecting frustration of some members of the public that factual records are not as effective as they might be, the JPAC and others have advised the Council to commit to some kind of follow-up to a factual record – for example, by requiring the Party whose enforcement is addressed in a factual record to report periodically to the Council on follow-up actions (JPAC 2001). To date, the Council has deemed follow-up to factual records to be a matter exclusively of domestic concern (Smith 2002).

Despite this limitation, however, submitters have found that the filing of a submission or publication of a factual record can have an effect. For example, the submitters of the Cozumel submission found that the submission "led to additional protection of coral reefs in the area, improvements to Mexican law on environmental impact assessment, and establishment of a trust fund for reef protection," among other benefits (Garver 2001). Likewise, the submitters of the BC Hydro submission have stated that substantive commitments that the Canadian and British Columbia governments made and that were recorded in the BC Hydro factual record have helped to keep on track a water-use planning process that responded to concerns highlighted in the submission (Bowman 2001). Indeed, the submitters found that the mere filing of the submission brought increased government attentiveness to their concerns regarding the impact of hydroelectric facilities on fish habitat (Bowman 2001).

Although Article 14 has features that limit to some extent the situations in which a person or NGO is likely to use it, the process can also serve as a means for bringing to the attention of the CEC matters that can be addressed with other tools in the CEC toolbox. For example, the Fort Huachaca submission showed that, in some situations, Article 13 might be a more suitable vehicle for addressing concerns that are brought forward in a submission. This is particularly likely where it turns out that a submission does not raise an appropriate issue for review under Articles 14 and 15. However, in considering whether Article 13, the CEC's cooperative programmes, or other mechanisms are suitable alternatives, the CEC must always assess whether it, the CEC (as opposed to another institution), is best placed to address the matter.

In the case of transboundary watershed management, the CEC must be particularly sensitive to the jurisdiction and comparative advantage of North American bilateral institutions such as the IJC, the IBWC, and the BECC. IJC and IBWC commissioners attended the annual session of the CEC Council in June 2002 and, through designated points of contact, the CEC maintains regular communication with all three entities with a view to cooperating on issues of common interest and minimizing duplication of effort. The CEC has also signed a letter of intent with the IJC to formalize cooperation in areas of common interest.

Conclusions

The CEC's citizen submission process and other public-participation mechanisms have the potential to contribute to transboundary watershed management. Whether a reactive mechanism focusing on environmental enforcement concerns, such as the citizen submission process, will prove valuable in regard to a particular matter will depend on the circumstances and the goals of potential submitters. The Cozumel and BC Hydro factual records have shown already that the process can prove useful in efforts to improve policies for protecting the environment, including water resources. However, even if the citizen submission process is not a good fit, the CEC's other programmes and initiatives might also play a useful role, particularly where (as in the case of the San Pedro River) management of a transboundary watershed raises concerns with continental significance. In all cases, the greatest promise of CEC's toolbox is in providing objective and rigorous factual information and analysis and involving a broad range of stakeholders with a view to untangling difficult problems of environmental or natural resource management.

Note

1. Director, Submissions on Enforcement Matters Unit, Commission for Environmental Cooperation. This chapter does not represent the views of the CEC, but instead solely represents the views of the author.

REFERENCES

Arizona Republic. 2003. "Riparian Rip-off" (editorial). 21 May, B10.

Bowman, Jamie. 2001. "Citizen Submission Process Proves Valuable in BC Hydro Case." *Trio* (Fall). Internet: ⟨http://www.cec.org/trio/stories/index.cfm?ed=5&ID=70&varlan=english⟩ (visited 25 June 2003).

CEC (North American Commission for Environmental Cooperation). 1996. "Fort Huachaca Submission." CEC Submission SEM-96-004. Internet: ⟨http://www.cec.org/files/pdf/sem/ACF158.pdf⟩ (visited 25 June 2003).

CEC (North American Commission for Environmental Cooperation). 1999. "Ribbon of Life: An Agenda for Preserving Migratory Bird Habitat on the Upper San Pedro River." Internet: ⟨http://www.cec.org/files/pdf//sp-engl_EN.pdf⟩ (visited 25 June 2003).

CEC (North American Commission for Environmental Cooperation). 2003. "Citizen Submissions on Enforcement Matters." Internet: ⟨http://www.cec.org/citizen/⟩ (visited 25 June 2003).

Fisheries Act. 1985. Canada. R.S., ch. F-14, sec. 1.

Garver, Geoffrey. 2001. "Citizen Spotlight is Beginning to Show Results." *The Environmental Forum* (March/April): 34.

Joint Public Advisory Committee of the Commission for Environmental Cooperation (JPAC). 2001. "Lessons Learned: Citizen Submissions under Articles 14 and 15 of the North American Agreement on Environmental Cooperation." June 6. Internet: ⟨http://www.cec.org/files/PDF/JPAC/rep11-e-final_EN.PDF⟩ (visited 25 June 2003).

Markell, David L. 2001. "The Citizen Spotlight Process." *The Environmental Forum* (March/April): 32.

NAAEC (North American Agreement on Environmental Cooperation). 1994. Internet: ⟨http://www.cec.org/pubs_info_resources/law_treat_agree/naaec/index.cfm?varlan=english⟩ (visited 25 June 2003).

NAFTA (North American Free Trade Agreement). 1992. Done at Ottawa, Mexico City and Washington, DC on 17 December 1992. Entered into force 1 January 1994. Reprinted in 32 I.L.M. 289 (1993).

Nature Conservancy. 2003. "Last Great Places: Upper San Pedro River." Internet: ⟨http://www.lastgreatplaces.org/SanPedro/conserve_efforts.html⟩ (visited 1 August 2003).

NEPA (National Environmental Policy Act of 1969). 1969. United States Code, Title 42, secs 4321–4347.

Smith, Norine. 2002. Letter to Jonathan Plaut, JPAC Chair. June 14. Internet: ⟨http://www.cec.org/files/PDF/JPAC/L-Coun-04.pdf⟩ (visited 25 June 2003).

Udall Center for Studies in Public Policy. 2003. "Dialogue San Pedro." Internet: ⟨http://www.udallcenter.arizona.edu/sanpedro/dialogue.html⟩ (visited 25 June 2003).

Varady, Robert B., Anne Browning-Aiken, and Margaret Ann Moote. 2001. "Watershed Councils Along the U.S.–Mexico Border: The San Pedro Basin." *Proceedings of the AWRA/IWLRI University of Dundee International Specialty Conference on Globalization and Water Resource Management: The Changing Value of Water.* Internet: ⟨http://www.awra.org/proceedings/dundee01/Documents/VaradyandMoote.pdf⟩ (visited 25 June 2003).

Part IV

Lessons from domestic watercourses

13

Improving sustainable management of Kenyan fisheries resources through public participation

Nancy Gitonga

Introduction

Since time immemorial, fishing has been regarded as a limitless food source and source of employment to fishers. However, the world is now aware that renewable, aquatic, living resources are threatened with over-exploitation and imminent collapse unless a deliberate effort is made to ensure sustainable utilization. As the fisheries sector continues to be a dynamic player in the world's food industry, investment in modern fishing fleets and fish-processing factories has grown to satisfy the growing international demand for fish and fishery products.

Consequently, fisheries resources worldwide are fast declining owing to increased fishing pressure as the demand for fish continues to outstrip supply. When the decline is realized through the reduction in fishers' catch per unit effort and sometimes corroborated by research findings, the policy makers (usually without consulting the users) take measures to control – and, in most cases, to restrict – the total amount to be harvested. This is usually done through various methods such as licensing, surveillance, total allowable catch (TAC), and quotas, among others.

Such measures do not usually work very well, and stocks often fail to recover. The TAC regime applied in Norway in 1977, for example, failed to reverse the stocks' decline in any significant way (Hannesson 1996). When the fishers are restricted, and also when they are unaware of the importance of a healthy ecosystem, they tend to resort to illegal fishing methods that destabilize the very environment that supports their liveli-hoods.

The need to reverse the alarming depletion of the world's fish stocks and the degradation of marine ecosystems has been recognized by the in-ternational community in many forums. For example, the World Summit on Sustainable Development (WSSD) pledged to maintain or restore fish stocks to levels that can produce the maximum sustainable yields. The forum stressed the need for these goals to be achieved as a matter of urgency and, where possible, not later than 2015 (WSSD 2002).

The conventional fisheries-management paradigm, which focuses on fishing activities and the target resource, is constrained in achieving sus-tainable utilization of the available resources and conservation of the environment and biodiversity. In contrast, the ecosystem approach to fisheries management, a relatively new management concept that identi-fies and defines the ecosystem to include human populations, offers a viable option for achieving sustainable fisheries utilization. In the new approach, stakeholders become the stewards of the resources and are, therefore, involved in the decision-making, implementation, and moni-toring processes. This new approach also provides a framework for man-aging fisheries that often are a transboundary or shared resource, for example in the case of Lake Victoria.

This chapter examines the roles of resource users, the fisheries man-agers, and scientific information in the management of fisheries and watersheds to achieve healthy and environmentally sustainable fresh-water and marine ecosystems in Kenya and East Africa. It focuses on the evolution of some novel approaches for engaging the public in do-mestic and transnational watercourses.

Evolution of techniques for managing fisheries

Conventional fisheries management

Conventional fisheries management is constrained in its delivery, mainly because of limited knowledge regarding the status and dynamics of the stocks, the tendency to give priority to the short-term social and eco-nomic needs at the expense of long-term benefits from sustainable fish stocks, and institutional weaknesses – particularly the lack of stake-holders' involvement in decision-making processes.

In conventional fisheries-management practice, a government typically formulates policies in the assumption that the users will comply with them even though they were not involved in the process. This manage-ment paradigm focuses on the fishing activity and target fish resources.

Since ownership of fish from a common resource occurs only upon capture, the resource users do not usually regard the unharvested resource as their own, but rather as belonging to the government which regulates their utilization. In this regard, the fishers do not usually see the value of conserving the stocks only for other fishers to harvest later. This attitude encourages fishers to capture as much as they can while they can, which in most cases results in the decline of fish stocks. The policies that are developed through the conventional top-down method, therefore, frequently result in low-level compliance by resource users who, because of their lack of involvement in the policy decision-making process, may not realize the benefit of their participation in the implementation.

The ecosystem approach to fisheries

The principles and concepts of the ecosystem approach to fisheries are not new, as they are contained in a number of international instruments, agreements, and conferences, either already adopted or in the process of being implemented. These include the 1972 World Conference on Human Environment (the Stockholm Declaration), the 1982 United Nations Law of the Sea Convention (UNCLOS 1982), the 1992 United Nations Conference on Environment and Development (the Rio Declaration) and its Agenda 21, the 1992 Convention on Biological Diversity, the 1995 United Nations Fish Stocks Agreement, and the 1995 FAO Code of Conduct for Responsible Fisheries (FAO 1995). More specifically applicable, however, are the 2001 Reykjavik Declaration and the 2002 WSSD Plan of Implementation. During the WSSD, the heads of state agreed to develop and facilitate the use of diverse approaches and tools, including the ecosystem approach, for sustainable management of fisheries.

The ecosystem approach to fisheries calls for strong stakeholder participation and decentralized decision-making structures. This highly participatory fisheries-management paradigm aims at achieving an integrated and realistic approach to fisheries that ensures that – despite variability, uncertainty, and natural changes in the ecosystem – the capacity to produce fish food, revenues, employment, and general wealth is maintained indefinitely for the benefit of the present and the future generations. The ecosystem approach to fisheries also provides scope for an increased involvement of regional bodies in establishing integrated management measures. The main implication of this paradigm is the need to cater for both human and ecosystem well-being, since people affect, and are affected by, their ecosystem.

Although the concepts underpinning the ecosystem approach to fisheries are not new, there has been little experience in its application. The success of this approach requires that an ecosystem is identified and described in order to manage it as a single interconnected system (FAO 2003). Once the ecosystem is defined, it is necessary to establish an effective consultation and decision-making process to facilitate legitimate and effective stakeholder involvement. The appropriate management measures to achieve these objectives are then developed through transparent stakeholder participation. Gathering relevant research information based on the gaps identified by stakeholders and establishment of regular effective review, monitoring, and enforcement systems are some of the ingredients for a successful ecosystem approach to fisheries.

Public participation in fisheries resource management in Kenya

The Lake Naivasha fishery collapsed in 2000. Many fishers from Lake Naivasha had voluntarily abandoned fishing by mid-2000, owing to scarcity of fish stocks; those who persisted and continued to fish could barely make a living. In 2000, the fish stocks in Lake Victoria also declined, and there were conflicts in the shrimp fisheries located in Kenya's marine territorial waters. Together, these collapses and strains have been the driving force behind the current reform in fisheries-management practice in Kenya, as elaborated below. These disastrous events prompted an urgent effort to restore lost and declining stocks rather than trial of a new fisheries-management approach. A number of factors have influenced the collapses: these include the open-access regime; the application of environmentally unfriendly fishing methods such as seining; excess fishing pressure, especially through the use of illegal gear; and corrupt practices by law enforcers.

The reforms involve extensive stakeholder consultation in policy decision-making and implementation processes. Public participation, mainly through stakeholder consultative meetings, has been instrumental in addressing the decline of fish stocks in Kenya. The most significant achievement of this consultative process with the public is the realization by the resource users that they have a fundamental role to play in the sustainable management of their resources and, therefore, the need to participate effectively. The participatory approach with the fisher community has, in effect, helped to significantly reduce the incidence of more destructive gear than when the government dealt with the situation without involving the public. The most significant output of this process was the removal of 22 trawlers in March 2001 (Kenya Fisheries Department 2001a) that had been fishing illegally in Lake Victoria since 1993, when the revised subsidiary regulation prohibited them from operating

(*Kenya Gazette* 1993). The banning of the destructive beach seines (*Kenya Gazette* 2001b) after consensus was reached at a stakeholder meeting (Kenya Fisheries Department 2001b), was another move towards sustainable fisheries management of Lake Victoria through public participation. The fishers were helpful in reporting the illegal fishers to the law enforcement agencies, and the joint surveillance by the government and communities thereby rapidly produced the desired results. The following sections outline how the Kenyan Government proceeded to address the concerns relating to the decline in fish stocks in a participatory process.

Stakeholder consultation in managing Lake Naivasha fisheries

In late 2000, it became obvious that something had to be done to restore and manage the Lake Naivasha fisheries. Stakeholders – including environmentalists, the lake riparian landowners, farmers who use lake water for irrigation, government agencies, politicians, wildlife and conservation agencies, research institutions, universities, and NGOs – were identified. A series of stakeholder meetings (Kenya Fisheries Department 2002a), convened by the Fisheries Department and attended by representatives of all the identified stakeholder groups, deliberated on the causes of Lake Naivasha fishery decline and came to the conclusion that the overfishing of the lake was due to the increased fishing effort, especially from the unlicensed illegal fishers. A consensus was reached to close Lake Naivasha to all fishing activities for six months. During this period, a research survey would be conducted to determine the extent of overfishing and the rate of stock regeneration. The information gathered would then be used to develop a strategy that would ensure sustainable resource utilization.

Management of the fishing ban

The government closed Lake Naivasha by enacting subsidiary legislation through a legal notice (*Kenya Gazette* 2001a), and a monitoring programme was put in place during the period of the ban, which was effected in February 2001. Because of the inadequate surveillance capacity of the Fisheries Department at the time the ban started, a task force composed mainly of the displaced fishers (i.e. those who had to stop fishing after the ban had been put in place), was put in place through an integrated approach, to augment the governmental enforcement effort. Stakeholders and friends of Lake Naivasha, through the Lake Naivasha Riparian Association, voluntarily supported the task force and worked closely with the government enforcement team, through a management committee that was established during the consultative meetings.

Although there were signs of recovery of the fishing stocks during the six-month ban period, as evidenced by survey data (KMFRI 2002a), researchers from the Kenya Marine and Fisheries Research Institute recommended that the ban be extended to allow appreciable recovery of the resource (KMFRI 2002a). The stakeholders agreed to extend the ban for another six months but demanded regular updates from researchers. During the waiting period, a subcommittee comprising all the stakeholders' representatives was established and mandated to develop a return-to-fishing formula, which would include new or additional regulations, a participatory surveillance system, closed areas and seasons, penalties for lawbreakers, and development of need-driven research protocols (Kenya Fisheries Department 2002b).

The survey data showed that the number of licensed fishers and boats had to be reduced from 300 and 130, respectively, by one-third. Moreover, in order to achieve sustainable utilization of the fisheries, a concerted effort to eliminate illegal fishing would be necessary.

Lifting the ban

After a one-year fishing ban, the fishery had shown encouraging signs of recovery to allow a cautious return to fishing. On the basis of the research results from the survey carried out during the ban period, and of previous information on the performance of fish stocks in relation to fishing effort (derived from data compiled by the research team), the Kenya Marine and Fisheries Institute researchers recommended that the number of boats, fishers, and gear be reduced to about one-third of the existing numbers (KMFRI 2002a). This would allow sustainable exploitation of the resource after the recovery brought about by the fishing ban.

The aim was to reduce the number of fishers who would be licensed to resume fishing upon reopening of the lake. All prospective fishers who had applied for a fishing licence were subjected to scrutiny using the criteria developed by a subcommittee that had been formed by the stakeholders during the consultative meetings. In developing the criteria, the committee took into account the need for information regarding those fishers who had other businesses besides fishing and those who depended solely on fishing. The criteria thus developed were used to award merit points in an effort to ensure that only the most deserving fishers would be licensed to return to fishing on a trial basis (Kenya Fisheries Department 2002c).

The stakeholders' meeting, through consensus, proceeded to approve 43 boats operated by three crew each with a maximum of ten nets per boat to resume fishing on a trial basis. It was also agreed that anyone permitted to return and who contravened any regulation would be permanently barred from fishing in Lake Naivasha.

Involvement of the fishers in the retrenchment exercise pre-empted any anticipated unrest by the fishers. Various stakeholders offered employment and other alternative sources of livelihood to alleviate any unemployment arising from the reduction of the number of fishers earning their living from Lake Naivasha.

Benefits from the process

The most significant achievement of the year-long closure of the lake and the stakeholder consultation process was the realization of secondary benefits, although these had not initially been anticipated. Such benefits included a significant recovery of birds and other wildlife, the growth of vegetation around the lake, general environmental improvement, acceptance of the resource-users' interests, appreciation of the different roles played by various stakeholders, and recognition of the importance of an integrative participatory approach to sustaining a healthy lake ecosystem.

Although many issues, including illegal fishing, still need to be resolved before truly sustainable management of Lake Naivasha is achieved, a firm and workable process for better management of the fishery has started. Environmental-restoration strategies, watershed management, and stock-replenishing programmes for the lake are some of the areas that need to be considered by stakeholders, in addition to fisheries management, in order to achieve a healthy and sustainable ecosystem.

The Ministry of Water has recently taken a lead role in addressing the Lake Naivasha water catchment intake, and utilization, through stakeholder consultations. The initial meeting was held in Naivasha, at which it was resolved that water usage by riparian horticulturalists would be controlled more stringently. The same meeting agreed on the need to monitor upstream water usage of the main rivers flowing into Lake Naivasha, and also to carry out a study on the extent of catchment destruction, with a view to developing a strategy for its restoration. It is expected that these consultative meetings (which have now been accepted by policy makers as a useful paradigm for sustainable management of shared ecosystems) will continue to be pursued by the Ministry of Water in order to develop a lasting solution in the management of water resources country-wide. The close collaboration between the Department of Fisheries, the Ministry of Water, and the stakeholders of Lake Naivasha is a good beginning towards sustainable management of the watershed. This consultation process has been a significant departure from previous approaches in which the Ministry of Water had approved water extraction for irrigation and other purposes without involving the Fisheries Department, downstream consumers of water, and other important stakeholders.

Lake Victoria (Kenyan section) fisheries

Lake Victoria and coastal fisheries play an important role in the socio-economic development of the peoples of East Africa by providing food, employment, and income, as well as hard currency for the nation, among other benefits. Fishing is also a way of life for many coastal and lakeside communities.

The current decline of fish stocks in Lake Victoria and the marine territorial waters of Kenya and Tanzania are influenced by the significant overinvestment in the fish-export business. Because of this overcapacity, the demand for fish is so great that, to satisfy this demand, the fishers are catching younger and smaller fish, thus not giving the fish a chance to mature and reproduce.

The fishers and the industrialists have thus fallen into what has been termed a "tragedy of the commons" (Hardin 1968). In such a common-property tragedy, each harvester continues to harvest until the resource is depleted, believing that they might as well do so before someone else does, because the common property (fish, in this case) is owned only upon capture and others are not holding back from fishing. In Hardin's words, the unilateral action is "self-eliminating."

Whereas the local fishers do not appear to be aware of the steep fall in fish stocks, the industrial fish processors that drive the exploitation of such resources have started to realize the danger of the current fishing pressure. Accordingly, they are beginning to be more proactive by holding meetings to discuss the future of the Nile perch fishery, offering remedial suggestions, and becoming more responsive to stakeholder consultative meetings. This is a positive step, an unusual initiative from major beneficiaries of the resources, and one that assists managers in promoting sustainable utilization.

The task before the East African Community (EAC) is to determine how to manage the fisheries of Lake Victoria and the Indian Ocean to ensure their sustainable utilization for the benefit not only of the present but also of future generations, while maintaining a healthy ecosystem for biodiversity conservation.

The decline in catches and the extra effort required to harvest adequate amounts of fish are signs that the fishery is currently managed unsustainably. This has been the case with Lake Victoria in recent years, highlighting the need for rapid intervention by the government. Kenya's Department of Fisheries organized a stakeholder meeting for the Kenyan section of Lake Victoria, but also invited the Lake Victoria Fisheries Organization (LVFO) Secretariat and the research coordinators of the Lake Victoria research and environmental management regional projects to participate as stakeholders.

During the meeting held in June 2001, the fishers expressed their willingness to participate actively in the management of the Lake Victoria fisheries. They suggested that closed seasons and areas would be a good beginning toward the restoration of stocks. Some of the measures suggested by the stakeholders have since been implemented through enactment of new subsidiary fisheries legislation (*Kenya Gazette* 2001b). Because of the involvement of stakeholders in the policy-making process from the outset, the compliance level with the closed season for fishing *Rastrineobola argentea* (small freshwater sardines, known as *omena*, *dagaa*, or *mukene* in East African local languages), which was effected in April 2002 was high (between 60 and 85 per cent), despite political interference with the ban at times. This has been an encouraging development, considering that closed seasons had not been imposed on Lake Victoria fisheries for the last decade.

At the end of the close season, the fishers were delighted by the abundance of good-quality target fish and the recovery of non-target fish such as tilapia. During the close season, the fishers themselves participated in surveillance, especially by reporting those who violated the new regulations. The resource users' sense of ownership was clearly demonstrable, during this period, by their full participation in the law enforcement. The second close season, on 1 April 2003, therefore met with less resistance and it is anticipated that a much higher degree of compliance will be achieved because, by general consensus, it was also decided that any trade in the target fish would be banned during the close season. In order to succeed, however, the government and the fishers must jointly participate in effective enforcement of the close season and ensure fishing does not take place. Besides ensuring that resource users obey the laws and regulations, this participatory management approach encourages and gives confidence to law-abiding resource users to participate, with the knowledge that the collective management is worth the effort and with patience, there would be gains in both the short and the long term. In Hardin's terminology, such a participatory approach facilitates "mutual coercion mutually assured." Nevertheless, there are still some surveillance weaknesses resulting from a lack of craft and of outboard motors for both the Department of Fisheries and the fishers, which must be urgently addressed in order to ensure rapid regeneration of fish stocks in Lake Victoria and other natural fishery resources in Kenya.

The current intensive exploitation of fisheries resources, especially the Nile perch in Lake Victoria, calls for the urgent development of a common strategy by the partner states of the EAC if resource depletion is to be avoided. The greatest weakness in the current system for managing Lake Victoria fisheries is the emphasis by partner states on restricting fishing by those from other states within their respective boundaries, despite the traditional amicable coexistence enjoyed by border communities, for whom cross-border fishing and trading has been the accepted way of life. This recent development is due to the increasing demand for fish to satisfy the expanding and lucrative Nile perch export industry. The increase in fishing efforts, exacerbated by the use of illegal fishing methods and gear as well as by an open-access regime (which encourages new entrants to fish for quick gains), needs to be urgently addressed by the EAC.

In developing a sustainable approach to managing the Lake Victoria fisheries, it is hoped that EAC partner states will be guided by an ecosystem approach. Such an approach would gather and review the available information on the lake and marine ecosystems, identify means by which ecosystem considerations can be included in fisheries management, and identify future challenges and relevant strategies. The establishment of the LVFO, a regional body whose Secretariat is already in place, is a good move to this end. The formation of the LVFO was based on the desire to manage Lake Victoria, which unites the three countries, as a single ecosystem. At the time of the establishment of the LVFO, Kenya (in spite of its small portion of the lake – 6 per cent) had a much higher exploitation capacity than Uganda and Tanzania until the mid-1990s. This has changed as all three countries realized the economic importance of the Nile perch, and now the processing capacities of each of the other two states are higher than those of Kenya. The competition for the target resource, due to the lucrative export market, has increased the level of fishing pressure to one that, if not urgently rectified, could result in the fisheries' collapse.

The current situation has brought into sharp focus the imminent danger of overexploitation. The increase in demand for Nile perch has started to cause conflict between fishing communities, as well as the resource managers. Such a conflict between Kenya and Uganda, which started in August 2001, had to be resolved by the Council of Ministers, the highest policy organ of the LVFO. The Council also directed that harvesting of Nile perch within a slot size of fish less than 50 cm and not more than 85 cm in length should be prevented by using nets with a mesh of 5–8 inches (≈ 12.7–20.32 cm) as recommended by the researchers. This is expected to expedite the fishery recovery, as the young fish would be allowed to grow to maturity (at 50 cm total length) and the spawners (over 85 cm total length) would also be spared from being harvested. The three partner states are implementing this directive, mainly through public participation by fish processors and fishers. The second survey mission around Lake Victoria, carried out by LVFO executive committee members, to evaluate the activities of the fisheries, showed better resource management where the public was involved than where the management was solely by the government (LVFO 2003).

The governments of Kenya, Tanzania, and Uganda on behalf of their people are mandated to manage the Lake's fisheries through the Fisheries Departments. To coordinate regional fisheries management, the three governments established the LVFO in 1997 (LVFO 2001), comprising the Fisheries Departments and Fisheries Research Institutions of Kenya, Tanzania, and Uganda. The LVFO is managed through several committees – namely, the Fisheries Management Committee, Executive Committee, Policy Steering Committee, and Council of Ministers (LVFO 2001).

At present, the Lake Victoria fisheries regulations, including cross-border fishing and fish trade, are not harmonized (Heck et al. 2003). Tanzania and Uganda, for instance, do not allow non-citizens (i.e. people from partner states) to fish without special licences, and harvested fish must be landed in their respective countries, not in Kenya. Kenya, on the other hand, has practised an open-border policy by allowing East African Community fishers to land fish in Kenya regardless of their nationality (Heck et al. 2003).

Although cross-border meetings have taken place on all international borders, mainly involving government officials, cross-border conflicts still persist. Conflict resolution has been complicated by various factors such as low compliance with existing regulations, weak patrolling and enforcement capacity, and changes in fishing technologies – which include night fishing, use of active fishing gear (such as drift nets, seine nets or trawl nets), and the use of large container boats to fish and to collect fish from the islands.

The partner states, through the LVFO management committees and policy organs described above, should seriously consider the application of an ecosystem approach to fisheries management in Lake Victoria as it offers a viable option to the prevailing conventional fisheries-management paradigm. One important weakness of the current management approach of those fisheries is its emphasis on restricting fishing capacity and effort instead of advocating a rights-based approach, which could encourage fishers to view their role as one of stewardship of the resource and the ecosystem rather than that of mere exploiters. Such a change in attitude, and the involvement of the public of the three states in the general management of the Lake Victoria fisheries, could expedite cross-border conflict resolution. The LVFO and the EAC policy organs have now realized the need for public participation in the management of shared resources and are encouraging the formation and strengthening of beach-management units (BMUs), including giving them legal status as smaller localized units.

Public participation in the management of Lake Victoria fisheries

The significance of public participation in management of Lake Victoria fisheries resources has been studied (Geheb and Sarch 2002), and the results have been shared with policy makers in the three partner states, through the auspices of the LVFO. Following realization of the immense fishing pressure, remedial decisions (such as net mesh and gear size) have been made and implemented through approaches that emphasize stakeholder participation.

The LVFO policy organs have played an important role in bringing the three states together to discuss the Lake Victoria fishery. At this level, however, stakeholders have not been involved in critical appraisal of the fishery and the decision-making processes for its management. As a result, enforcement by individual governments is usually not as effective as one would wish, and the costs of preventing the use of illegal fishing methods are exorbitant.

In this regard, BMUs, which originated in Tanzania, have been adopted informally in the other two partner states. The BMUs have been in place in Kenya and Tanzania for a long time. In Kenya, the government has used BMUs to disseminate information to fishers, through a process originally referred to as "beach leadership." It was expected that the policies would be complied with without question, and no opinion was sought from the resource users during the policy-formulation process. The fishers thus viewed the resource as belonging to the government and, therefore, would comply only in order to avoid punishment by enforcement agencies. The government (owing to ineffective surveillance and, occasionally, corruption by enforcement officers) failed to curb illegal fishing or to stop overexploitation. The major role of beach leadership at that time was to collect revenues from fishers, without giving much service to them in return.

As previously stated, the BMUs' approach to fisheries management was pioneered in Tanzania, which has a decentralized political system that empowers village governments to manage their own resources independently as long as there is no infringement of national laws. The BMUs are small, localized units at the fish-landing sites and consist mainly of the fishers themselves. The members usually elect a management committee, which oversees the activities of a landing site.

The fundamental role of BMUs, now adopted by the three East African countries, is to ensure that fisheries regulations are complied with; in the case of Tanzania, offenders are fined. The BMUs are also channels for transmission of fishers' views to the government, and through which information from the government is received. Because of the success of some of the BMUs, the EAC has recognized these grass-roots units as a viable option for sustainable management of Lake Victoria fisheries, including law enforcement and enhancement of security in the lake. The EAC policy organs have, therefore, directed that the BMUs be given legislative powers to assist in apprehending any fishers who contravene the regulations – for instance, by using undersized nets. The BMUs are also expected to determine and recommend those fishers eligible for fishing licences, depending on their compliance regarding the use of correct fishing methods and gear. Given such powers, the fishers would assume full stewardship of their resource. The BMUs in the three East African states are at different stages of development, but the countries have agreed to harmonize the roles of these localized units in an effort to facilitate sustainable management of Lake Victoria as a single unit.

It is essential for the EAC partner states to view the lake as a single ecosystem for management purposes. The prevailing management system, in which each state manages its portion of the lake separately, is a self-defeating strategy. A change of attitude is necessary, especially now that the Nile perch has become an extremely popular fish in the world market.

It is not possible to separate aquatic "pastures" by fences, as can be done with terrestrial units. At the same time, aquatic ecosystems are complex, dynamic, and still poorly understood when compared with terrestrial ecosystems (Keen 1988); therefore, human and natural aspects need to be taken into account when designing an optimal ownership system. Ownership requires the support of political systems with effective and readily enforceable laws, responsive to the needs of fishery resource management, and the involvement of resource users. The EAC partner states are in the process of harmonizing fisheries legislation, and it is hoped that the stakeholders will be involved before the process is finalized. Because the three states have different views on public participation in fisheries management, the LVFO [in collaboration with the IUCN – the World Conservation Union and the European Union through the Lake Victoria Fisheries Research Project (LVFRP)] carried out studies on cross-border fisheries management in Lake Victoria (Heck et al. 2003) and a three-beach study (Geheb and Sarch 2002). Both studies clearly demonstrated that resource users wished to be included in decision-making processes as well as in the implementation of the collectively agreed management measures. In the three-beach study, there was a high level of compliance among those fishers who were members of BMUs that were allowed to manage their resources independent of the governments.

Owing to the economic importance of the Nile perch, some governments are reluctant to agree to harmonized regulations, fearing that other partner states might harvest their resources. Nevertheless, both the LVFO and the EAC are in favour of harmonization of the fisheries legislation of the three countries; accordingly, the process was started in 2002, when a task force was established to review existing fisheries laws and to draw up recommendations for harmonization. The task force (composed of fisheries personnel and lawyers from the three countries) has completed the study; the recommendations have been discussed and consensus reached at the LVFO Executive Committee. The harmonized fisheries legislation has now been approved by the Council of Ministers, and the EAC has taken up the process of enactment into legislation, although, in the interim, individual countries have incorporated the harmonized legislation into their national legislation instruments.

Prawn fisheries in Kenya

Prawn-fishery exploitation in Kenya, which started in the mid-1970s, has been riddled with controversies and conflict from other resource users in the same environment. The conflict was so great in 1999 and 2000 that it became necessary for the government to intervene and call for a consultative stakeholder meeting in October 2000 consisting of artisanal fishers (i.e. small-scale or traditional fisherfolk), conservationists, trawlers, researchers, and fisheries managers. The first meeting was very contentious, with artisanal fishers complaining that trawlers destroy their fishing environment and fishing gear without compensation. The government was also accused of siding with trawler owners. Conservationists complained that the trawlers endangered the sea turtle population. The meeting resolved that the information available was not adequate to enable the government to take a policy decision that differed from the current approach, and therefore a scientific committee was formed to carry out a research survey for a year. The information gathered from such a research programme was expected to assist in determining the fisheries-management options that would lead to resource recovery and sustainability.

Another stakeholder meeting was convened in April 2003 to evaluate the results of the research and surveys, with the aim of developing an appropriate management strategy that was acceptable to the majority of participants. (The composition of stakeholders who attend the meetings has been consistent.) The result of surveys, for example, showed that prawns are abundant between the coast and 5 nautical miles off shore (KMFRI 2002b), but the current legislation restricts prawn trawling to

taking place only in waters more than 5 nautical miles from the shore. Although the scientific committee recommended that trawlers should operate more than three nautical miles from the shoreline, the information
from the 1-year research survey was too sparse to enable the government
to change the legislation from the 5-mile to a 3-mile limit (Kenya Fisheries Department 2003).

The second meeting, which again aroused strong feelings, decided that
the available information was not adequate to enable enactment of new
legislation and, therefore, further research was necessary. Conservationists (and a few fishers) came to the meeting in the conviction that trawlers
should not be allowed to operate in Kenyan waters; some of them walked
out of the meeting when they realized that trawlers would be allowed to
continue to operate (albeit under strict observation) until there was adequate information for the government to decide otherwise. The scientific
committee was mandated to develop a research protocol in order to collect more information on environmental degradation and on the economic viability of trawlers operating more than 3 nautical miles from the
coast.

Conclusions

One practical lesson learned from the experience of stakeholder involvement in the decision-making processes regarding management of East
African fisheries is that various stakeholders hold high expectations that
the meetings will see their points of view. Although, therefore, participants may ultimately be disappointed, in most cases stakeholders accept
the outcome of the decisions made through consensus after consultation
and are therefore more likely to participate in the implementation
process.

Improved stakeholder consultation in Lake Naivasha has demonstrated that public involvement can greatly improve livelihoods through
the sustainable management of fisheries, especially where a collective
and integrated approach to management is pursued. Lake Victoria, a
shared fishery resource, could benefit from a similar approach. The resource users need to be made aware of their fishery resource and the
role they can play in its sustainable management so that they can participate effectively. In this regard, therefore, it is important that research results should be widely disseminated in order to reach all stakeholders.

The "tragedy of the commons" – which includes the feeling that it is
imperative to exploit the resource before someone else does, the need
to want to harvest the most valuable resource (or species) first, and the

lack of interest in investment that would improve productivity of the resource – is the major cause of the decline in the world's fisheries (Keen 1988). It is imperative to recognize and address this tragedy, which can in part be done by effective advocacy for full ownership by resource users and therefore for property rights, amounting to a right to enjoy the fruits of such users' own efforts and restraint. The challenge is to harness such property rights through an ecosystem approach to fisheries management that promotes efficient, appropriate, conservation of resources, while at the same time using such resources more economically. When the resource users are fully accustomed to regarding the fisheries resource as their own, they become involved voluntarily in both the decision-making process and the implementation of collectively agreed fisheries-management measures. The public thus becomes proactive and participates in ensuring that all resource users obey regulations that protect the resource, with localized management units (e.g. BMUs) being elected by fishers to ensure such compliance.

The changes required in a transition from the conventional management, with its are top-down policies, to the ecosystem-based fisheries-management paradigm, with policies developed through public participation and that encourage voluntary restraint, may entail considerable sacrifices by resource users and costs to the fisheries sector in the short term. However, these are likely to be compensated in the medium term and would definitely accrue huge benefits for all the interested and participating parties in the long term. The deliberate efforts by the Fisheries Department to involve the public in fisheries resource management in Kenya has had a number of positive outcomes, such as the formation of many BMUs, which continue to be well supported by fishers through membership enrolment in their respective beaches. The BMUs have been recognized by the EAC as cohesive groups with a full mandate from fishers and as institutions that can be used effectively in management of the shared resources. In this regard, the EAC and LVFO policy organs have directed that the BMUs should be strengthened and be given legal mandates to assist in implementing those decisions made through public participation and consensus.

REFERENCES

Convention on Biological Diversity. 1992. Internet: ⟨http://www.biodiv.org/convention/articles.asp⟩ (visited 7 February 2005).

FAO. 1995. United Nations Fish Stocks Agreement. Agreement Relating to Conservation and Management of Straddling Fish Stocks and Highly Migratory Fish Stocks. Rome: FAO.

FAO. 2003. *The Ecosystem Approach to Marine Capture Fisheries.* FAO Technical Guidelines for Responsible Fisheries. Rome: FAO.

Geheb, K., and M.T. Sarch (eds). 2002. *Africa's Inland Fisheries: The Management Challenge.* Kampala: Fountain Publishers.

Hannesson, Rognvaldur. 1996. *Fisheries Mismanagement. The Case of the North Atlantic Cod.* Oxford: Fishing News Books, 113–125.

Hardin, Garrett. 1968. "The Tragedy of the Commons." *Science* 162:1243–1248.

Heck, S., J. Ikwaput, C. Kirema-Mukasa, C. Lwenya, D.N. Murakwa, K. Odongkara, P. Onyango, J. Owino, and F. Sobo. 2003. *Cross-border Fishing and Fish Trade on Lake Victoria.* Fisheries Management Series No. 1. Nairobi: IUCN – The World Conservation Union, Eastern Africa Regional Programme.

Keen, Elmer A. 1998. *Ownership and Productivity of Marine Fishery Resources. An Essay on the Resolution of Conflict in the Use of the Ocean Pastures.* Blacksburg, Va.: MacDonald and Woodward Publishing Company.

Kenya Fisheries Department. 2001a. *Fisheries Department Executive 2001 Report.* Nairobi, Kenya: Agriculture Information Resource Centre.

Kenya Fisheries Department. 2001b. Report of Stakeholder Meeting of L. Victoria, Kisumu, Kenya, 21 June.

Kenya Fisheries Department. 2002a. Minutes of Stakeholder Meetings for Sustainable Management of L. Naivasha, 2001–2002.

Kenya Fisheries Department. 2002b. Reports of the Lake Naivasha Ban Management Sub-Committee, 2001–2002.

Kenya Fisheries Department. 2002c. Report of the Lake Naivasha Return-to-Fishing Sub-Committee.

Kenya Fisheries Department. 2003. Minutes of Stakeholder Meetings for Sustainable Management of Prawn Fishery, held in Malindi, Kenya, April 2003.

Kenya Gazette. 1993. Legal Notice No. 351, Fisheries Act (Cap 378) of 5 November. Fisheries (General) (Amendment) Regulations,1993. Amendment of Regulation 43(1) (a). *Trawling is a Prohibited Fishing Method in the Waters of Lake Victoria.* Nairobi, Kenya: Government Printer.

Kenya Gazette. 2001a. Legal Notice No. 50, Fisheries Act (Cap 378) of 8 February. *Closure of Fishing in Lake Naivasha.* Nairobi, Kenya: Government Printer.

Kenya Gazette. 2001b. Gazette Notice No. 7565, Fisheries Act (Cap 378) of 9 November. *Imposition of Management Measures.* Nairobi, Kenya: Government Printer.

KMFRI (Kenya Marine and Fisheries Research Institute). 2002a. "Status of Fisheries and Water Quality of L. Naivasha with a discussion on Water Abstraction and Socio-economic Issues." Final Report to Lake Naivasha Stakeholders.

KMFRI (Kenya Marine and Fisheries Research Institute). 2002b. Current Status of Trawl Fishery of Malindi–Ungwana Bay. Report to the Malindi–Ungwana Bay Stakeholders.

LVFO (Lake Victoria Fisheries Organization). 2001. The Organs of the Organization in the Convention for the Establishment of the Lake Victoria Fisheries Organization, as amended in November 1998. Jinja, Uganda: LVFO and IUCN.

LVFO (Lake Victoria Fisheries Organization). 2003. Second LVFO Monitoring and Outreach Mission around Lake Victoria 20–31 January 2003. Jinja, Uganda.

United Nations Convention on the Law of the Sea (UNCLOS) of 10 December 1982. *Doc. A/RES/48/263.*

World Summit on Sustainable Development (WSSD). 2002. Plan of Implementation. 4 September. Johannesburg. Internet: ⟨http://www.johannesburgsummit. org/html/documents/summit_docs/2309_planfinal.htm⟩ (visited 15 July 2003).

14

Public participation in a multijurisdictional resource recovery: Lessons from the Chesapeake Bay Program[1]

Roy A. Hoagland[2]

Introduction

Perhaps no estuarine system in the United States has received as much political attention and federal and state funding as that of the Chesapeake Bay. Beginning in 1976, with a Congressional directive to the United States Environmental Protection Agency (USEPA) to conduct a study of the Bay and the problems then facing it, the United States invested approximately $30 million in research and evaluation, releasing in 1983 a report highlighting the pollution and overharvesting problems threatening this historic estuary.

Today, over 25 years later, a multijurisdictional, intergovernmental initiative called the Chesapeake Bay Program (hereinafter referred to as the Program), funded by the US Congress at a level of approximately $20 million per year, works to implement strategic goals and objectives designed to restore and preserve the health of the Chesapeake Bay. However, as discussed below, this sum is a fraction of the overall budget.

With a priority placed on partnerships, consensus, and high levels of public participation, many question whether the Program, with its current annual federal investment and supplemental dollars of state investments, is saving the Bay. One recent analysis argues that the Program's progress has stalled, as has the restoration of the natural resource itself (CBF 2002a). The Chesapeake Bay Foundation (CBF), a 35-year-old non-governmental organization (NGO) equipped with experts in a variety of technical, legal, and political arenas – from water-quality scientists and lawyers to land-use planners and restoration managers – argued in an October 2002 release highlighting its 2002 State of the Bay Report that the Program was "floundering on the reefs of bureaucracy" and that the restoration of the resource had "stalled" (CBF 2002a). On a scale of 1 to 100, with 100 representing the Chesapeake Bay in a pristine condition, the 2002 State of the Bay report rated the Chesapeake's condition at 27, a score nearly identical to the score it received when CBF first began releasing its annual report five years ago in 1997 (CBF 2002b). (CBF estimates that the Chesapeake's condition bottomed out in 1983, at a score of 23.)

This chapter examines the state of the Chesapeake Bay restoration effort as well as the state of the resource itself, investigating the role of public participation in the success or failure of the restoration effort.

The Chesapeake Bay: A primer

The Chesapeake Bay spans approximately 200 miles, from the reaches of the Susquehanna River in New York to south-eastern Virginia, where it meets the Atlantic Ocean (fig. 14.1). With a watershed of 64,000 square miles extending through six states (New York, Pennsylvania, Delaware, Maryland, West Virginia, and Virginia) and Washington, DC, and an average depth of only 22 feet, the Chesapeake Bay exhibits the largest watershed to water-volume ratio of any estuary in the world. As figure 14.2 illustrates, this ratio is five times larger than that of the world's next-largest estuary, the Gulf of Finland.

The shallow Chesapeake Bay estuary is the recipient of run-off from any and every land use from its watershed – from agricultural cropland in Lancaster County, Pennsylvania; from urban streets in Washington, DC; from forestlands and feedlots in the Shenandoah Valley of Virginia. With about 50 major tributaries and thousands of streams, creeks, and ditches carrying fresh water laden with run-off, the Chesapeake's watershed is a drainage basin of enormous size, funnelling pollution into a receiving estuary of relatively shallow depth and little volume.

Home to more than 3,600 species of plants, fish, and animals, the Chesapeake Bay is North America's most biologically diverse and productive estuary. Species such as crabs, oysters, shad, and rockfish (striped bass) are a part of its history and culture. Native American food-gathering lifestyles were dependent on the productivity of the bay and its tributaries. Today, watermen still provide half of the US blue crab harvest from the waters of the bay, as well as 90 per cent of the nation's soft-shell crabs.

Figure 14.1 Map of the Chesapeake Bay (the heavy dashed line delimits the Chesapeake Bay watershed)

Scenic views and extensive waterways have delighted inhabitants for centuries and given commerce an opportunity to flourish. Captain John Smith, during his voyage into the Chesapeake in 1608, wrote that "[t]he land was beautiful, and one of the most pleasant in the whole world for large and useful navigable rivers. Heaven and earth never agreed to better frame a place for man's habitation ... here are mountains, hills, valleys, rivers and brooks, all running most pleasantly into a fair bay ... In summer no place affords more plenty of sturgeon, nor in winter more abundance of fowl ..." (Smith 1612). Today, shipping terminals, residences, industrial plants, marinas, and more developments line the shores of the Chesapeake and its tributaries. "Man's habitation" and man's occupations consume the land and use the waters of the Chesapeake.

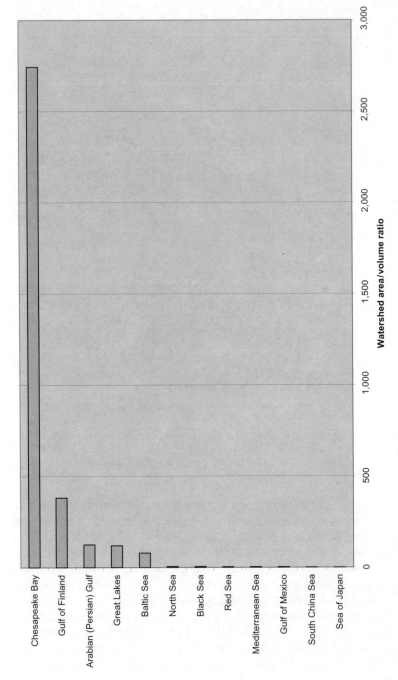

Figure 14.2 Area-to-volume ratios for various watersheds

Nevertheless, all is not well with the Chesapeake Bay. Its qualities of biological diversity, of magnificent beauty, and of navigable channels and waterways have suffered a precipitous decline:

- Underwater grasses, one of the primary indicators of the health of the Chesapeake, cover a mere 12 per cent of their historic acreage. In addition, the diversity of species (thirteen in total) has declined at some historic locations from five or six to one or two.
- Maryland, Pennsylvania, and Virginia have experienced a 58 per cent decline in wetland acreage (both tidal and non-tidal) since colonial times.
- Although sustainable fisheries of rockfish and summer flounder exist, bluefish, weakfish, American shad, and other fish populations remain depleted or depressed. Stocks of reproductive-age female crabs are down by 80 per cent.
- Once known as "Chesapeake Gold," the native oyster powered a major commercial industry. Today, the oyster is commercially extinct, existing at only 2 per cent of its historic levels. Harvests in the 1880s hit historic highs of 20 million bushels landed annually; today, the harvest is less than 500,000 bushels annually [1 bushel = 8 gallons = 36.4 litres (UK) or 35.3 litres (US)]. This decline has environmental impacts, as scientists estimate that, at historic levels, the native oyster population could filter the entire water column of the Bay within three to six days.

Moreover, excess levels of nitrogen and phosphorus, or nutrient pollution, threaten the very existence of the Chesapeake Bay's living resources. These excess nutrients feed algae that grow profligately, or "bloom," in the bay and its tributaries. The algal blooms prevent sunlight from penetrating the bay's waters, thereby inhibiting the growth of underwater grasses. When the algae die, they rob the water of oxygen, thereby choking the life out of aquatic living resources. In the warm summer months, this process of excess nutrient loading and algal decomposition creates a "dead zone," an area devoid of necessary oxygen levels, in the bay's mainstem. Nutrient pollution is, by scientific consensus, the Chesapeake Bay's primary water-quality problem.

Nutrient pollution is due, in large part, to the dramatic changes occurring across the watershed. Population growth is converting forests, wetlands, and farmlands into homes, shopping malls, and industrial parks, removing the natural filters that once protected the bay. There is, according to the Program, "a clear correlation between population growth and associated development and environmental degradation in the Chesapeake Bay system" (Chesapeake Bay Agreement 1987). It is estimated that, by 2020, an additional 3 million people will join the 15 million people already living in the watershed. The consequent, detrimental, impacts, (both direct and indirect) on the water quality and living resources of the bay may be incalculable.

The Chesapeake Bay Program: A primer

In 1983, the USEPA; the states of Maryland, Pennsylvania, and Virginia; the District of Columbia; and the Chesapeake Bay Commission, all entered into a historic compact. The Chesapeake Bay Commission (CBC) is a statutorily created interstate agency composed of elected legislators from the states of Maryland, Pennsylvania, and Virginia, and several citizens. The Chesapeake Bay Agreement of 1983 established the Chesapeake Bay Program (Chesapeake Bay Agreement 1983).

The 1983 Agreement was the first of a series of multijurisdictional, intergovernmental agreements designed to save the bay; these agreements have defined the agenda for the Program, setting forth an evolution of strategic plans containing measurable objectives. The first agreement, however, set no measurable goals or objectives and few commitments: it was a brief, terse document, committing to little more than dialogue among different jurisdictions and different levels of government.

Nevertheless, this was not an inconsequential accomplishment. The 1983 Agreement launched a major cooperative partnership among state and federal governments and jurisdictions, with the aim of restoring and preserving Chesapeake Bay. In the 1983 Agreement, the signatories (termed the Executive Council) agreed (a) to coordinate and assess plans to improve and protect the water quality and living resources of the Chesapeake Bay; (b) to create an implementation committee to coordinate the development and evaluation of management plans; and (c) to create a liaison office to staff the new partnership.

It was not until the second (1987) Chesapeake Bay Agreement that the signatories set clearly defined and measurable environmental goals and objectives. Foremost among the 1987 commitments was a specific, measurable goal to reduce the levels of nitrogen and phosphorus entering the bay by 40 per cent by the year 2000. The 1987 Agreement also included other goals, some measurable and some not. Some of these goals were as follows:

- reducing the discharge of untreated or inadequately treated sewage into Chesapeake Bay waters from such sources as combined sewer overflows, leaking sewage systems, and failing septic systems.
- commissioning a panel of experts to produce a report by December 1988 on population growth and land development patterns.
- providing curricula and field experiences for students.

- by March 1988, developing a state and federal communications plan for public education and participation.

The Executive Council supplemented the goals of the 1987 Agreement in subsequent years through directives, initiatives, and amendments: for example, in 1995, the signatories signed the Riparian Forest Buffers Initiative, setting a goal for restoring 2,010 miles of riparian buffers along streams and shorelines in the watershed by 2010. Nevertheless, reduction of nutrients by 40 per cent remained the pre-eminent and overriding goal.

In 2000, the Executive Council took a dramatic and aggressive step with the signing of a third agreement – Chesapeake 2000 (Chesapeake 2000 2000). In this agreement, the Executive Council acknowledged that "[w]hile the individual and collective accomplishments of our efforts have been significant, even greater effort will be required to address the challenges that lie ahead." Chesapeake 2000 contains over 100 commitments in areas of Living Resource Protection and Restoration; Vital Habitat Protection and Restoration; Water Quality Protection and Restoration; Sound Land Use; and Stewardship and Community Engagement. Goals and objectives incorporated into Chesapeake 2000 include, for example, the following:

- by 2010, achieve, at a minimum, a 10-fold increase in native oysters in the Chesapeake Bay, based upon a 1994 baseline;
- by 2005, develop ecosystem-based multi-species management plans for targeted fish species;
- by 2010, work with local governments, community groups, and watershed organizations to develop and implement locally supported watershed management plans in two-thirds of the bay watershed;
- by 2010, achieve a net resource gain in tidal and non-tidal wetlands by restoring 25,000 acres ($\approx 101 \times 10^6$ m^2);
- by 2010, correct the nutrient- and sediment-related problems in the Chesapeake Bay and its tidal tributaries sufficiently to remove the bay and the tidal portions of its tributaries from the list of "impaired waters" under the federal Clean Water Act;
- by 2010, permanently preserve from development 20 per cent of the land area in the watershed;
- by 2010, rehabilitate and restore 1,050 brownfield sites to productive use;
- by 2005, increase the number of designated water trails in the Chesapeake Bay region by 500 miles;
- beginning with the class of 2005, provide a meaningful bay or stream outdoor experience for every school student in the watershed before graduation from high school.

Although Chesapeake 2000 states that it "reflects the Bay's complexity" and "responds to the problems facing this magnificent ecosystem in a comprehensive, multifaceted way," it acknowledges that "[i]mproving water quality is the most critical element in the overall protection and restoration of the Chesapeake Bay and its tributaries." Notwithstanding the proliferation of measurable targets, the reduction of nutrient pollution remains the key factor in improving water quality under Chesapeake 2000.

Since 1983, not only have the agreements that govern the agenda of the Program grown in complexity, but so has the Program itself. Its mission is more complex, covering all the Chesapeake Bay restoration opportunities and needs – from toxic-discharge reduction to preservation of open space, and from restoration of fish passageways to engagement of community watershed organizations. Ann Pesiri Swanson – the Executive Director of the Chesapeake Bay Commission and a recognized expert on the Chesapeake Bay, who provided key leadership in the authorship of Chesapeake 2000 – observed that:

Over the last 25 years, the bay program has gone through its own evolution. What began as a water-quality program designed to address the decline of the bay's living resources has grown to involve integrated management of land, air, water and living resources, including man. (Swanson 2001)

The bureaucracy that administers this integrated Program has also grown. It now includes not only the Executive Committee (the Agreement's signatories), but also a Principals' Staff Committee, Implementation Committee, Citizens Advisory Committee, Local Government Advisory Committee, Scientific and Technical Advisory Committee, Fisheries Steering Committee, Water Quality Steering Committee, Water Quality Technical Workgroup, Federal Agencies Committee, Budget Steering Committee, Nutrient Subcommittee, Trading and Offsets Workgroup, Toxics Subcommittee, Monitoring and Assessment Subcommittee, Modeling Subcommittee, Living Resources Subcommittee, Land Growth and Stewardship Subcommittee, Communications and Education Subcommittee, and Information Management Subcommittee. These are but a few of the (over 50 such) workgroups that now comprise the Program's bureaucracy.

Such a bureaucracy and the initiatives it pursues need money to function. Thus, the Program's budget has grown also, from an annual Congressional appropriation of approximately $3.5 million to one of approximately $20 million. The current figure of 20 million US dollars represents only the Congressional line-item funding for the Program; appropriations from Maryland, Pennsylvania, and Virginia supplement this budgetary line item by millions of dollars. While the Program estimates this supplemental income at over $100 million annually, Virginia alone has estimated its additional contribution (through state appropriations to state agencies pursuing implementation of the 2000 Agreement's commitments) at $191 million (for the fiscal year 2001). In addition, Congressional funding to bay-restoration initiatives separate from the line-item appropriation to the Program is not included in the $20 million figure: for example, the US Senate Appropriations Committee, for the federal fiscal year 2003, included an additional $57.7 million.

Lessons learned from the Program

Swanson recently described 12 lessons that the Chesapeake Bay Program has taught its leaders (Swanson 2001). These learned lessons, she concludes, are as follows:

1. *Begin with comprehensive scientific studies that combine theory, detailed knowledge, monitoring, and modelling.* Science did, in fact, serve as the foundation upon which the Program was conceived, and science has remained a core function. The wealth of scientific and technical information now found within the expansive framework of the Program is nothing short of phenomenal. The Program's website (http://www.chesapeakebay.net) is a doorway into a multitude of studies, analyses, data, and research focused on the Chesapeake Bay. In addition, the staff of the liaison office now comprises nationally and internationally recognized scientists, natural resource managers, and policy makers.

2. *Involve the highest levels of leadership possible.* The Executive Council consists of state-level political leaders and the USEPA Administrator – those whose political decisions provide a direct link to the Chesapeake Bay restoration efforts. It is interesting to note that neither the US President nor the political leaders of the federal legislative branch – the US Congress – are part of the Council, even though the seat of the federal government sits squarely in the watershed and the Chesapeake Bay is (in the words of the 1987 Agreement and former President Reagan) "a national treasure."

3. *Embrace clear, strong, specific, comprehensive, and measurable goals.* The goals and deadlines set in the three agreements are of a breadth and complexity unmatched anywhere in the United States for a natural resource restoration effort.

4. *Encourage the participation of a broad spectrum of participants.* The Program has evolved to include among its ranks governors, state

legislators, local government officials, regional and local environmental organizations, businesses, and more.

5. *Provide incentives and methods for institutional cooperation.* A fundamental premise of the Program is its focus on voluntary agreements and consensus-based decision-making. As a result, incentives are common, mandates few. The Program often pursues incentives that provide for leveraging of funding: for example, of the current annual federal appropriation of nearly $21 million, the Executive Committee annually provides "implementation grants" to the signatory states and Washington, DC, to utilize in Chesapeake Bay restoration priorities selected by the receiving jurisdiction. The recipients supplement these funds through their own existing budgets.

6. *Inform and involve the public.* Although the Program has recognized the importance of an informed and involved public to effective policy-making and improved natural resource management, organizations outside the Program bureaucracy have, in fact, provided the key leadership on public information and involvement. NGOs such as the CBF, with 110,000 members across the United States, have led the public-awareness campaign through advocacy efforts in the political, legislative, and media arenas. The CBF website (http://www.cbf.org) gives evidence of its critical role in educating and engaging the public through electronic and print correspondence, middle- and high-school education programmes, restoration activities, pollution-reduction campaigns, and ongoing advocacy efforts.

7. *Develop a balanced set of management tools.* "No one approach works best in all ecological, political, and economic situations," Swanson concludes. In fact, the signatories to the agreements have, through independent or cooperative actions, taken implementation steps ranging from legislative mandates to tax incentives to grant programmes.

8. *Choose pollution prevention before restoration or mitigation.* Few environmental strategists would debate this conclusion, but prevention is not always an available option.

9. *Test scientific theories and management approaches on a small scale.* Demonstration projects can lead to broader application. Currently, for example, there are several pilot projects or small-scale tests within the Chesapeake Bay watershed that focus on nutrient trading, which seeks to accomplish nutrient reductions in the water-pollution arena.

10. *Focus on integrating government agencies.* Swanson argues for the need to cross governmental agency lines within a jurisdiction, for integration among governmental agencies with differing substantive responsibilities. For example, a state agency making siting decisions

on the location of industrial facilities should not act wholly independently of another state agency making permit decisions on whether such facilities receive water pollution-discharge permits. There is also a need to integrate governmental agency action across jurisdictions. The flow of excess nutrients plaguing the Chesapeake Bay and its tidal tributaries does not stop at the borders of Pennsylvania, but continues on into the waters of Maryland and Virginia. Although the Program seeks to effect voluntary integration across jurisdictional lines, such integration is often incomplete or unsuccessful.

11. *Conduct regular assessments of goals and progress.* This lesson is straightforward and an essential element of any strategic plan for improving natural resources. The bottom-line question is whether the resource displays the desired and projected improvement.

12. *Demonstrate and communicate results.* A programme without success and without public knowledge of results simply cannot obtain or maintain the funding necessary to continue. Demonstrated success and communication of success is key.

Swanson's lessons are perceptive and instructional; however, there are four additional, critical lessons missing from this list. This author would add the following: (13) demand courageous political leadership; (14) engage aggressive public and NGO advocacy; (15) create and utilize strategic public and private partnerships; and (16) pursue accurate implementation budgets.

13. *Demand courageous political leadership.* The Program's underlying strategic plans – the three agreements – are merely pieces of paper in the absence of the political will of elected officials to ensure their implementation. The onset of the Program in the 1980s, with its fresh ideas and mandate, brought with it a high level of political commitment to the implementation of the 1983 and 1987 agreements. US Senator Paul Sarbanes, former Maryland Governor Harry Hughes, and former Virginia Governor Gerald Baliles, for example, dedicated time, funding, personnel, and political capital to advancing the agreements' goals and objectives. However, political commitment to these goals/objectives began to wane by the mid-1990s as the Program matured and leadership changed. Executive Council meetings grew less important to the signatories: former Virginia Governor James Gilmore attended only one meeting during his four-year tenure as a member of the Council. The lack of political commitment remains an issue today: at the meeting of the Council in October 2002, not only did Pennsylvania's governor fail to attend but also he failed to send any Pennsylvania governmental official as his representative.

The most dramatic change occurred in 1992, with the election of former Virginia Governor George Allen. With his anti-environment philosophy demonstrated by the appointment of a controversial property-rights advocate as his senior environmental cabinet official, Virginia's progress on some of the major commitments set by the Program ground to a halt. For example, Virginia's strategies designed to address the 40 per cent nutrient-reduction goal on a tributary-by-tributary basis encountered such a deliberate delay in development and implementation that the Virginia legislature had to mandate their completion by statute in 1996. With these declines in political leadership, the Program plodded along without its previous drive and direction.

At the same time, affirmative political leadership can be essential. For example, many would argue that former Maryland Governor Parris Glendenning did much within Maryland to move that state's Chesapeake Bay restoration initiatives forward, even if these efforts in Maryland did not translate into regional leadership by the Program.

14. *Engage aggressive public and NGO advocacy.* Swanson's sixth lesson deals with the need for educating and involving the public; however, education and involvement are insufficient: there must be present a higher level of activism and advocacy that extends beyond what most consider as education. The Program does have a structural vehicle for public participation: the Program's Citizen Advisory Committee (CAC) is charged with providing advice directly to the members of the Executive Council. With membership consisting of those appointed from Maryland, Pennsylvania, Virginia, and Washington, DC, as well as several at-large appointees (i.e. appointed not by the state but by an independent third party), the professional and political backgrounds of the CAC members are varied, ranging from energy-industry officials to environmental advocates, to city land-use planners, to homemakers, to retired corporate executives. CAC has played numerous roles, both as critic and participant. It was, for example, the only advisory committee in the Program to serve on the drafting committee for Chesapeake 2000. Nevertheless, as an organization internal to the Program, CAC has limits on its ability to "push the envelope" (i.e. do something new). For the Chesapeake Bay restoration effort, the critical role of watchdog comes from CBF, a prominent and persuasive NGO. The ongoing advocacy efforts of CBF have played a key role in developing and implementing the agreements. As non-governmental "outsiders," NGO advocacy organizations such as CBF can hold governments accountable for promises made and monies allocated. Moreover, they force governmental initiatives such as the Program to reach beyond the safe confines of political conformity and mediocrity.

With 110,000 members, CBF is a force of substantial size and effectiveness. For example, during the development of Chesapeake 2000, both CBF and CAC sought to ensure that the agreement addressed, through measurable objectives, the threats resulting from population growth. No agreement prior to Chesapeake 2000 experienced as much haggling or bargaining among the staffs of the agreement's signatories, and the issue of growth and development heightened the levels of contention. In spite of the recognition that land conversion and population growth was a determining factor in the future restoration of the Chesapeake, more than one signatory sought to minimize the provisions of the Sound Land Use section of the agreement. CBF publicly challenged the lack of Program leadership on this issue and aggressively lobbied the representatives of the Executive Council to include measurable goals for land conservation and reductions in land conversion. CAC, too, argued for such, seeking and receiving an audience with the Executive Council's top-level Cabinet officers on this issue. Had it not been for the participation of these advocates, Chesapeake 2000 would probably have been silent on measurable goals and objectives in the area of population growth.

15. *Create and utilize strategic public and private partnerships.* As the Bay's oldest, largest, and strongest advocate, CBF has long recognized the need for an active and engaged constituency – one that demands progress and change. Consequently, CBF has worked cooperatively and strategically with a number of public and private organizations to advance the restoration of the bay, to leverage resources, and to recruit additional advocates. These alliances range from other non-profit groups (such as Ducks Unlimited and the Nature Conservancy), to corporate interests (such as Toyota and MasterCard), to state and federal government agencies (such as the US Army Corps of Engineers and the Virginia Marine Resources Commission), to public and private schools, as well as individual citizens. These partnerships have met with varying degrees of success. Key among the reasons for success are strong leadership, mutuality of purpose ("win–win" situations), clear and measurable goals, and adequate funding. Examples of successful partnerships include the following:

(i) *Ducks Unlimited (DU).* This is a much larger non-profit organization than CBF, with members including thousands of waterfowl hunters across North America. DU raises millions of dollars for waterfowl-habitat conservation and restoration. In 1997, DU and CBF entered into a $20 million partnership ($10 million from each organization) to restore 125,000 acres ($\approx 506 \times 10^6$ m^2) of wetlands and 1,500 miles ($\approx 2,400$ km) of streamside buffers in the Chesapeake Bay watershed by 2010. The two organizations also donated more than \$3.4 million to the Conservation Reserve Enhancement Program (CREP), a governmental incentive programme that pays farmers to restore stream buffers and wetlands. The funding provided by DU and CBF has encouraged farmers in bay states to enlarge the buffers funded by CREP and to protect them permanently with conservation easements. This well-funded partnership was extremely successful. Virginia soil- and water-conservation district leaders credited the CBF/DU partnership with making the CREP a success (Talley 2002; Whitescarver 2002). To date, the expenditure of over \$15 million has led to the restoration of more than 9,567 acres ($\approx 38.7 \times 10^6$ m^2) of wetlands and 2,621 miles ($\approx 4,220$ km) of riparian forested buffers.

(ii) *Toyota.* In partnership with the automotive giant Toyota Motor Services Corporation, CBF in 1996 adapted its award-winning hands-on outdoor environmental education programme to train similar conservation organizations in three diverse US cities to replicate the CBF programme. The three-year project, named CLEAN (Children Linking with the Environment Across the Nation), used a \$2 million grant from Toyota and successfully reached more than 20,400 ethnically and economically diverse students in California, North Carolina, and Alabama. This partnership led to the creation of three successful, popular, and self-sustaining programmes. From CBF's perspective, one of the keys to this partnership's success was the hands-off approach by Toyota: the company provided funding but left programming and implementation decisions exclusively to the experts, CBF.

(iii) *Rappahannock River Project.* This five-year partnership with citizens and local governments in the Rappahannock River area of Virginia sought to engage, educate, and motivate local protection of the Rappahannock River watershed, one of the subsidiary watersheds of the Chesapeake Bay. The effort included dedicating a full-time CBF staff member to the effort; establishing a temporary satellite office; sponsoring issues workshops, restoration projects, and special events; and ultimately nurturing a "friends of the river" group of citizens to become advocates for the Rappahannock River. These efforts led not only to the creation of a regional watershed planning council of local citizens, government representatives, businesses, and other stakeholders but also to the identification of environmentally significant land tracts and the establishment of the Rappahannock River National Wildlife Refuge, currently consisting of more than 5,000 acres ($\approx 20.2 \times 10^6$ m^2).

(iv) *Virginia Oyster Heritage Program.* One of the most promising oyster-restoration projects in the United States, this effort partners US government agencies, Virginia State Government, university researchers, the seafood industry, and CBF. The federal government helps with funding; the state matches the federal funds and constructs artificial oyster reefs using empty oyster shells; university researchers help to develop disease-resistant strains of oysters; the waterman and seafood industries help produce seed in hatcheries, move shells, and plant oysters; and CBF "grows out" the seed and stocks the reefs with over a million native oysters per year, using volunteers to stock the reefs. This collaborative formula for restoring native oysters – stocking massive quantities of mature, disease-resistant oysters on protected "sanctuary" reefs in key areas of Chesapeake Bay – has become a model for oyster restoration (Allen, Brumbaugh, and Schulte 2003).

16. *Pursue accurate implementation budgets.* The CBC calculated the cost of implementation of Chesapeake 2000 at a whopping $19.1 billion. With this projected cost and a projected income of only $6.1 billion, there is a projected shortfall of $13.0 billion (or $1.6 billion per year, calculating from 2003 to 2010). The CBC has stated: "on our current course, it's doubtful that we will meet our C2K [Chesapeake 2000] goals by 2010" (CBC 2002, 2003). CBF, in response to these figures, has noted that the Chesapeake Bay is worth every penny spent on it. CBF President Will Baker – a nationally recognized leader on Chesapeake Bay issues, with more than 25 years of experience – argues, "The price tag, $19.1 billion, sounds high, yet it is an indispensable investment in our future.... The cost of [Chesapeake] Bay restoration is a relatively small investment to make in the region's most important natural and economic resource" (CBF 2002c). The reality is that the Program lacks the funding necessary to implement the strategic plan embodied in Chesapeake 2000 (and, one can surmise, in the prior Agreements). Without the funds, Chesapeake 2000's goals and objectives will not be met.

Conclusions

This chapter has sought to examine the Chesapeake Bay Program and the role of public participation in the success or failure of Chesapeake Bay restoration efforts. It is appropriate to return to the bottom-line question posed above: is the resource displaying the desired and projected improvement?

As noted at the outset, CBF's 2002 "State of the Bay Report" challenges all who are involved with (or who are watching) the Program to question if the actions taken to date are improving the resource or are merely preventing additional degradation. Baker argues that "[f]or three decades we have been fighting to save the [Chesapeake] Bay. We have made progress, but pollution is still winning" (CBF 2002a). The majority of natural resource indicators utilized by CBF in its 2002 "State of the Bay Report" to evaluate the health of the Chesapeake Bay remain at unhealthy levels.

This is not to say that the Program (partly because of the targeted goals and objectives set by the 1987 Agreement) has not driven improvements in natural resources: for example, the Program notes that collaborative efforts have (a) removed dams and blockages to achieve 849 miles opened to migratory fish and an additional 143 miles to resident fish between 1988 and 2001; (b) restored the striped bass (rockfish) fishery; and (c) improved the overall condition of the Chesapeake Bay and its rivers since the signing of the first Agreement in 1983 (CBF 2002b).

Nevertheless, these accomplishments touch only the edges of the Chesapeake Bay restoration goals and objectives. Poor water quality, degraded by nutrient pollution, still remains the primary – and unresolved – problem facing the Chesapeake Bay. The Program has failed to achieve the 40 per cent nutrient-reduction goal identified 15 years ago in the 1987 Agreement. Recent analyses show that the 40 per cent reduction will not now be enough: there is a need to reduce nutrient loadings dramatically further (Chesapeake Bay Program 2003). The CBC sounds a more frightening warning in light of years and years of effort: "In 2001, water quality monitoring data from the Chesapeake Bay's largest tributaries revealed no discernable trends in nutrient loads [i.e. nutrient pollution], despite modeling results showing a 15 per cent reduction in the amount of nitrogen entering the Chesapeake Bay from 1985–2000" (CBC 2001).

The Program's greatest strength, however, is its commitment to embrace the active involvement of the public at all levels. Voices of local community and non-profit advocates are heard along with those of governors, legislators, and agency staff. This commitment ensures a transparency and accountability of an unprecedented nature, providing a dramatic contrast to the typical opaque government bureaucracy. Naturally, public participation does not ensure success; the restoration of the Chesapeake Bay is a Herculean task. None the less, allowing watershed residents and a multitude of interest groups to engage as partners with the government – critically challenging political agendas, examining failures, and celebrating achievements – fuels a creative energy that is essential to pursuit of such a large restoration effort. Only with this open dialogue can the

Chesapeake Bay hope to return to the full, productive health that has inspired Captain John Smith and so many others.

Notes

1. This chapter represents the personal opinions of the author and is not an authorized statement of the Chesapeake Bay Foundation (CBF) or the Citizen Advisory Committee to the Executive Council (CAC).
2. At the time of writing, Mr Hoagland was the Virginia Executive Director of the Chesapeake Bay Foundation, Inc. (CBF), a non-profit environmental advocacy organization dedicated to the restoration and preservation of the Chesapeake Bay. He serves as the Chairman of the Chesapeake Bay Program's CAC.

REFERENCES

Allen, Standish K., Jr, Robert Brumbaugh, and David M. Schulte. 2003. "Terraforming Chesapeake Bay." *Virginia Marine Resources Bulletin* 35(1):2–8. Spring. Internet: ⟨http://www.vims.edu/GreyLit/SeaGrant/vmrb35-1.pdf⟩ (visited 16 November 2003).

CBC (Chesapeake Bay Commission). 2001. Annual Report.

CBC (Chesapeake Bay Commission). 2002. PowerPoint Presentation Slides. Nov. 15.

CBC (Chesapeake Bay Commission). 2003. "The Cost of a Clean Bay: Assessing Funding Throughout the Watershed." January.

CBF (Chesapeake Bay Foundation). 2002a. "Press Release." Oct. 15.

CBF (Chesapeake Bay Foundation). 2002b. 2002 "State of the Bay Report." Internet: ⟨http://www.cbf.org/site/PageServer?pagename=sotb_2002_index⟩ (visited 16 November 2003).

CBF (Chesapeake Bay Foundation). 2002c. "Press Release." Nov. 15.

Chesapeake 2000. 2000 Internet: ⟨http://www.chesapeakebay.net/agreement.htm⟩ (visited 16 November 2003).

Chesapeake Bay Agreement. 1983. December 9. Internet: ⟨http://www.chesapeakebay.net/pubs/1983ChesapeakeBayAgreement.pdf⟩ (visited 16 November 2003).

Chesapeake Bay Agreement. 1987. December 15. Internet: ⟨http://www.chesapeakebay.net/pubs/1987ChesapeakeBayAgreement.pdf⟩ (visited 16 November 2003).

Chesapeake Bay Program. 2003. Internet: ⟨http://www.chesapeakebay.net/⟩ (visited 16 November 2003).

Smith, John. 1612. "A Map of Virginia with a Description of the Countrey, the Commodities, People, Government and Religion." Internet: ⟨http://etext.lib.virginia.edu/etcbin/jamestown-browse?id=J1008⟩ (visited 16 November 2003).

Swanson, Ann Pesiri. 2001. "Chesapeake Bay, USA: Lessons Learned from Managing a Watershed." *International Newsletter of Coastal Management*. Fall.

Talley, Stephen E. 2002. Letter to the Chesapeake Bay Foundation from the Director and Chair, Forestry and Riparian Buffer Committee, Headwaters Soil and Water Conservation District. June 26.

Whitescarver, Robert N. 2002. Letter to the Chesapeake Bay Foundation from the District Conservationist, Headwaters Soil and Water Conservation District. May 31. (On file with author.)

15

Chesapeake Bay protection: Business in the open[1]

Rebecca Hanmer

Introduction

The Chesapeake Bay is the largest and most productive estuary in North America. Even after centuries of intensive use, the bay remains a highly productive natural resource. It supplies millions of pounds of seafood; it functions as a major hub for shipping and commerce; it provides natural habitats for wildlife; and it offers a variety of recreational opportunities for residents and visitors.

The watershed includes parts of New York, Pennsylvania, West Virginia, Delaware, Maryland, and Virginia and the entire District of Columbia. Close to 16 million people live in the bay watershed, each one living just a few minutes from one of the more than 100,000 streams, creeks, and rivers that drain into the bay. Their daily activities have had, and continue to have, direct impacts on local and bay water quality.

The Chesapeake Bay Program has evolved over the past 25 years into a transparent and participatory institutional and legal framework to respond to these effects on the bay. Because the extent of the bay's watershed is massive compared with the volume of water, land management is a critical component of the restoration, requiring the involvement of partners from all of the watershed states. The Program is a partnership among all of the entities that have a stake in the restoration of the bay, including representatives of federal, state, and local governments; non-governmental organizations (NGOs); universities; private industry; and citizens ranging from scientists and environmental advocates to farmers and watermen.

The Program operates using a blend of regulatory and voluntary processes. It takes advantage of the benefits of a well-established federal and state regulatory system. However, by using a consensus-based approach to adopt voluntary goals, the Program has attempted to achieve better and faster results than might have been achieved using only regulatory processes. The voluntary approach allows the partnership to take full advantage of regulatory flexibility and market-based mechanisms.

Public access to information and public participation in decision-making and management have resulted in an informed and involved public who have supported the development of numerous voluntary commitments expected to bring about the bay's restoration. Indeed, the transparent and participatory process has contributed to the development of the Program into what is now considered a national and international model for estuarine research and restoration programmes. Through their participation in the various committees, all of the Chesapeake Bay Program partners have an opportunity to influence policy and develop plans to carry out the policy decisions.

This chapter examines some of the experiences in engaging a broad range of public and private actors in watershed governance. It identifies lessons learned and draws conclusions, particularly relating to public involvement in the process.

Public access to information and participation

Public access to scientific and other bay information

The Chesapeake Bay is one of the most carefully monitored bodies of water in the world. Because concern for the bay dates back to the 1970s, and implementation of restoration efforts has been ongoing for nearly two decades, there is a considerable body of scientific information and data on environmental conditions in the bay. Consistent and comparable data on all traditional water parameters have been taken at over 130 sites in the watershed and the open bay since 1984. The data and trend analyses available from this monitoring programme are some of the best in the United States. A major strength of the Chesapeake Bay Program's monitoring programme is that it does not rely solely on data generated by the US Environmental Protection Agency (EPA) but also leverages (i.e. through cost share requirements, memoranda of understanding, and other partnership agreements is able to acquire goods and services above and beyond those possible via its initial investment) and accesses many other reliable information sources maintained by cooperating state and federal agency partners.

Although the environmental data were critical to programme development, prior to 1991 they were not used systematically to inform Bay Program partners and the public of the bay's condition, environmental problems, and progress being made in the restoration. Moreover, information on environmental outcomes was not used to make or justify management decisions. Early in 1991, EPA leadership decided to make the programme more accountable to the public on a day-to-day basis by defining and communicating the environmental results achieved by the restoration programme. While EPA staff began this effort, states and other stakeholders became involved early in the process. The Bay Program began to develop a set of environmental indicators and measures to support goal setting and to serve as targets for the restoration effort.

Although the indicators were used successfully in presentations to managers and scientists, there were concerns that the materials were too technical to be useful for the general public. Through the efforts of an NGO partner, the Alliance for the Chesapeake Bay, workshops were held in 1994 and 1995 to build stakeholder involvement in the design and refinement of the indicators and the communication products. The stakeholders included representatives of citizen groups and the press. The goal of the workshops was to reach consensus on clear messages that could be used with key indicators to help convey a story to the public about the overall health of the bay and how the water quality and living resources were responding to restoration efforts.

The products developed through this and other outreach efforts, along with the wealth of monitoring data and technical information available at the EPA Chesapeake Bay Program Office, helped to provide the foundation for developing the Bay Program's first website in 1995. In 1996, the Chesapeake Executive Council adopted a Strategy for Increasing Basin-wide Public Access to Chesapeake Bay Information. As a result, the Chesapeake Information Management System (CIMS) evolved into an organized, distributed library of information and software tools designed to increase basin-wide public access to Chesapeake Bay information. Website improvements were made to enable more efficient delivery of governmental services throughout the watershed, as well as enhanced opportunities for the public to engage in bay policy development and to more fully understand the activities of the restoration programme and how individuals and organizations can contribute to it (Chesapeake Bay Program 2003).

Ongoing improvements to the Bay Program website and the accessibility of information (ranging from fact sheets and press releases to indicators and raw data) have provided an excellent educational resource for students and teachers of all ages. They also have been a useful resource for other organizations and countries interested in environmental restoration.

Public involvement in Bay Program decision-making

The involvement of NGO partners in the Bay Program has been a critical driver for the transition from data to messages, and subsequently the translation of messages into public policy. One of the key partners in these efforts has been the Alliance for the Chesapeake Bay. Founded in 1971, the Alliance has focused on improving the bay watershed through collaboration and consensus building. The Alliance works to bring government, business, academic, and non-profit players to the table, serving as a neutral facilitator of bay issues.

Since 1984, the Alliance has staffed the Bay Program's Citizens Advisory Committee, which provides a non-governmental perspective on the bay clean-up effort and on how Bay Program policies affect citizens who live and work in the Chesapeake Bay watershed. The Alliance works with the business and development community to promote the Businesses for the Bay and Builders for the Bay programmes, two initiatives that promote sustainable development and business practices.

In 1999, the Alliance led the Chesapeake Renewal Project in response to a request from the Bay Program to report on the status of the bay-restoration efforts and to find out what the public thought should be in the next bay agreement. The Alliance solicited, organized, and reported the input from 95 stakeholders, 22 focus groups, and 750 questionnaires in order to develop a consensus on issues that needed to be addressed in the Chesapeake 2000 Agreement.

Through the efforts of such partners as the Alliance, the public has been actively engaged in Bay Program decision-making. This public involvement and support has been critical for the development of numerous voluntary commitments expected to bring about the bay's restoration.

Results

Environmental benefits of voluntary goals and accountability using indicators

Experience in the Bay Program shows that environmental benefits are gained through participatory, voluntary, goal setting and indicator development. For example, in 1993, Bay Program partners voluntarily committed themselves to removing stream blockages and reopening 1,357 miles of bay tributaries for migratory fish by the year 2003. Resources have been targeted, and progress is reported annually. As a result, hundreds of miles of historic spawning habitat have been reopened. Although the interim goal was not attained on time, it can be argued that, without the goal and indicator, resources would not have been targeted for this purpose and very few miles, if any, would have been reopened. In fact, the long-term goal is expected to be met in December 2004.

The Bay Program utilizes a variety of indicators to measure and track the status and trends of living resources and water quality in the bay, as well as to track the progress made in specific water-quality restoration efforts. This subsection considers a few of these indicators, which also are applied in various ways to monitor responses to restoration efforts, drive management decisions, and hold managers accountable to the public. They also are useful for demonstrating linkages among various indicators used in the Bay Program.

One of the key measures of success in achieving improved Chesapeake Bay water quality is the restoration of bay grasses (also known as submerged aquatic vegetation or SAV). SAV is one of the most important biological communities in the bay – producing oxygen, nourishing a variety of animals, providing shelter and nursery areas for fish and shellfish, reducing wave action and shoreline erosion, absorbing nutrients such as phosphorus and nitrogen, and trapping sediments. Although recent improvements in water quality have contributed to a resurgence in SAV [from a low of 38,000 acres (15,000 ha) in 1984 to more than 89,000 acres (36,000 ha) in 2002], even more improvements are necessary.

Bay grass recovery is linked to improvements in water clarity, a second indicator. Although there are extensive areas with adequate water clarity, several areas in the bay fail to meet the standard necessary for a bay grass habitat.

One of the factors affecting water clarity is excess nutrient loads (a third indicator), which stimulate algal blooms that cloud the water and lead to depleted oxygen levels. The Bay Program partners made a voluntary commitment to reduce nutrient loads by 40 per cent by 2000. Although nutrient loads have declined significantly since 1985, the goal of a 40 per cent reduction has not yet been achieved: additional reductions in nutrients, as well as in sediment, will be needed to restore water clarity to all areas of the bay, and to achieve bay-grass restoration goals.

One of the key elements by which the nutrient load has been reduced has been the installation of nutrient removal technology (NRT) at wastewater-treatment facilities (a fourth indicator). Currently, 55 per cent of the flow from significant facilities is treated using NRT; 63 per cent of the flow will be treated using NRT by the year 2005, and 79 per cent of the flow will be treated using NRT by 2010.

In addition to nutrient reductions, the Bay Program is concerned about other factors that affect water clarity, such as sediment loads and shoreline erosion. One of the on-the-ground practices used by Bay Program partners to reduce nutrient and sediment loads and to prevent shoreline erosion is the restoration of riparian forest buffers (a fifth indicator). In 1996, Bay Program partners voluntarily committed themselves to restoring 2,010 miles of streambank and shoreline by planting trees in the buffer zone by 2010; this goal was achieved well ahead of schedule, and efforts are under way to establish a new goal.

There has been increased concern about blue crabs since the indicator used to track spawning stock abundance showed levels at or near historical lows. In addition to efforts to restore bay grasses, Program partners recently adopted a bay-wide threshold for the blue crab spawning stock and, in 2001, agreed to reduce the harvest by 15 per cent over three years in order to achieve this threshold.

The indicators are used extensively in outreach with Bay Program partners and stakeholders, and many have been used in public outreach products (materials) for over a decade. The public's familiarity with these measures of progress in the bay-restoration effort has significantly affected management decisions made by programme partners. These measures, which are used to monitor environmental conditions and responses to restoration efforts, have driven the development of Bay Program goals and commitments as well as implementation and management strategies; furthermore, by tracking restoration progress (or lack of it), they hold managers accountable to the public.

Enhanced public understanding, concern, and support

Achievement of reductions of some pollutant loads and encouraging trends in some environmental measures have led to continuing public support. It is not uncommon for 100 people to attend a meeting focused on the restoration of a single, small watershed within the greater bay watershed. Close to a thousand people and organizations provided input during development of the Chesapeake 2000 Agreement.

The results of a recent survey of Chesapeake Bay watershed residents' knowledge, perceptions, and attitudes toward the bay and its restoration show that nearly 90 per cent are concerned about the health of the bay and of its rivers and streams (McClafferty 2002). However, the survey also found that nearly half the watershed's residents do not understand that their daily actions have a direct impact on water quality both locally and in the bay. As restoration efforts continue, the Bay Program plans to use information gained from the survey to communicate better with citizens and encourage them to become more involved in "bay-friendly" activities that reduce residents' impact on the bay and its rivers.

Budgetary support

Experience has shown that growing public support for, and financial investment in, the Bay Program have been associated with the development and communication of bottom-line environmental results. Additionally, Bay Program Office staff believe that the increased support given to the Program in recent years reflects the enthusiasm for supporting effective federal–state–local partnerships to address problems. Unlike many other EPA programmes, the Bay Program does not have independent regulatory authorities, and strong support by state and local governments and other institutions is key to its success. Coincident with vigorous efforts to develop goals and environmental indicators, federal funds appropriated for the EPA Chesapeake Bay Program increased from approximately $13 million in the financial year (FY) 1990 to nearly $21 million in FY 1996, before levelling off to the current level of roughly $20 million. When matching funds and other "leveraging" options [e.g. the states' use of additional federal funding opportunities (both grants and loans) and funds from local and state taxes and fees] are considered, the total amount spent by all federal agencies, state and local governments, and other entities is significantly higher than this total alone: estimates are that approximately $150 million are spent each year by a combination of the EPA, other federal agencies, and the states.

Conclusions and lessons learned

At the beginning of the bay-restoration effort, the focus was on mobilizing the public by appealing to their emotions: during this phase, the public was activated by their concern about the degradation of the bay and their love of a treasured resource. In the next phase, the restoration effort focused on mobilizing the public by appealing to their reasoning capacity: efforts were geared to providing information about the results of the restoration effort and to engaging the public in the decision-making process.

The Bay Program has been effective in these public-mobilization efforts, but the environmental results have been mixed. There have been marginal improvements in some areas, and the fact that degradation did not increase during a period of continuing population growth is encouraging. However, significant improvements are still needed, and reliance on regulatory processes is becoming a larger component of operations that were previously guided primarily by voluntary and consensus-based decisions. It will continue to be a challenge to move the partnership through this changing paradigm.

Moving to a more regulatory mode has not lessened public involvement. On the contrary, the Bay Program has engaged stakeholders in a two-year process to develop revised water-quality criteria, and has publicized the new criteria well beyond the normal Chesapeake Bay constituency. This is, without a doubt, the most open criteria-development process that USEPA has ever instigated. During 2004, the Chesapeake Bay watershed states and the District of Columbia will establish strategies to implement the new criteria and involve tributary-specific stakeholder groups. It can be awkward sometimes for government authorities to know that they are sitting at the table, in a completely transparent government decision-making process, with representatives of organizations who may bring legal action against the government; indeed, that is a significant difference between voluntary and regulatory goal-setting processes.

From more than two decades of engaging the public in restoring the Chesapeake Bay, a number of lessons may be drawn. Eight of the key lessons from the Bay Program are highlighted below:

1. *Keep the public engaged.* It is essential to inform and involve citizens in setting and achieving goals.
2. *Public participation takes time and money.* Some members of the public have more time and money than others. This raises concerns about who is actually participating and whether all stakeholders are fairly represented. Sometimes governmental authorities have to go to the stakeholders instead of relying on them to have the time and/or money to be able to come forward. In addition to "face to face" meetings with these stakeholders, the Internet has also become a very useful tool for "levelling the playing-field."
3. *Environmental restoration takes time and money.* Public support is essential because people are being asked to support the use of tax dollars for environmental restoration, to support regulations that may increase business costs, and – a difficult step in the United States – to support land-management measures that may restrain how landowners develop their property. The public needs to understand that environmental improvements often take a long time: political leadership often changes faster than environmental results can be realized. Steadfast public support and patience is needed to ensure that environmental restoration stays on the agenda of new political leaders.
4. *Begin interactions with a core public group.* It can be more efficient to

work first with those who are well educated about the problems and potential solutions, and who have a vested interest in achieving success.

5. *The Internet is a powerful tool for reaching out to the public.* Although the Bay Program has made great strides involving the public through its website, it has only begun to utilize the full potential of the Internet.

6. *Managers have to consider carefully what they are seeking to achieve.* Managers need to answer these questions each time they interact with a different stakeholder: what is the behaviour that we are trying to achieve and reward; who benefits; who pays; are we reaching out in the right way?

7. *Consensus building requires a great deal of effort.* Consensus building can be valuable, but it takes time and resources. In this respect, there are a few steps that can facilitate the consensus-building process:
   - focus on reaching consensus among the "players" with the greatest impact;
   - focus on what you can contribute to solutions, not on what you think others did to cause the problem;
   - decide what your share of the solution will be, not what you think others' shares should be;
   - do not worry about someone else's relative power; focus on your combined power;
   - do not argue over whether current conditions are good enough; let one person's "restoration" be another person's "preservation;"
   - give the partnership process a chance to work, and avoid statements critical of others.

8. *Seek challenging and measurable goals.* There are two aspects to this. First, seek simple, measurable goals. These should address matters of how much, by when, and based on what baseline? Second, it is important to avoid letting the fear of *not* attaining voluntary goals prevent parties from setting them. Any progress is good, and probably will not occur unless challenging goals are set. If the results do not meet the goals, celebrate the progress and admit that more work needs to be done.

Editors' note

1. Parts of this chapter that overlapped with the previous chapter (by Roy Hoagland and also on the Chesapeake Bay) have been excised. For further background on the Chesapeake Bay and the Chesapeake Bay Program, please refer to that chapter (14).

REFERENCES

Chesapeake Bay Program. 2003. Internet: ⟨http://www.chesapeakebay.net⟩ (visited 16 November 2003).

McClafferty, Julie A. 2002. "Final Report for A Survey of Chesapeake Bay Watershed Residents: Knowledge, Attitudes and Behaviors Towards Chesapeake Bay Watershed Water Quality Issues." Report Prepared for the Chesapeake Bay Program. Internet: ⟨http://www.chesapeakebay.net/info/pressreleases/survey2002/Report.pdf⟩ (visited 23 November 2003).

16

A cooperative process for PCB TMDL development in the Delaware Estuary

Tomlinson Fort III

A TMDL is not just a technical issue, rather a combination of the disciplines of philosophy, chemistry, biology, economics and a little bit of theology.

Marasco Newton Group 2002

Introduction

The Delaware River Estuary spans three states and two United States Environmental Protection Agency (USEPA) regions (fig. 16.1), and has been the subject of a concerted effort by the Delaware River Basin Commission (DRBC) to develop a scientifically credible total maximum daily load (TMDL) for polychlorinated biphenyls (PCBs). Levels of PCBs in fish flesh, and resulting fish-consumption advisories, have caused the estuary to be listed as impaired, which in turn is driving development of the TMDL. Fish-consumption advisories are currently in effect from Trenton, New Jersey to the mouth of the Delaware Bay. The stakeholder group involved in setting the TMDL is large, diverse, and vocal, owing to the high environmental and financial stakes. The task is hindered by inconsistent state regulations, extremely low detection and action levels for PCBs, relatively large base-loads of PCBs already in the environment, and a lack of guidance or precedent from similar studies.

Although most parties involved agree that fish-consumption advisories need to be addressed on a priority basis, permitted dischargers are concerned that they could be required to bear a disproportionate share of

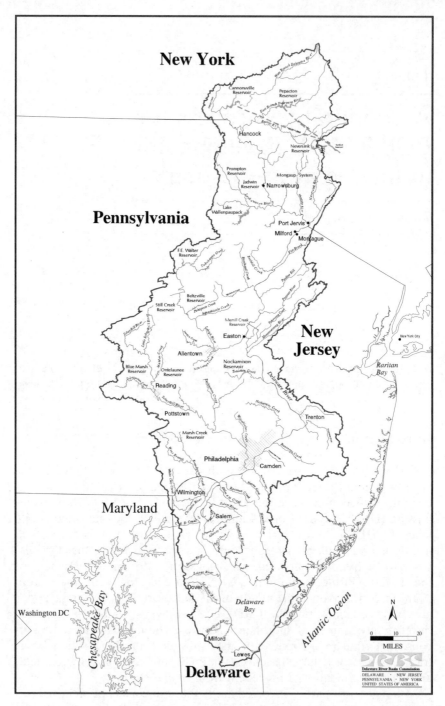

Figure 16.1 Map of the Delaware River Estuary (Source: Delaware River Basin Commission)

the burden of PCB reduction. Most permitted dischargers, including industries and municipal sewage plants, neither manufactured nor stored PCBs, but may have used products containing PCBs in the past. Generally, PCBs in most of these permitted discharges have not been historically regulated. Permitted dischargers are concerned that regulation of PCBs in their effluents to extremely low levels now could cause them significant expense and also may cause unavoidable permit violations, because the sources of PCBs are poorly understood and, in many cases, are not controlled by the permittee. For example, some industrial permit holders pump river water for cooling purposes. If the water they pump from the river contains PCBs, then it is easy to understand their concern that they could be required to build expensive treatment facilities to remove PCBs from their effluent. Some researchers have estimated that contributions from the vast array of non-point sources may be so large as to render point-source controls meaningless in solving the problem of PCBs in fish tissue.

New Jersey had a 2003 deadline for producing a TMDL, established in a memorandum of understanding with USEPA Region 2; Delaware had a court-ordered deadline contained in a consent decree; Pennsylvania lists the estuary as impaired, but has no definite deadline for a TMDL. These deadlines do not offer any flexibility for implementation delays, which translates to a limited capacity for conducting additional studies or analysis that otherwise might improve the database upon which decisions are being made.

A coalition of municipal and industrial dischargers (hereinafter termed Coalition) was formed in 2000 to interact more effectively with other stakeholders and regulatory officials in defining both the data needs and terms of the PCB TMDL. Coalition members and others are cooperating to share the burden of analysis and evaluation of alternatives for reducing the PCB load to the estuary. Members include municipal sewage-treatment authorities and corporations involved in transportation, petrochemicals, and power generation. What follows are general observations made over several years as the process has unfolded. These observations are not unique to this particular case, and have been purposefully generalized to apply to a wide variety of situations. The views expressed are the author's and do not necessarily represent those of the Coalition.

Background

Total maximum daily loads (TMDLs)

TMDLs have their origins in Section 303(d) of the Clean Water Act and seek to reduce loadings of a particular compound to required levels by limiting contributions from all controllable sources. The idea is to deter-mine the relative degree of health (or impairment) of a receiving water-body, impacts on that waterbody from past or current discharges of specific constituents, and safe loads of specific constituents that the waterbody can readily assimilate. The USEPA, charged with determining impairment status and with establishing and enforcing TMDLs where necessary, may delegate authority to state agencies. Waste load alloca-tions for permitted discharges and load allocations for discharges from non-point sources are established by the agency in such a way that the total loading is reduced below the TMDL and so that the formerly im-paired waterbody may be reclassified as unimpaired.

The statute recognizes that not all load sources are necessarily control-lable, which is tantamount to acknowledging that TMDLs and unim-paired status may, in some cases, be unattainable. However, in these cases there is a presumption that the loading must be reduced to the maximum extent possible, in order to approach the TMDL as closely as feasible. This raises the real possibility that controllable sources could be regulated to zero allowable discharge of targeted compounds, regardless of the relative contribution of those discharges to the overall loading problem. This is a key point of concern for some parties. The most easily controlled and best understood sources are the permitted point sources, such as discharges from factories and municipal sewage-treatment plants. In a scenario where non-point sources (such as run-off) are uncontrolla-ble or poorly understood, a disproportionate burden of reduction could be placed on the point sources. The cost and feasibility of achieving the reductions specified by TMDLs is uncertain, because technology to reli-ably reduce PCBs to levels of parts per quadrillion (ppq) has not been tested and may not exist. Further, if the uncontrolled loading from non-point sources remains high, the impaired waterbody may experience only limited benefits from any point-source reductions that are made.

The problem with PCBs

PCBs are a class of synthetic compounds not found in nature. Structur-ally, they consist of phenolic rings linked together with various numbers of chlorine atoms bonded to the rings in different configurations. There are over a hundred different variations, or congeners. They were manu-factured and used extensively in the United States, primarily as oil-based insulators for the electronics industry, from the onset of their commercial exploitation in 1929 until production was banned in 1979. They had ex-cellent dielectric and fire-resistance properties, making them ideal for use as insulators in transformers, capacitors, and switches. The vast majority of PCB-containing equipment has been replaced, or flushed and refilled with non-PCB insulating oil. However, limited continued use in older industrial equipment, large numbers of historical spills, and past mixing with organic oils in reuse/recycling programmes has caused widespread distribution of PCBs in the environment at low concentrations.

Once released to the environment, PCBs accumulate in sediments and animal tissue. PCBs generally have a low volatility, but the congeners of lower molecular weight are more volatile than heavier ones. These light congeners may volatilize into the vapour phase to be transported by wind and subsequently deposited back to the earth in rain, sorbed to dust, or via direct atmospheric diffusion into waterbodies. PCBs have low aqueous solubilities and high octanol–water partition (K_{ow}) coefficients, meaning that they tend not to dissolve in water in high concentrations; instead, they attach to soil or organic carbon particles, which may be carried by the wind as dust or become entrained in turbid run-off and enter streams and rivers attached to the sediment load. In quiescent areas such as estuaries, eddies, and point bars, the soil particles settle out as mud deposits. These PCB-containing sediments are periodically reworked by storm scour, tidal flow, downstream currents, and bioturbation.

Benthic (bottom-living) organisms may ingest contaminated sediment, as will bottom-feeding fish. Predator species, in turn, feed on organisms that may have consumed sediments containing PCBs. PCBs are highly fat soluble, or lipophilic; once consumed by an organism, they are readily partitioned from the gut into fat deposits, bioaccumulating within the organism over time. They are not readily broken down by metabolic or natural degradation processes, and continue to accumulate in the tissues of the organism over the course of its life. If that organism is subsequently consumed by a predator species such as a striped bass or fish-eating raptor, the PCB is transferred to the predator's fat tissues. PCB concentrations in predator species that routinely feed in areas containing bioavailable PCBs can develop higher PCB concentrations in their flesh through a process termed biomagnification. The primary focus regarding PCBs, therefore, has been on concentrations in (and potential effects on) predator species. Potential human-health issues arise by virtue of our position at the top of the food chain. Adverse health effects such as cancer and effects on the immune, reproductive, nervous, and endocrine systems have been noted in animal species at high concentrations (USEPA 2002). Studies in humans provide additional evidence for potential carcinogenic and non-carcinogenic effects associated with PCBs (USEPA 2002).

Stakeholder participation

Diverse input is important to the creation of lasting solutions. Whereas discussion of environmental issues was previously limited to regulators and regulated parties, it now routinely includes a broad cross-section of the community. Participation may be voluntarily solicited by a responsible party in order to generate new ideas or to obtain public support for a desired regulatory action. Participation may be required by law, as in a public comment period for review of a draft industrial-discharge permit. It may be driven by political or popular demand, as when neighbourhood coalitions are formed in order to influence a process or plan. Regardless of the reason, increased participation in environmental decisions illustrates the importance that communities and individuals place on finding what they believe are the correct answers and approaches to environmental problems. Many of these stakeholders are capable of derailing or delaying a process if they have been excluded from it.

Stakeholder diversity

All stakeholders share one key trait: they are there for a reason. An effective stakeholder group must address each person's reason for being there (Fisher and Ury 1983; Nazzaro and Strazzabosco 2003; Sandelin n.d.; Wertheim n.d.). Stakeholders include municipal and industrial dischargers, politicians, neighbours, attorneys, special-interest groups, environmental advocates, consultants, regulators, watershed managers, researchers, teachers, and students. Some of these people are looking for work; others seek to influence the outcome; some attend primarily to learn. They may provide resources to help with resolving issues, or may be there primarily to express concern. Some are well informed and supported by technical experts who assist them in understanding highly technical issues, whereas others do not have such resources. They may be well-balanced and pragmatic, or one-sided in their views. Some participate on principle; others may have tangible concerns regarding cost, property value, or quality-of-life issues. Participants may or may not have specific agendas. Regardless of their purpose, all participants provide input by stating positions or asking questions, and process coordinators must then assimilate that input.

Stakeholder differences

Stakeholder differences are often rooted in the relative importance that people assign to project drivers. Four of the most basic drivers are (a) speed, (b) accuracy, (c) cost-effectiveness, and (d) public acceptability. Participants generally agree that all four drivers are important, but frequently differ in how they rank their relative importance. For the setting of the PCB TMDL in the Delaware Estuary, agencies are highly concerned with speed and cost-effectiveness of TMDL development, because of the imposed TMDL deadlines and limited budget for completing it.

Dischargers are concerned with the accuracy of science used to develop the TMDL and with the cost-effectiveness of complying with the standards. Managing public acceptance is a goal of both groups.

To resolve the differences, the TMDL working group has remained committed to an open, public process, thereby focusing on a value that all participants share. The Coalition has remained sensitive to regulators' deadline concerns and has worked to fund technical experts who can rapidly contribute to the knowledge base, thereby relieving some budget pressures while strengthening the technical defensibility of TMDLs under development. Out of this work has come a recognition among some participants that there may be a real, justifiable need to develop an interim solution and to delay final waste-load allocations for a period of time so that additional agreed tasks (such as sediment-transport modelling and decadal scale-model verification) can be completed first.

Stakeholder similarities

Similarities among people are helpful in bringing together diverse stakeholder groups, and must be employed with skill and frequency in complex situations (Wertheim n.d.). Broad similarities, regardless of peoples' backgrounds or affiliations, are the most useful initially. Focusing on these aspects helps the group to feel more like a team. For example, most people agree that preventing impact from pollutants on the environment is a worthy goal, and that pollution prevention is preferable to remedial activities to clean up. People can agree to focus first on the most toxic pollutants, or on those having a disproportionately negative impact, as long as the toxicological differences are clear and easily understood. People can usually agree to focus first on achieving an early win on an easy, intermediate issue in order to build momentum and make early progress toward the ultimate goal. People often place a high priority on protecting wildlife, elevating its importance almost to the human-health level. We trust objective scientific analysis to help us to understand impacts and alternatives, as long as the input data are accurate, precise, and immune to manipulation. People value a permanent solution, as opposed to one that is temporary or vulnerable to challenge. When the working group is faced with fragmentation over a difficult issue, returning to these agreed points can help coordinators to refocus.

The most influential stakeholder similarity for the PCB TMDL project has been common recognition of the need for good, reliable data upon which to base decisions. In 2002, the Coalition convened a two-day technical seminar to present and discuss the data, uncertainties, priorities, and ongoing efforts. Presenters included university researchers working in the estuary, the Academy of Natural Sciences, consultants, the

DRBC, state regulators, and USEPA, as well as Coalition members. Organizers specifically focused on technical presentations, and political or partisan statements were not admitted. The seminar was immensely successful in that it defined the state of knowledge to all parties and aligned them with each other as to what needed to be done next. It created trust and respect, as data developed by various participants were verified or supported by data from independent researchers and from historical databases developed by academia in the estuary over decades of study. The seminar represented a turning-point at which all participants firmly grasped the complexity of the issues and began to reconcile the remaining tasks with the project schedule.

A second significant similarity among PCB TMDL participants is a desire for an open, participatory process. When faced with concerns regarding trust among participants, focus of the overall effort, and problems of implementation, DRBC retained a consultant to evaluate problems with the collaborative effort and propose solutions. This consultant interviewed 71 stakeholders representing industry, municipalities, environmental groups, various Coalition representatives, and members of all involved state and federal regulatory agencies. The consultant then produced a detailed report in two volumes, the first outlining and analysing comments received and the second making recommendations and drafting a charter for a TMDL-implementation advisory committee (Marasco Newton Group 2002). This report was widely regarded by stakeholders as valuable to the effort and a major factor in bringing the group together. This report is discussed later in greater detail.

A third generally agreed point is that all involved regulators from the Commission, three states, and two USEPA regions need to cooperate in defining requirements and avoiding inconsistency. The involved regulatory authorities have acknowledged this and have been working to eliminate any issues. DRBC, as a commission that is sponsored largely by other involved agencies, plays an important role in accomplishing this coordination.

Roles of dischargers

The unique position of dischargers allows them to contribute technical resources, financial resources, and their time to help ensure the best possible outcome for the project. In this case, municipal and industrial representatives participated actively on DRBC's Toxics Advisory Committee, the Water Quality Subcommittee, the Implementation Subcommittee, and the Tidal Subcommittee. In addition, the Dischargers' Coalition sponsored the two-day technical symposium (described above), retained TMDL and PCB modelling experts to collaborate with DRBC, contributed funds to an Academy of Natural Sciences food-web study, provided screening-level PCB sampling data, and assisted DRBC's consultant with decadal-scale calibration of PCB models.

Successful cooperation

To achieve lasting results, experience with the Delaware Estuary TMDL has shown that a stakeholder group must consistently press for truthful, objective communication among members. This finding is common in studies of group dynamics, conflict resolution, and high-performance team building (Nazzaro and Strazzabosco 2003). The following discussion draws upon experiences with participatory problem-solving in this particular context.

A forum is needed in which different views may be aired informally. Similarities among members need to be fully explored and voiced, so that they may refocus the group when discussions become divisive. More importantly, the differences in stakeholder requirements, concerns, and motivations must be honestly discussed and understood by all participants. Hearing others' concerns helps each member to recognize small "win–win" strategies that may not have appeared before (Wertheim n.d.).

New working groups usually start with individual participants expressing their individual positions. No one is inclined to compromise on the first day, nor are they particularly inclined to listen. A more mature group with an evolved structure based on trust can focus more on listening, problem analysis, solidarity, and goal attainment (Nazzaro and Strazzabosco 2003). This means that benefits from group listening take time to accrue, often over the course of numerous meetings. Although this is an oversimplification, it is helpful to envisage working groups traversing four basic phases, as they move from defining the problem to developing the solution. Each phase is important and must be allowed to occur. In order, the general phases of a dynamically maturing group are described in figure 16.2. Nazzaro and Strazzabosco (2003) take this simplistic view further by integrating a group-trust dynamic to arrive at their four phases, described as (1) come together, (2) "norming," (3) performing, and (4) transforming.

As dynamics of the group mature, positions of individuals also evolve in response to the changing group context. Long-term stakeholder relationships give each participant an opportunity to think through his or her reasons for being there. This takes place in the "Listen" and "Analyse" stages. Listening to other stakeholders leads each individual to consider which of his desired outcomes from the process are reasonable, based on group support or aligned issues (Wertheim n.d.). Each person

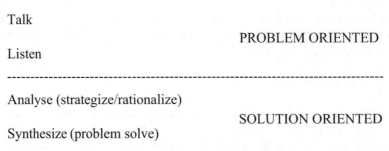

Figure 16.2 Four-phase schematic for participatory problem solving

forms an opinion, during this stage, of what he or she might reasonably get out of the process. With time, individual positions are moderated by the group. Consciously or unconsciously during this stage, each individual places his various desired outcomes into one of three categories:

- things that I must cause or prevent from happening;
- things that I value highly, but could compromise on;
- things that I could easily give up to get something else.

When people begin to listen to other stakeholders' priorities, they get ideas for bargaining, alliances, strategy, and acceptable compromise. In this way, understanding priorities creates the currency required to negotiate "win–win" solutions that preserve the points most important to each person (Wertheim n.d.).

Once this process has started with a few individuals, group dynamics make it very difficult for any participant to claim that all of their issues are non-negotiable. Particularly if the issues are unique to that person, it becomes obvious that an inflexible position risks losing credibility with the group and even losing the relevance of those issues. This represents a critical turning-point, at which the group focus changes from individualistic problem definition to a solution-oriented, group dynamic.

Challenges in implementing public participation in the Delaware Estuary PCB TMDL

Forums in the PCB TMDL process

In the PCB TMDL process, several forums have been established to facilitate communication. Coalition meetings provide a venue for airing differences of opinion between the municipal and industrial membership and for building consensus. The Coalition has two subcommittees – a technical committee and a steering committee – which meet separately or concurrently. The steering committee gives direction to the technical committee, and the technical committee provides recommendations, or may request that certain policy decisions be made in order to clarify or focus its efforts. This is all within the discharger Coalition itself. At a broader level, the Toxics Advisory Committee (TAC), appointed by the DRBC Commissioners and co-chaired by members of the Coalition and DRBC, includes technical representatives from most stakeholder groups and has the responsibility of advising the DRBC on appropriate handling of technical issues within the TMDL development process. The TAC includes members from public health, fish and wildlife, environmental, agriculture, municipal, academic, industrial, and state and federal regulatory sectors. An Expert Panel was created to advise the TAC and DRBC on developing a model for calculating TMDLs. Similarly, the Implementation Advisory Committee (IAC) advises the DRBC on process and strategy for implementing TMDLs and, until they are complete, also advises on adjusting current procedures in order to plan most effectively for the ultimate conclusion of the process. Improving communication among stakeholders is a major part of the IAC's charter.

Late in 2001, during a period of significant stakeholder criticism of the process, the DRBC contracted with the Marasco Newton Group to help facilitate development of a TMDL implementation plan (Marasco Newton Group 2002). This consultant was charged with identifying barriers and recommending solutions. A broad group of 71 stakeholders were interviewed, giving priority to those with large stakes or ideas that might be applied to the problem. Interviewees represented industry, municipalities, public-interest groups, environmental groups, and regulators. Interviews resulted in a large number of constructive comments on identifying issues, improving communication, improving the scientific basis for TMDL development and data quality, need for solidarity among regulators, need for more clarity regarding the fish-consumption guidelines, need for an open and cooperative processes, and clarification of roles for the various advisory groups. Concerns were voiced regarding lack of trust or misunderstanding of motives, funding, delays in the schedule, differences of opinion over the appropriate level of study prior to setting TMDLs, appropriate use of screening-level data, and reliability of analytical methods.

The Marasco Newton interviews illuminated the division between point sources and non-point sources as perhaps the most controversial technical issue, and one at the heart of the PCB TMDL development. The potential impact of TMDLs on discharger cost is significant because of the connection with PCB-load (non-point source) and waste-load (point source) allocations. At the same time, point sources are relatively easy to understand and regulate, whereas non-point sources are less so.

Regulatory concerns over the TMDL-implementation schedule forced an initial focus on the point sources, because of their capacity to provide an early and tangible result. When combined with the recognition that non-point sources are significant (if not the primary) contributors of PCBs entering the estuary today, the concern held by point-source dischargers becomes clear. Indeed, the report found that, "identifying non-point sources is critical to getting the regulated community to cooperate" (Marasco Newton Group 2002); however, this is only partly true, because the regulated community had been cooperating for a long time. Still, it is probably a fair assessment that few, if any, point-source dischargers would accept a final TMDL result that failed to take non-point sources into account.

A good indicator of an evolved, trusting, and solution-oriented process is direct communication among the retained technical experts for the different parties. At this stage, the parties focus more on ascertaining the status and possible solutions than on politicking. In the early stages of stakeholder group interaction, most discussions among the various parties are handled by the most outspoken strategists or attorneys. These discussions tend to be partisan and may be defensive; there is usually an underlying agenda.

Once this posturing is over and trust begins to prevail, more of the initiative and direction for the group's activities flow from technical or business representatives, as opposed to legal or strategic ones. In the final stage, experts (retained or otherwise) from each side may be encouraged by their respective employers to interact directly with one other to reach the correct solution as efficiently as possible. Data are regularly shared in real time; restrictions on sharing qualified or unverified data may be relaxed; in the case of the PCB TMDL process, computer simulation code and assumptions for modelling sediment distribution were shared. Additionally, the tasks of running the model and performing decadal-scale calibrations on its output were divided and shared by technical experts from both regulatory and discharger groups.

Equal footing

It may be important in some cases to recognize and address problems relating to an uneven balance of resources. Although this has not proved to be necessary in the case under discussion, this is frequently an issue for environmental stakeholder groups. For example, a vocal, concerned, and affected residents' group may have insufficient technical expertise to understand technical data upon which decisions are being made. If participants are not able to understand and interact at the required level to affect the decision, then it may be argued that they do not have meaningful input. Models exist in federal legislation, such as Superfund, that allow for technical representatives to be funded and appointed to assist less technically sophisticated participants, thus enabling them to provide meaningful input to the process. Another option is the designation of a technical consultant, agreeable to all parties, to advocate for non-technical stakeholders, but who may be funded by another party. Without such technical assistance, the stakeholder process may ultimately fail because a significant participant, who may feel railroaded, could decide to challenge the decision in court. True, meaningful consensus helps to ensure the permanence of a solution (Sandelin n.d.).

Risk perception and uncertainty

How participants perceive risk is a primary factor in determining the complexity of the stakeholder process. Risk does not have to be actual, but need only be perceived as possible, to cause concern. Possible risk or uncertain severity can cause the same complications in a stakeholder process as actual loss, damage, or impact from an environmental issue. The most prevalent risks addressed by environmental stakeholder groups are human-health risks, financial risks, and environmental risks. The more severe a risk is perceived to be, typically the more groups become involved and the more vocal and adamant the parties become.

Uncertainty implies possible inaccuracy or imprecision. Uncertainty will complicate any environmental process or effort involving stakeholders. If a party is unable to prove that risk does not exist, others tend to assume that it does exist. Likewise, if risk is expressed as a range of probability that spans both the acceptable and the unacceptable, then people tend to assume that an unacceptable risk could exist. These tendencies demonstrate the well-established principle of conservative assumptions applied to managing risk uncertainty. This works well for filling-in isolated data gaps; however, where uncertainties are numerous, multiple conservative assumptions become compounded in such a way that the output may grossly overpredict risk and have little utility. In this way, uncertainty can drive process complexity and magnify perceived risks. Thus, the more that uncertainty (regarding effects, level of conservatism, and appropriate methods) can be resolved or identified as an insignificant factor in the overall analysis, the more smoothly and efficiently the process will flow.

In the PCB TMDL process, there have been significant issues surrounding uncertainty of regulatory requirements. This is compounded by having three states and two USEPA regions involved with the commission in regulating the issue. Points of policy and standards differ markedly: in several instances, solutions or approaches suggested during meetings by one agency were criticized as inadequate by another. This leaves the regulated community frustrated and fuels angst among the stakeholders. Differences revolve around standards, priorities, schedules, water-quality criteria, and acceptability of a phased implementation. Differences in fish-consumption advisories among the states cause confusion and difficulty. The fact that allowable tissue concentrations in fish are higher for those fish available in supermarkets (regulated by the Food and Drug Administration) than for those caught in the estuary further complicates matters.

Voluntary and involuntary risks are perceived in different ways. We do not live in a risk-free world, and we are all used to living with a certain, inevitable, amount of risk. Every day, we voluntarily accept risks (such as driving to work, or smoking) that carry health risks in excess of most environmental-exposure scenarios. However, involuntary risks, and often poorly understood ones, are the usual subjects for environmental debate. When an involuntary environmental risk is thrust upon people, even if it is very small it is often rejected as unacceptable. The problem is compounded by the fact that risk is described in statistical terms – such as one cancer death in 100,000 defining the threshold for "acceptable" risk. What if that one person is someone in your family? In general, people are more likely to reject risks that they cannot control.

Staying current on external research

Because every TMDL case will be unique in significant ways, including the physical and regulatory settings, other PCB TMDL efforts are unlikely to be sufficiently similar to serve as a template for the Delaware Estuary. However, much research that can be useful has been conducted on elements of PCB toxicity and TMDL development. An example would be work conducted by General Electric and other stakeholders relative to the Hudson River, which is also affected by PCBs (Environmental Media Services 2001). The stakeholder group for the Delaware Estuary has drawn extensively from work done by external parties on other projects, to the extent that such results might be relevant. In December 2002, a live webcast on non-cancer health effects of PCBs was jointly sponsored by the National Institute for Environmental Health Sciences (NIEHS) and the USEPA Office of Emergency and Remedial Response. The webcast highlighted the research being conducted by two NIEHS scientists into the non-cancer endpoints of exposure to PCBs, with particular emphasis on findings related to growth and neuro-cognitive development in infancy and later childhood. Another focus of this same study involved evaluation of whether exposure of pregnant rats to PCBs could reduce the length of gestation. These are just two of the numerous external projects that have been studied.

Further work

It is estimated that several more years of work may remain to achieve a reasonable understanding of the whole PCB "budget" in the Delaware Estuary. Among major tasks left to complete are the following:

• quantify existing PCBs in sediments already in the estuary;
• focus on conservatism inherent in the fish-consumption advisories and reconcile them from state to state, to similar watersheds, and to federal protection levels;
• address bioavailability, bioaccumulation, and biomagnification issues;
• examine tributary loading;
• improve sediment-transport modelling;
• characterize contributions from non-point sources and atmospheric sources;
• develop PCB modelling; and
• complete the long-term, decadal-scale, model calibration.

Agreement as to water-quality goals and related allowable concentrations in fish flesh, and the resulting impact on fish-consumption advisories, are key issues to be confronted in the near future. Whereas all parties want goals that are adequately protective, there is also the need to be certain that the goals established are feasible and beneficial. High levels of uncertainty, goal conservatism, the ubiquitous occurrence of PCBs in the environment, and extremely low detection limits, contribute to differences of opinion and complications in the negotiation process. This is a critically important point, both to dischargers (who must ultimately pay to accomplish reductions) and to regulators (as stewards of public safety).

The need to characterize the atmospheric loading illustrates the dilemma. Early indications are that further study may show PCB occurrence to be ubiquitous in the environment at the extremely low levels specified. The present regulatory goal is to reduce targets for PCBs in fish tissue to 2 parts per billion (ppb) through lowering the water-quality standard from the current 44 ppq to 3–10 ppq in the Delaware Estuary. Although this is an admirable goal, 2 ppb in fish is below levels detected in fish in remote areas far from residential, commercial, or industrial developments (Washburn 2003). The Agency for Toxic Substances and Disease Registry (ATSDR) believes that the source of these remote PCBs is probably atmospheric deposition, because there are no other known sources (ATSDR 2000). If this is true, then attempts to attain the

2 ppb goal through point and non-point source control may be futile. For example, the following mean PCB concentrations have been reported in fish from remote areas (Dewailly et al. 1993; Wilson et al. 1995; Kidd et al. 1998; Datta et al. 1999; ATSDR 2000; Kannan et al. 2000; Lewis et al. 2001):

- Char, Lake Ellasjoon (Arctic): 55–2,500 ppb
- Char, Arctic Quebec: 152 ppb (mean of nine muscle samples)
- Lake trout, Sierra Nevadas: 18–430 ppb
- Kokanee, Sierra Nevadas: 13–44 ppb
- Lake trout, Siskiwit Lake: 40–460 ppb
- Lake trout, Schrader Lake (Arctic): 1–18 ppb (muscle only)
- Grayling, Schrader Lake (Arctic): 1.3 ppb (mean of five muscle samples)
- Whitefish, Mackenzie River: 2–11 ppb (muscle only)
- Lake trout/char, Peter Lake: 10–82 ppb.

Summary PCB concentration results, from sampling of fish tissue from remote areas, were extracted from these references (Washburn 2003) and are consistent with the findings of EPA's National Sediment Quality Survey (EPA 1997). In that survey, PCBs were detected at 2,370 (73 per cent) of 3,232 total fish-flesh sampling locations located in fresh water in the United States where collected samples were analysed for PCBs (EPA 1997). Of these 2,370 detections, the survey concluded that approximately 99 per cent had PCB concentrations exceeding 2 ppb, with more than 95 per cent exceeding 14 ppb and more than 70 per cent exceeding 140 ppb (EPA 1997).

Adaptive implementation

Phased implementation of a PCB TMDL is being utilized as a way to reconcile the large amount of work yet to be done with the implementation deadlines. Given the incomplete status of data collection and analysis, phased implementation allows for near-term reductions of those PCB loads that are reasonably well understood and controllable through best management practices (BMPs), while deferring those reductions that are less well defined until the remaining analysis is concluded. While interim BMPs are in place, the proposal is to continue monitoring and collecting data to supplement the existing database and also to measure any beneficial effects from implementing the BMPs.

Members of the regulated community have generally supported a phased approach, but regulatory agencies and environmental groups were initially more mixed in their perceptions. Concerns of the latter group tended to hinge on an overriding concern that real, meaningful reductions needed to occur by the deadline, and they were concerned that a phased approach could allow delay of some significant, possible, reductions. The regulated community remains reluctant to spend large sums on treatment and control equipment before all the research is done that shows that such controls are both necessary and cost-effective. Ultimately, a plan for phased implementation emerged as the most pragmatic approach.

The dischargers are currently working with the DRBC to develop an adaptive implementation model (including BMPs and monitoring) that is acceptable to the environmental and regulatory stakeholders. The present proposal under discussion calls for the remaining analysis to be completed over the course of the next two and a half years, with the next phase of TMDL implementation in place by 2005. Whether this would be the final phase has yet to be determined. When the final phase is in place, regulators envisage waste-load allocations for point-source dischargers and load allocations for non-point sources. The waste-load allocations for point sources could be implemented through National Pollution Discharge Elimination System (NPDES) permit renewals, although the actual mechanism has yet to be determined.

Conclusions

Building on the cooperative approach now in place, continued discussions are expected to resolve the question of adaptive implementation in a manner satisfactory to all stakeholders. In spite of our differences, our similarities are well known. We are all seeking the right answer and best approach. Our goals are similar: we are motivated to produce a defensible TMDL with a solid technical foundation that is resistant to challenge because of our rigorous methods. We are talking and listening to each other.

REFERENCES

ATSDR (Agency for Toxic Substances and Disease Registry). 2000. *Toxicological Profile for Polychlorinated Biphenyls. Update. (Final Report)*. Atlanta, Ga.: ATSDR, Public Health Service, US Department of Health and Human Services. 935 pp. NTIS Accession No. PB2000-108027.

Datta, S., K. Ohyama, D.Y. Dunlap, and F. Matsumura. 1999. "Evidence for Organochlorine Contamination in Tissues of Salmonids in Lake Tahoe." *Ecotoxicology and Environmental Safety* 42:94–101.

Dewailly, E., P. Ayotte, S. Bruneau, C. Laliberte, D.C.G. Muir, and R.J. Norstrom. 1993. "Inuit Exposure to Organochlorines through the Aquatic Food Chain in Arctic Quebec." *Environmental Health Perspectives* 101(7).

Environmental Media Services. 2001. "General Electric, PCBs and the Hudson River." Internet: ⟨http://www.ems.org/pcb/general_electric.html⟩ (visited 23 November 2003).

Fisher, R., and William Ury. 1983. *Getting to Yes: Negotiating Agreement Without Giving In*. New York: Penguin Books.

Kannan, K., N. Yamashita, T. Imagawa, W. Decoen, J.S. Khim, R.M. Day, C.L. Summer, and J.P. Geisy. 2000. "Polychlorinated Naphthalenes and Polychlorinated Biphenyls in Fishes from Michigan Waters Including the Great Lakes." *Environmental Science and Technology* 34:566–572.

Kidd, K.A., R.H. Hesslein, B.J. Ross, K. Koczanski, G.R. Stephens, and D.C.G. Muir. 1998. "Bioaccumulation of Organochlorines through a Remote Freshwater Food Web in the Canadian Arctic." *Environmental Pollution* 102:91–103.

Lewis, M.A., G.I. Scott, D.W. Bearden, R.L. Quarles, J. Moore, E.D. Strozier, S.K. Sivertsen, A.R. Dias, and M. Sanders. 2001. "Fish Tissue Quality in Near-coastal Areas of the Gulf of Mexico Receiving Point Source Discharges." *Science of the Total Environment* 284:249–261.

Marasco Newton Group. 2002. "Summary and Analysis of Stakeholder Comments and Consultant Recommendations Regarding the Formation and Functioning of an Implementation Advisory Committee for the Total Maximum Daily Load (TMDL) for Polychlorinated Biphenyls (PCBs) in the Delaware Estuary." Report prepared for the Delaware River Basin Commission (DRBC).

Nazzaro, Ann-Marie, and Joyce Strazzabosco. 2003. *Group Dynamics and Teambuilding. Hemophilia Organization Development Monograph Series, No. 4.* Montreal, Quebec: World Federation of Hemophilia. Internet: ⟨http://www.wfh.org/ShowDoc.asp?Rubrique=99&Document=366⟩ (visited 23 November 2003).

Sandelin, Rob. n.d. "Basics of Consensus." (Summary compiled by Intentional Communities.) Internet: ⟨http://www.ic.org/nica/process/Consensusbasics.htm⟩ (visited 23 November 2003).

USEPA (US Environmental Protection Agency). 1997. *The Incidence and Severity of Sediment Contamination in Surface Waters of the United States. Volume 1: National Sediment Quality Survey.* 2nd edn, (EPA 823-R-97-006). Washington, DC: Office of Water. Internet: ⟨http://www.epa.gov/OST/cs/congress.html⟩.

USEPA (US Environmental Protection Agency). 2002. "Polychlorinated Biphenyls (PCBs)." Internet: ⟨http://www.epa.gov/opptintr/pcb/effects.html⟩ (visited 23 November 2003).

Washburn, S. 2003. "Response to Regulatory Proposal for Lowering Water Quality Standards in the Delaware Estuary." Letter to Delaware River Basin Commission and obtainable from the DRBC, 25 State Police Drive, PO Box 7360, West Trenton, NJ 08628-0360, USA.

Wertheim, E. n.d. "Negotiations and Resolving Conflicts: An Overview." Boston, Mass.: College of Business Administration, Northeastern University. Internet: ⟨http://web.cba.neu.edu/~ewertheim/interper/negot3.htm⟩ (visited 23 November 2003).

Wilson, R., S. Allen-Gil, D. Griffin, and D. Landers. 1995. "Organochlorine Contaminants in Fish from an Arctic Lake in Alaska, USA." *Science of the Total Environment* 160/161:511–519.

17

Public participation in the resettlement process of dam-construction projects: A post-project survey of the Saguling and Cirata dams in Indonesia

Mikiyasu Nakayama

Introduction

Despite the fact that involuntary resettlement has been regarded as a major issue in dam-construction projects, best practices are still not yet established. This stems in part from the fact that only a few, limited, detailed post-project surveys have been conducted thus far on issues of involuntary resettlement. Such post-project reviews of the implemented resettlement schemes associated with dam-construction projects are essential in order that a better methodology to deal with the issue may be developed.

Experiences have shown that participation of those who are forced to resettle due to inundation of their residence (termed "resettlers" for the purposes of this chapter) leaves much to be desired. The ways in which resettlement processes are planned and implemented do not reflect the desires, needs, or priorities of these resettlers. In fact, to date, very limited post-project analysis has been carried out in this regard.

The aim of this chapter is to learn from the post-project surveys of two dam-construction projects in Indonesia in order to identify ways in which public participation in resettlement decisions could be better planned and implemented for similar projects.

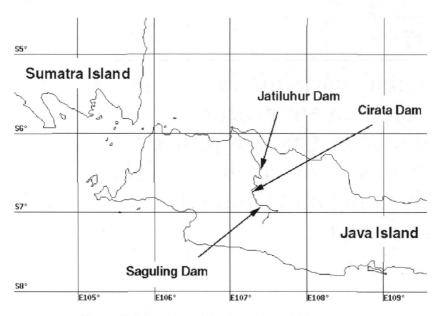

Figure 17.1 Location of the Saguling and Cirata dams

Analytical framework

This study was conducted by analysing dam-construction projects in Indonesia – specifically, the Saguling and Cirata dams on the heavily populated island of Java. These dams are located in the Citarum River system (fig. 17.1). The Citarum River originates in the Bandung Plain and flows into the Java Sea; it has a catchment area of 6,590 km² and an average annual rainfall of 2,232 mm (PLN 1990). It is the largest river in West Java and ranks third among the major rivers in the whole island. The Saguling and Cirata dam sites are located 30–50 km from the city of Bandung, which is the third-largest city in Indonesia and is within the catchment of these dams. Construction of the Saguling Dam started in 1983 and the dam was completed in 1985. The Cirata Dam was built shortly thereafter, between 1983 and 1988. The purpose of these dams was power generation to meet the then-increasing demand for power in the island; the Saguling and Cirata dams have power-generation capacities of 700 MW and 500 MW, respectively. Both dams were planned and constructed by the Indonesian National State Electric Company (Perusahaan Umum Listrik Negara, or PLN) with loans provided by the World Bank and Overseas Economic Cooperation Funds (OECF) of Japan. Both the Saguling and Cirata dam projects were (owing to the funding from the World Bank) among the initial projects in Indonesia in which serious attention was paid to their environmental impacts.

The environmental impact assessment (EIA) for the Saguling and Cirata dams was carried out by the Institute of Ecology (IOE) of Padjadjaran University in Bandung, Indonesia under contract by PLN (Nakayama 1998, 1999). As noted above, these two dam projects were among the first projects in Indonesia in which serious attention was paid to their impacts on the environment. For the Saguling Dam, possible impacts of the planned dam were identified by the flow-diagram method (Bisset 1987), whereas no specific flow diagram was developed for the Cirata Dam. In the latter case, provision for the effects of the dam's construction and operation merely mirrored that of the Saguling Dam, implying that the effects of both dams would be similar.

For this study, details of the Saguling and Cirata dam-construction projects were examined mainly through literature survey and interviews. The former made extensive use of the EIAs, both before and after dam construction, carried out by the IOE for the two dams. The interviews were conducted several times between 1991 and 2000. Interviews were also conducted with those in the World Bank and OECF at Washington, DC and in Tokyo, respectively. Targeted individuals included those who either (a) used to be involved in the Saguling or Cirata dam projects, or (b) worked on environmental aspects of dam-construction projects, particularly resettlement issues.

Issues related to public participation

Destination of resettlers

Indonesia may be characterized by the concentration both of populations and of various socio-economic activities on the island of Java, whereas the other islands are much more sparsely populated. Many dams have been constructed on Java for agricultural, industrial, and domestic purposes. Although the construction and operation of these dams forced a number of residents to relocate, the compensation schemes applied have been criticized by non-governmental organizations (NGOs) and researchers because the resettlers were, in general, worse off after relocation.

Because both the Saguling and Cirata dams were constructed in the densely populated region of West Java, PLN was of the opinion that it was not feasible to secure alternative lands on Java for all the resettlers. Instead, PLN proposed the creation of jobs (in sectors other than agricul-

Table 17.1 Mode of living of resettlers, as assumed by the PLN at the planning stage of the Saguling Dam Project

Mode of living	No. of families
Transmigration into other areas	2,000
Resettlement by agricultural development schemes	625
Mixture of agriculture and aquaculture	1,500
Relocation into the southern Bandung area	250
Employment in project-related activities	600
Monetary compensation	5,689
Total	10,664

Source: IOE 1984.

ture) as essential for the smooth implementation of the resettlement schemes for these dam-construction projects (Soemarwoto 1990).

The resettlement scheme initially planned for the Saguling Dam is shown in table 17.1 (PLN 1988). About 73 per cent of the resettlers were supposed to have alternative lands after relocation, either on the island of Java (by "Agricultural development in Java island" and "Agricultural development near Bandung") or on another island by the transmigration scheme. It should be noted that the World Bank regards Saguling Dam as a success in terms of mitigating the impacts of the dam on the human environment: reducing the height of the dam by five metres during the design stage had the mitigating effect of reducing the number of resettlers by nearly 50 per cent (Scudder 1997). Nevertheless, the Saguling Dam project involved the relocation of a number of families.

According to the research conducted by the IOE in 1979 (IOE 1980), only 11 families out of 316 families to be relocated wished to migrate to other areas by the transmigration scheme. The PLN, however, assumed that some 79 per cent of the resettlers would be transmigrated to other areas (i.e. not on Java), as shown in table 17.1.

Table 17.2 shows the destinations of the 3,078 families resettled as a result of the Saguling Dam (IOE 1984). This research (IOE 1980) ad-

Table 17.2 Destinations of resettlers from the Saguling Dam

Destination	No. of families	Percentage
The same village	2,416	78.5
Other villages in the same district	197	6.4
Neighbouring districts	391	12.7
Transmigration	74	2.4
Total	3,078	100

Source: IOE 1980.

dressed practically all of the families relocated. This table clearly shows that few families wished to migrate by the transmigration project or other agricultural-development schemes. In light of the PLN's assumptions of most resettlers participating in transmigration schemes, the reality of almost 80 per cent of the resettlers staying in the same village implies that the participation of resettlers in formulating the resettlement programme was almost non-existent. The outcome of the survey carried out by the IOE, before the resettlement scheme was designed, was not taken into consideration. Those who would be obliged to resettle were not consulted in terms of their preference about their destinations. Accordingly, the resettlement scheme planned for the Saguling Dam leaves much to be desired, particularly in terms of public participation regarding destination of the resettlers (Nakayama 1998).

Distrust among residents

Receiving compensation can be a sensitive issue. In some Asian and African cases, the community spirit was lost, owing to a development project. The villagers tended to mistrust the community leaders, particularly with regard to compensation: many feared that the leaders would use their influence with the executing body of the project in order to receive more compensation than other villagers would receive. This same phenomenon was observed with the Saguling Dam. Table 17.3 indicates the mode of payment by which the resettlers wished to receive compensation money, as surveyed in each village. The table shows the overwhelming preference of most resettlers to receive compensation money directly from the executing body, namely PLN, not through the community leaders nor the local committee that had been established especially for the project.

The reasons behind the villagers' preference are shown in table 17.4. More than half of the resettlers wished to avoid possible fraud by media-

Table 17.3 Villagers' preferred mode of payment for compensation from the Saguling Dam

Mode	Percentage of villagers			
	Sindangkerta	Cililin	Batujajar	Padalarang
Through leaders	6	6	0	0
Through committee	0	1	0	5
Directly from executive agency	94	93	100	95

Source: IOE 1980.

Table 17.4 Reasons for the villagers' preferences

Reason	Percentage of villagers			
	Sindangkerta	Cililin	Batujajar	Padalarang
To avoid fraud	80	76	88	37
Faster payment	13	22	4	58
Less complicated procedures	7	2	8	5

Source: IOE 1980.

tors (namely, the community leaders). The villagers apparently distrusted community leaders, and they suspected that these leaders received more compensation than they were entitled to by bribing officials (although this was, in fact, not happening with the Saguling Dam). The lesson learned was that "leaders" in a community may be relied upon and functional for many aspects of daily life, but in emergency situations (such as inundation of the village by dam construction) involving money, the villagers wish to be directly involved and to receive their payments directly.

In the case of the Saguling Dam, distrust of community leaders seems to have been amplified by the fact that not all the resettlers obtained compensation money at the same time, even in the same village. Some villagers thus suspected that the community leaders had given bribes to the executing agency in order to have priority in receiving compensation (Achmad 1986). The EIA conducted during project formulation foresaw the possibility of tension among villagers as a result of "rumours." However, the EIA failed to anticipate the distrust among community members toward their "leaders" – which, in fact, prevented those leaders from exercising their leadership.

The lack or loss of village governance mechanisms was reported in some villages within the Cirata Dam project area. In these instances, the chief of a village or "community leader" used to play a pivotal role in building consensus among villagers. However, once such people had left the village after having obtained compensation, the remaining villagers suffered from the lack of a mechanism by which to communicate with local government or other institutions involved in implementing the project (Nakayama et al. 1999).

The distrust observed in the Saguling Dam case, and the non-functioning of community leaders observed in the Cirata Dam case, suggest that traditional ways of consultation among village people (which relies on leadership by community leaders) may not function properly for resettlement issues following dam construction. Resettlers apparently wished to be able to communicate directly with the implementation body of the project in the case of the Saguling Dam. This should definitely be borne in mind when planning public participation in future dam-construction projects.

Extension services provided versus desired

The resettlement plan for the Cirata Dam project anticipated that about 20 per cent of the resettlers could be absorbed by construction work for the dam or by secondary development. However, it turned out that only a very limited number of resettlers could obtain a job at the construction site (IOE 1985), because of (a) their inadequate educational backgrounds, (b) their lack of relevant skills and experience, and (c) their lack of connections with the contractor.

The resettlers in general had disadvantages in this respect. For example, 35.3 per cent of the resettlers had only an elementary school education; only 2.4 per cent enrolled in secondary school; only 1.3 per cent had a high-school education; and only 0.13 per cent of resettlers studied at a university. Drop-outs from elementary school amounted to 29.3 per cent, and 31.3 per cent of resettlers had never received any education (IOE 1985). The resettlers generally did not have a sufficient level of education or set of skills to gain employment as dam-construction workers.

Nevertheless, the government tried to improve the people's economy by providing extension services. The major institutions that provided such extension services included the Fishery Office (49.1 per cent of services), the Agriculture Office (22.6 per cent), the local village government (10.2 per cent), and the IOE (5.3 per cent). As these figures show, most of the efforts concentrated on aquaculture and agriculture.

Table 17.5 shows the content of extension services provided, as surveyed by IOE in 1992 (IOE 1992). In fact, most of the extension services concentrated on fishery and aquaculture, and only limited efforts were made to enable resettlers to rebuild their livelihood as construction workers or through secondary development. The expectations of resettlers from extension services are shown in table 17.6. Although fishery- and aquaculture-related topics are significant, resettlers also hoped to enhance their knowledge in such fields as home industry, security, trading and stallholding, furniture making, and running a cooperative. These hopes and aspirations of the resettlers were not fully met by the extension services provided.

These shortcomings of the extension service for the Cirata Dam project illustrate the lack of effective public participation. The needs and priorities of the resettlers should have been learned at the planning stage of the extension service, so that they could be planned for and properly reflected in the topics of the extension service provided.

Table 17.5 Extension services provided

Service	Percentage
Fish-trap fishery	32.6
Fish processing	26.1
Home industry	9.8
Floating net fish management	6.5
Welding	6.0
Fish fodder	3.3
Fence system fish management	2.7
Animal husbandry	2.2
Cooperative	1.6
Agriculture	1.6
Rice-fish culture	1.6
Floating net construction	1.1
Trade and vending	0.5

Source: IOE 1992 (percentages as given therein).

Table 17.6 Extension services desired by resettlers

Service	Percentage
Less capital floating net aquaculture	18.2
Agriculture	18.2
Home industry	13.2
Trading and stallholding	12.1
Fish processing	11.6
Security	6.1
Furniture making	5.5
Running a cooperative	3.3
Knowledge of tourism	1.1
Skill in electronics	0.6
Other	7.2

Source: IOE 1992 (percentages as given therein).

Conclusions and lessons learned

The cases of the Saguling and Cirata dams show that efforts to promote public participation were not apparent in these projects. This was particularly true with regard to the fate of resettlers in the Saguling Dam project: the lack of previous public participation, and the subsequent disconnect between the resettlement plans and the resettlers' preferences, led to many difficulties for the resettlers in regaining their livelihood after relocation (Nakayama 1998). The preferences of resettlers should have been more carefully surveyed and considered during the planning stage of the resettlement scheme, so that these people would not have had to be forced to accept unwelcome locations and occupations after relocation.

The ways and means of communicating with resettlers should differ from those generally believed appropriate for approaching village people under normal circumstances. Resettlement is an emergency situation for most resettlers and the village in which they live. The "order" within a village, which has been functional for a long time, may become non-functional in such a situation so that, for example, villagers distrust their leaders on key issues. Special attention should be paid to dealing with such villages, for example in communicating with resettlers. The cases examined show that resettlers tend to prefer direct communication with the implementing project body, rather than through traditional information-flow mechanisms within the village.

Educational needs, as assumed by the project-implementation body of the Cirata Dam project, showed major discrepancies from the actual wishes of the resettlers. This stemmed from the fact that the former re-garded aquaculture as almost the only viable option for the resettlers to earn a living, whereas the resettlers wished to explore many other ave-nues. These intentions had not been fully taken into account by those who designed the extension services provided for resettlers, and this led to a lack of the necessary vocational training among those resettlers who wished to re-establish their livelihoods in sectors other than aquaculture.

The projects examined were planned and implemented in the early to mid-1980s, when the importance of public participation had not been fully recognized by the project proponents and funders – partly because the concept of "public participation" was not sufficiently mature at that time. Nevertheless, listening to the resettlers should have been given greater priority by those who designed and implemented the resettlement scheme; the lack of such efforts created difficulties both for the resettlers and for the implementing body of the projects, and increased the costs of the project. Currently, public participation is recognized as essential to planning processes, including those for dam construction; experience in these two projects is educational in terms of specific aspects of public par-ticipation, particularly in resettlement. Accordingly, the lessons gained through the post-project survey of these projects should be carefully ex-amined by those who will be planning and implementing resettlement schemes for future dam-construction projects.

Acknowledgements

The research for this chapter was partly funded by the Japan Society of Promotion of Science (JSPS) in Tokyo, Japan. The research was also funded by the Core Research for Evolutional Science and Technology (CREST) of the Japan Science and Technology Corporation (JST). The author wishes to thank these institutes for the support provided. The author is also indebted to the Institute of Ecology, Padjadjaran University, in particular Oekan S. Abdoellah PhD, Director, and the late Dr Nani Djuangsih, former Vice-Director, for their assistance.

REFERENCES

Achmad, H. 1986. "Adaptation Strategies of Saguling's People to Forced Resettlement." MA Thesis. Boston, Mass. USA: Boston University.

Bisset, R. 1987. "Methods for Environmental Impact Assessment – A Selective Survey with Case Studies." In A.K. Biswas and Qu Geping (eds). *Environmental Impact Assessment for Developing Countries*. London: Tycooly.

IOE (Institute of Ecology). 1980. *Environmental Impact Analysis of the Saguling Dam: Mitigation of Impact – Report to Perusahaan Umum Listrik Negara Jakarta Indonesia*. Bandung, Indonesia: Institute of Ecology, Padjadjaran University.

IOE (Institute of Ecology). 1984. *Environmental Impact Analysis of the Saguling Dam*. Bandung, Indonesia: Institute of Ecology, Padjadjaran University.

IOE (Institute of Ecology). 1985. *Environmental Impact Analysis of the Proposed Cirata Dam*. Bandung, Indonesia: Institute of Ecology, Padjadjaran University.

IOE (Institute of Ecology). 1992. *Socio-Economic Monitoring Impact Study of the Cirata Project*. Bandung, Indonesia: Institute of Ecology, Padjadjaran University.

Nakayama, M. 1998. "Post-Project Review on Environmental Impact Assessment Methodology Applied for Saguling Dam for Involuntary Resettlement." *International Journal of Water Resources Development*. 14(2):217–229.

Nakayama, M., B. Gunawan, T. Yoshida, and T. Asaeda. 1999. "Involuntary Resettlement Issues of Cirata Dam Project." *International Journal of Water Resources Development*. 15(4):443–458.

PLN (Perusahaan Umum Listrik Negara). 1990. *Proyek Pusat Listrik Tenaga Air Cirata*. Jakarta: Perusahaan Umum Listrik Negara (Indonesian National State Electric Company) [in Indonesian].

PLN (Perusahaan Umum Listrik Negara). 1988. *Completion Report on Saguling Hydroelectric Power Project*. Jakarta: Perusahaan Umum Listrik Negara (Indonesian National State Electric Company).

Scudder, T. 1997. "Social Impacts of Large Dams." In T. Dorcey (ed.). *Large Dams*, pp. 41–67. Gland: IUCN and Washington, DC: World Bank.

Soemarwoto, O. 1990. "Introduction." In B.A. Costa-Pierce and O. Soemarwoto (eds). *Reservoir Fisheries and Aquaculture Development for Resettlement in Indonesia*. ICLARM Technical Report 23. Manila: International Center for Living Aquatic Resources Management.

Part V
Emerging tools

18

Internet-based tools for disseminating information and promoting public participation in international watercourse management

Carl Bruch

With the dramatic proliferation of telecommunications and computer technology, decision makers, advocates, and the broader public are turning to the Internet to obtain information, share information, coordinate campaigns, and engage in policy and project-specific discourses. The Internet is changing how decisions are made and who is involved in the process. This revolution has spread to the management of international watercourses.

Internet-based tools focusing on international watercourses are undergoing rapid developments. Almost all are less than a decade old; accordingly, there remains a great deal of variability in the type, quality, and completeness of information. Within this dynamic growth, however, there is great promise for improving governance of international watercourses by enhancing transparency and public dialogue on watercourse management. This chapter surveys a variety of approaches from watercourses around the world – from the Mekong River, to the North American Great Lakes, to the Danube River. Some of the innovative approaches at the sub-national level (for example in the multi-state Chesapeake Bay) are also considered in the light of how their experiences may inform further development of Internet-based tools for international watercourses.

The first main section of this chapter (pp. 332–337) starts with an overview of the types of Internet-based tools that promote public access to information and public participation in international watercourse management. This section also briefly considers why these tools seek to utilize the Internet, as opposed to other media. The next main section (pp. 337–344) reviews the different Internet-based approaches for disseminating information regarding international watercourses. Having reviewed the various approaches, it examines elements of effective Web pages as well as some limitations. The section that starts on page 345 examines some of the Internet-based approaches that are starting to be used to facilitate public participation in watercourse management. This examination also considers some of the initial attempts to facilitate public access to administrative complaint mechanisms to redress grievances (often termed "access to justice"). The chapter closes with some thoughts on future developments of Internet-based tools to engage civil society in the management of international watercourses.

Introduction

Overview of tools

This chapter considers a wide range of Internet-based tools that have facilitated, or are poised to facilitate, public involvement in international watercourse management. The primary tool for disseminating information is the Web page. Web pages may have a variety of different emphases (for example, watercourse- or issue-specific); they may be hosted by a range of public or private institutions or individuals; and they include different types, quantities, and qualities of information. Web pages constitute the most frequently used and well developed Internet-based tools.

Although not as advanced as Web-based tools for disseminating information, the range of tools to engage the public in decision-making processes is more diverse. These tools range from announcements (frequently by e-mail or listservs – a listserv being a computer application that manages electronic mailing lists: when someone submits a query, comment, or response by e-mail to the listserv, the submission is automatically forwarded to all members of, or subscribers to, the listserv), to "virtual" discussion forums (such as chat rooms and listservs), to decision support systems DSSs, to mechanisms for soliciting and processing public comments and grievances.

Electronic mail, or e-mail, is perhaps the most commonly used Internet-based tool. At an early stage, environmental organizations took advantage of the possibilities to communicate and exchange information rapidly and affordably via e-mail. For example, the Environmental Law Alliance Worldwide (E-LAW) was established in 1989, three years before the establishment of the World Wide Web, for public-interest environmental lawyers and scientists to exchange experiences, resources, and skills (E-LAW 2003; Organization Summary 1997). Since then, e-mail has become an indispensable medium for communicating and sharing information for non-governmental organizations (NGOs), governmental institutions, and international organizations working in the field of international watercourse management.

Benefits of Internet-based tools

The benefits of transparency, public participation, and accountability in general, and in international watercourse management have been examined in detail elsewhere (Bruch 2002, 2003); however, the specific benefits of Internet-based approaches merit some attention here. The primary benefits of Internet-based tools are their speed of delivery, affordability, and ease of use.

The Internet allows people to log on to the Internet, to search for the information they require, and to download it. Meta-search engines – such as Google (http://www.google.com) – enable people to find information and relevant websites and to filter the results using Boolean logic. Internal search engines allow people to identify the resources available on a particular website. Often, Web pages provide links to related sites that may contain useful information.

Once the information has been located, people do not have to wait for days (or even for weeks) for printed material to be located and mailed; instead, data and reports can be downloaded in a matter of seconds or minutes. Often, this is free; even when there is a fee for the material, various e-business mechanisms allow users to purchase the materials using a credit card.

The Internet constitutes an affordable means of disseminating and accessing information. For institutions generating data, reports, and other information, once the material is ready for dissemination it is easy and inexpensive to post the information on an existing Web page. Because there are limited production (and no printing, handling, or shipping) costs for such an arrangement, many institutions are able to make the information available electronically to users free of charge. In some instances, a modest fee is charged for downloading electronic versions of the manuscript.

Increasingly, Internet-based tools also facilitate public participation. As Kazimierz Salewicz describes in the next chapter (19) of this volume, the ongoing emergence of Web-based decision support systems (DSSs) stands to empower decision makers, advocates, and members of the public alike in decision-making processes. In many instances, people are able to learn about potential actions through the Internet, and they can coordinate and submit their responses by e-mail, listservs, or chat rooms. As such, these Internet-based tools can allow people to participate when it is convenient for them: they can learn about an action, formulate a response, and submit the response from their office, home, or Internet café. It does not matter whether they are travelling or in another time zone (or continent). The Internet, and particularly e-mail, thus facilitates public participation with a delay of hours and without requiring everyone to meet at the same time, whether by telephone or in person. As such, Internet-based tools can allow more people to participate.

For these reasons – speed of access, affordability of dissemination and access, ease of use, and flexibility – the Internet has become a preferred mode of sharing and obtaining a wide range of information. At the same time, it has its limitations, and thus remains only one of the tools in the toolbox.

Limitations of Internet-based tools

Although Internet-based tools can be fast, accessible, affordable, easy, and flexible, they are not necessarily so for all people. Lack of access to computers and the Internet (including the lack of functional access due to the cost of Internet access) means that significant portions of most populations lack access. Internet connectivity (that is, access to the Internet) varies greatly by nation and by community, with developed nations and urban communities typically having greater access.

A 2002 survey found that, worldwide, a minority of people (1 of every 12 people in the world) have access to the Internet (Worldwatch Institute 2002). In some countries, connectivity is quite high: in nine developed nations, more than half the population used the Internet. However, the situation is different in developing nations: whereas the vast majority of people live in developing nations, only 1 in 5 Internet users lives in a developing nation (approximately 100 million people). In Africa, there are only approximately 4 million people with Internet access (or approximately one-half of 1 per cent of the continent's population). The growth in Internet connectivity globally continues to be brisk, with an annual increase of approximately 8 per cent from 2001 to 2002 (UN Wire 2004).

In addition, differences in computer and telecommunications technologies can impede access. Older computers often do not have the processor speed or memory capable of running the newer computer programs necessary to take full advantage of the Internet. Modems, common in developing countries, do not have the bandwidth capacity to transmit the data that DSL, T-1 lines, cable lines, and other fast technologies in the developing worlds can transmit. The new technologies facilitate the development and use of graphics-intensive Web pages that rely on larger programs and faster computers. However, these new pages can be inaccessible, or of limited use, to users who have technology that is even a few years old. At worst, certain Web pages may require recent versions of programs (browsers and associated programs called "plug-ins"); more often, the heavy use of graphics can make websites functionally inaccessible to older computers using slower communication technologies.

In addition to the cost, there are other functional barriers to access. Literacy is often limited in developing nations, where computer literacy is even lower. This is particularly pronounced in Africa, which has numerous transboundary watercourses and policy priorities on development that can affect the management of transboundary watercourses. For example, in Niger (one of the riparian nations of the Niger River), only 17.6 per cent of the population is functionally literate and there are only an estimated 12,000 Internet users (approximately 0.1 per cent of the population) (CIA 2003).

Language is another barrier to the effective access of information or the participation in decision-making processes through the Internet. As Ashton and Neal observe in their chapter on the Okavango River Basin (chap. 9), the Okavango Basin includes populations speaking 13 different indigenous languages and 5 "official" languages; India has 15 official languages (CIA 2003), and the Niger River Basin has many more [Nigeria alone has 505 living languages (SIL 2002), although not all of those are found in the basin]. "Official" national languages and shared trading languages (such as Swahili) can provide a common lingua franca and reduce the number of languages into which Web pages, documents, and other information need ideally be translated. However, translation can be expensive, particularly for developing nations. Even if a country can translate the information into only the official language(s), many people may not be able to understand it. This constraint is not unique to Internet-based tools – the same applies to printed material – but it remains one of the limitations to effectively engaging the public via the Internet.

Assuming that a person has functional access to the Internet, it can be difficult to find the desired information. In using search engines such as Google, random websites often showed up early in the search results, whereas the more relevant sites showed up later. Variations in spelling (particularly common in transliterated information from one written script to another) can mean that the name of the search object is spelled "wrongly" for the purposes of the search. For example, raw data for the Mekong River tend to be focused on specific stations; however, spellings in English can vary for the particular Vietnamese spelling, so the search might not yield all (or any) of the data sought.

Sorting through the search results can be difficult, with fragmentation of information a particular challenge. This is due, in part, to the multiplicity of information generators and, in part, to an apparent desire on the part of many institutions to establish themselves as the definitive source of information. Accordingly, it can be difficult for someone (especially a lay person) to sort through the different websites to figure out how the different pieces of information relate to one another.

The information itself may be limited, out of date, or of uncertain or mixed quality. Generally, raw data can be difficult to find, and the data might be available at irregular intervals, making it difficult to use to draw defensible conclusions. Few of the Web pages cited in this chapter had much data on water quantity and water quality from 2003 or late 2002. Frequently, there are gaps in the water-quality or -quantity data for particular months or years. Moreover, information may be copyrighted, limiting its use.

Availability of information can also be irregular. Servers hosting Web pages can go down, making the information inaccessible (although this usually is a short-term inconvenience). Information can also be removed from the Web page. Perhaps the most common inconvenience is reorganization of Web pages, so that the user follows a link or a search result only to receive a "404" message that the Web page cannot be accessed. (Indeed, for this reason, the references in this chapter are to the overarching website, rather than the pinpoint citation.) To a certain extent, these difficulties are minor. Search engines such as Google increasingly store Web pages in a "cache," so that even if the page is not accessible on its host site the user can access the cache. Increasing use of internal search engines also makes it possible for users to track down the information they seek on the reorganized site.

A related problem is the abandonment of Web pages, which leaves an obsolete address and out-of-date (and perhaps no longer accurate) information. These virtual "corpses" affect the efficiency of Internet searches and undermine the trust of users.

These various limitations mean that institutions should be careful not to rely too heavily on Internet-based tools. Internet connectivity is growing rapidly in developing countries, and most advocacy organizations (including many grass-roots institutions) have access. Accordingly, these organizations can act as a valuable link between those without access and those providing the information or making decisions. In this context, matters of organization representation become important (Ribot 1999). At the same time, governmental and intergovernmental institutions increasingly rely on the Internet to disseminate information, to communicate amongst themselves, and to communicate with outside organizations and individuals.

Following terrorist attacks on civilian infrastructure, national debates have arisen in many countries seeking to determine which information should be made publicly available, to whom, and in what manner. Security advocates argue that making such information easily accessible aids potential terrorists in selecting targets that would have the greatest impact; however, environmental and governance advocates tend to argue that the benefits of public review can reduce vulnerabilities and improve environmental performance (Echeverria and Kaplan 2002; OMB Watch 2002; Baker et al. 2004).

Internet-based tools are not (and should not be considered to be) the only tool – or even the best tool – for involving the public in international watercourse management. They represent one class of tool for disseminating information and engaging the public – but only one class. Internet-based tools can be fast, affordable, flexible, and easy to use. The limitations discussed above simply mean that other tools cannot be ignored, and that Internet-based tools (as noted above) remain one of the set of tools in the toolbox, not the only one. For that matter, anecdotal experiences suggest that the Internet can be an incredibly powerful tool (Organization Summary 1997).

Disseminating information through the Internet

The most common way to disseminate information regarding international watercourses through the Internet is by using Web pages. There is a great range of types of Web pages hosted by different individuals and institutions. This section highlights some of the different types of information that are made available regarding aspects of transboundary watercourse management, concluding with some observations on what makes an effective Web page.

Types and hosts of Web pages

A wide range of websites disseminate information on international watercourses. As shown in table 18.1, these include the following:
- watercourse-specific websites (for example for the Mekong River, the North American Great Lakes, and other international watercourses);
- issue-specific websites (for example, addressing the topics of dams or drinking-water) that canvas many watercourses;
- Websites addressing a specific issue in a specific watercourse, often for campaigning purposes;
- general websites that include relevant information.

Table 18.1 Selected websites addressing international watercourse management

Type of site	Subject and website(s)
Watercourse specific	Danube River: ⟨http://www.icpdr.org/pls/danubis/danubis_db. dyn_navigator.show⟩ Mekong River: ⟨http://www.mrecmekong.org⟩, ⟨http://www.mekonginfo.org⟩ Nile River: ⟨http://www.nilebasin.org⟩ North American Great Lakes: ⟨http://www.great-lakes.net⟩, ⟨http://www.glerl.noaa.gov/data/⟩ Rio Grande: ⟨http://www.ibwc.state.gov/wad/rio_grande.htm⟩ Ganges River: ⟨http://www.thewaterpage.com/ganges.htm⟩ Aral Sea: ⟨http://www.thewaterpage.com/aral.htm⟩, ⟨http://www.ce.utexas.edu/prof/mckinney/papers/aral/ aralhome.html⟩ Africa: ⟨http://www.africanwater.org/⟩ Latin America and the Caribbean: ⟨http://www.cathalac.org/⟩
Issue specific	Dams: ⟨http://www.dams.org⟩, ⟨http://www.unep-dams.org/⟩ Drinking-water: ⟨http://www.nesc.wvu.edu/ndwc/ndwc_index. htm⟩, ⟨http://www.cyber-nook.com/water/⟩ Irrigation: ⟨http://www.wiz.uni-kassel.de/kww/projekte/irrig/ irrig_i.html⟩
Addressing a specific issue in a specific watercourse	Three Gorges Dam and the Yangtze River: ⟨http://www. thewaterpage.com/yangtze.htm⟩ Lesotho Highlands Water Project: ⟨http://www.sametsi.com/⟩
General	Environmental: ⟨http://www.unep.net⟩, ⟨http://www. earthportal.org⟩ Environmental monitoring: ⟨http://www.gemswater.org/⟩ Environmental law: ⟨http://www.transboundarywaters.orst.edu/ publications/atlas/⟩ Water (many of which have watercourse-specific information): ⟨http://www.iwlearn.org/⟩, ⟨http://www.wrds.uwyo.edu/wrds/ wwwsites.html⟩, ⟨http://www.thewaterpage.com/⟩, ⟨http:// www.iwra.siu.edu/⟩, ⟨http://www.worldwater.org/links.htm⟩, ⟨http://www.waterweb.org/⟩, ⟨http://www.webdirectory.com/ Water_Resources/⟩

Table 18.1 enumerates many of the websites referred to in this chapter. Because of the volume of relevant sites, however, this is necessarily a selective list.

An equally wide range of institutions and individuals host the websites. These include:

- watercourse authorities, such as river basin organizations (RBOs);
- governments, either in their capacity as members of a particular watercourse authority or as a service to their citizens;

- research institutions, such as universities and other academic centres;
- NGOs and private individuals;
- international institutions working in the field, such as multilateral development banks;
- private businesses.

Although most of these hosts emphasize information dissemination and offer the information free of charge, some private businesses, as well as research institutions and NGOs, offer raw data, reports, and other information at a price.

Data on the status of international watercourses

An increasing number of websites offer raw data on international watercourses. This include information on the water level, flow, and precipitation; water quality; and biological aspects, such as invasive alien species. As discussed below, the practice and methodologies for making such information available are still evolving; accordingly, the availability of such data varies from watercourse to watercourse, and there frequently are gaps in the available data. Nevertheless, organizations increasingly are collecting data on the status of international watercourses and posting that information as a matter of practice on the Internet.

Information on water level, flow, and precipitation is available for many watercourses. For example, it is possible to obtain information on discharge, precipitation, flood-extent maps, and flood-depth change for the Mekong River (MRC 2003). Some isolated geophysical information is available for the Rhine River (Hachenberg n.d.). Similarly, websites provide data on water level, outflow, and precipitation for many points along the North American Great Lakes (ACE 2003; GLERL 2003; GLIN 2003). The Rio Grande has daily and historical (dating back more than 70 years) reports on flow conditions at various points along the river, as well as storage conditions (IBWC 2003b). The Colorado River also has reports on average daily flow and historical flows dating back 40 years (IBWC 2003a).

To a lesser extent, websites also provide information on water quality and factors affecting water quality of international watercourses. For example, the US Geologic Survey (USGS) provides data on the temperature, dissolved oxygen content, pH, and specific conductance for waters throughout the United States, including parts of the North American Great Lakes (USGS 2003). An emissions inventory is available for the Danube River, viewable by river basin or country, by pollutant, and in user-selected units (ICPDR 2003). Along the Rio Grande, it was previously possible to obtain data on pesticide levels in the river (http://wq.water.usgs.gov/ccpt/pns_data/data.html), although these data have since been moved or removed. Nevertheless, some relevant reports on pesticides in the Rio Grande can be obtained (Levings et al. 1998).

The newest, and least developed, realm of data to be made available is that regarding biological aspects of international watercourses. For example, the (North American) Great Lakes Environmental Research Laboratory posts data on the status and spread of invasive alien species (GLERL 2003).

Although most laypersons might not be inclined (or have the technical capacity) to fully process or interpret the raw data made available through these websites, the raw data are particularly useful for governmental, academic, business, and advocacy-oriented individuals engaged in research. The data allow such researchers to identify opportunities for development, to identify emerging or existing problems, and to consider possible solutions. Moreover, this data availability enables modellers and scientists to develop, improve, and validate models. Making the data available over the Internet can facilitate such analyses. As such, there appear to be at least three priorities, namely:

- developing consistent formats for presenting the data so that they can be processed and analysed by programs either on the site (a computationally intensive endeavour) or by user-selected and structured programs;
- enhancing the completeness of existing data; and
- extending approaches for collecting and posting data to the Internet in real time to other watercourses.

The main section that starts on page 345 expands upon these priorities and places them in the broader context of improving public involvement in international watercourse management.

Information on projects and activities

A wide range of projects affect international watercourses: these include dams, irrigation schemes, diversions, power plants that use river water for cooling or steam (and power) generation, industrial facilities that discharge their waste into the river, and so on. Institutions involved in managing international watercourses increasingly use the Internet to disseminate information regarding proposed and existing projects and activities that could affect international watercourses.

As a matter of national law and/or the policies of international financial institutions providing project support, most proposed projects that could affect international watercourses must first conduct an environmental impact assessment (EIA) of the project (Bruch 2002). Usually, draft EIAs must be subject to public review, with an opportunity for members of the public (particularly those who could be affected by the project) to comment on the analysis in the EIA. Thus far, few EIAs have been posted on the Internet, although institutions such as the World Bank often use their websites to indicate how an interested person may review or obtain a copy of a particular EIA (World Bank 2003a). The InfoShop website of the World Bank makes the EIAs freely available for download, while the hard-copy versions are available for a fee (World Bank 2003b). The Bank also posts summaries of EIAs on its website (IFC 2003). More often, organizations post analytical or advocacy information, reporting on or arguing for or against a particular proposed project. Such documents are common and typically include news releases, articles, updates, and campaign materials.

Websites addressing ongoing projects similarly tend to lack critical information. This may entail significant gaps (such as an appraisal of the environmental impacts) or more administrative data, such as project start and end dates. To the extent that there is ongoing monitoring of activities, such as monitoring and reporting on emissions, such data may be made available at the water basin level (as in the case of the Danube River; ICPDR 2003) or at the national (domestic) level through pollutant release and transfer registers, where available, which track emissions from specific facilities, including those along international rivers (RTKNet 2003; USEPA 2003b; Kemp 2004).

Analytical reports

Many websites make available reports on water quantity, water quality, and factors affecting international watercourses. Sometimes these analyses are accessible free of charge, and sometimes there is a fee. In most cases, however, the underlying raw data are not available on the website.

Information on institutions and instruments governing watercourses

For many international watercourses, websites provide information on watercourse governance. This typically includes information on the relevant governing documents (treaties, agreements, policies, and guidance) and operational information (such as announcements regarding the time and place of forthcoming meetings, as well as Minutes of past meetings).

For the Rio Grande, the website of the Border Environment Cooperation Commission (BECC) (which funds projects along the Rio Grande), includes treaties, rules of procedure, information on the organization and its management, annual reports, a virtual library of certified project documents, and meeting information and documents (BECC 2003a). The International Boundary and Water Commission (IBWC) website, also relating to the United States–Mexico border, provides text of the relevant governing treaties dating to the 1800s, organizational information, and procedures for resolving boundary and water disputes (IBWC 2003c). Also in North America, the website for the International Joint Commission (IJC) has extensive information on the relevant law, procedures, rules of procedure, institutions and their meetings, and mechanisms for members of the public to participate (IJC 2003). Along the Danube River, the International Commission for the Protection of the Danube River (ICPDR) website provides information on the permanent secretariat, and expert groups; annual reports; conventions and relevant domestic legislation; and various action plans, research programmes, and task forces (ICPDR 2003). The Regional Environment Center (REC) website also provides information on the legal and institutional structures governing the Danube (REC 2003). The Nile Basin Initiative has information on its organizational structure, meetings, policy guidelines, programmes, and projects on its website (NBI 2003). The Mekong River Commission (MRC) has a particularly detailed site, with information on the relevant legal instruments (treaties, plans, and programmes, as well as the MRC public participation policy) and institutional information (MRC 2003). In addition to over 2,000 documents, MekongInfo posts guidelines, information on meetings, contact information, and news (MekongInfo 2003).

As with other websites discussed above, the functional accessibility and ease of use of these sites varies. For example, the BECC and IBWC sites are easy to navigate because they are logically organized and laid out, whereas information on some other sites is more difficult to find.

Campaigns and public education

As an increasing number of people rely on the Internet to answer their questions, a wide range of governmental, intergovernmental, nongovernmental, and private institutions are using the Internet to educate the public. This may be towards the end of conducting a campaign on a particular issue, ongoing activity, or proposed action (GLU 2003; IRN 2003), or it may be more broadly to educate people about water issues. For example, the US Environmental Protection Agency's "Surf Your Watershed" site provides environmental information to the public on watersheds, and is one of the most frequently accessed EPA sites (USEPA 2003a).

These sites tend to be written in a plain language that is designed to be accessible by a wide range of people, of diverse ages from diverse backgrounds. In addition to providing briefings on particular issues, the sites often include reference documents for those seeking additional information, means of contacting the organization to ask questions by e-mail (and sometimes by telephone), and links to other relevant websites.

Other information

In addition to the specialized websites that focus on watercourses or related issues, such as dams, there are a number of general sites that have relevant information. For example, if someone is interested in information on the environmental laws in a particular watershed, ECOLEX would be a natural starting point (ECOLEX 2003). In addition to national environmental legislation from around the world, ECOLEX has online databases on multilateral treaties, international soft law and related documents, instruments of the European Union, court decisions, and law and policy analysis. Although the databases are not restricted to international watercourses, or even to water, ECOLEX has a broad range of relevant materials. IUCN – The World Conservation Union has launched a website to collect judicial decisions on environmental law, including those affecting watercourses (IUCN 2002).

There are also numerous sites on various aspects of general water resource management that may be relevant in different contexts. In addition to those listed in table 18.1, above, a few of these include:

- the Global Environment Monitoring System, which includes inter alia a global water-quality database and freshwater assessments (GEMS 2003);
- the Global Water Partnership Toolbox for Integrated Water Resources Management, which seeks to exchange experiences in IWRM (GWP 2003);
- the International Water and Sanitation Centre (IRC), which provides information on affordable water supply and sanitation (IRC 2003);
- the IUCN Web page on "Wetlands and Water Resources" (IUCN 2003).

There are many other relevant websites. This sampling is simply intended to illustrate some of the relevant information that is available but does not especially relate to international watercourses.

Elements of effective Web pages

There is a rich body of academic literature setting forth criteria that define the effectiveness of Web pages (Abrams 1999; Anderson 1999; AWRA 1999; Fernández-Jáuregui 1999; Halverson and Burton-Radzely 1999). Typical criteria include the following:

- ease of use;
- clear understanding of the site's ultimate purpose, goal, or message;
- artistic style, colour, typography, and other visual aspects, so that the site has a consistent look and feel;
- orientation [which may be linear, hierarchical (horizontal and vertical), or non-hierarchical];
- ease of navigation;
- cross-browser functionality.

In addition to these established criteria, while conducting research for this chapter, as a practical matter it became clear that some Web pages were more effective than others in organizing and conveying information.

The remainder of this subsection examines some of these lessons, which draw upon the numerous searches and surveys of different websites.

Internal search engines are essential. It helps to have drop-down menus and key words, but being able to enter a name, word, or other search term can greatly facilitate finding the desired information, as well as making such a search faster. Ranking the search results by relevance is also useful. Such search mechanisms are becoming increasingly important as websites are reorganized and renamed, and old bookmarks and references become outdated. The information is still on the site, but finding it without a search engine can be challenging.

Formatting affects the ease (or difficulty) of finding information. For example, use of appropriate programs and formats can mean that text comes through clearly on many browsers and does not overlap with other text or extend beyond the boundaries of the screen. Within a particular Web page, bullets and internal links can make it easy to find information. With hyperlinked bullets, it is easier to move through the website quickly, rather than sorting through large amounts of information on each page.

As there often are many related websites, it can be challenging to sort through the different providers and the information. Accordingly, websites that provide numerous links and place them in the overall context are particularly useful. Nevertheless, such an endeavour has its costs, as it requires frequent updating to ensure that the links are still live.

Even with search engines, internal and external links, and well-structured pages, it can sometimes be difficult to find specific information. In such instances, opportunities to obtain rapid responses to questions can be helpful. Most websites, especially for watercourse institutions, provide the opportunity to submit comments and questions by e-mail; however, in practice, the response times vary. A few websites [such as that for the IBWC (US section)] posted toll-free telephone numbers for helpdesks that people could call to obtain assistance with the Web page, or more generally.

In many countries, functional access to the Internet is constrained by the telecommunications capacity. In order to be accessible, then, it is necessary to be strategic about using graphics that can take a long time to load, by limiting their number and size. For those graphics that are necessary but large, it may be possible to put them on subsidiary Web pages that carry a warning that they may take some time to load.

Finally, websites that took into account the diversity of languages in a particular basin promised to be more accessible and thus more utilized. In many instances, where there were multiple languages in a watershed, English was the lingua franca; however, when there were only two languages, for example along the United States–Mexico border, the website tended to be posted in both of the major languages.

Internet-based tools for engaging the public in decision-making

Whereas information dissemination via the Internet has developed rapidly and dynamically, Internet-based tools for engaging the public have emerged more slowly. Although there are a number of these tools and applications, the number of watersheds that have applied them is still relatively limited in comparison to Internet use for information dissemination.

Announcements

Many websites relating to different aspects of international watercourse management post announcements of upcoming meetings or events so that interested persons can participate. Increasingly, NGOs are also using websites, e-mail, listservs, and chat rooms to announce upcoming events or decisions to be made with the aim of educating and engaging people in the decision-making process. This might be in the form of a letter or e-mail campaign, bringing people to a meeting, or another form of participation. In the context of the North American Great Lakes, Great Lakes United has used such announcements effectively and, at the domestic level, the Chesapeake Bay Foundation and the Sierra Club frequently announce and coordinate campaigns through Internet-based tools. E-mails have also been used to announce Web-page developments – for example, to the African Water Page (Abrams 1999).

Soliciting public comment

In a similar manner, the Internet can be used to solicit public comments. For example, the World Bank website announces how interested people can obtain a copy of an EIA for a proposed project and provides opportunities for public comment (World Bank 2003b). In a broader policy context, the Third World Water Forum (3WWF), held in March 2003 in Japan (the "Water Voice" Project and Virtual Water Forum) sought to engage people at the grass-roots level – including many who could not attend the actual event – on a wide range of water issues (3WWF 2003).

Decision support systems (DSSs)

As Kazimierz Salewicz explains in the next chapter (chap. 19) of this volume, Internet-based DSS constitutes an important emerging tool in building capacity for the decision makers and the affected public alike to engage in an informed dialogue on proposed policy or project decisions. In particular, the Internet offers an opportunity to popularize the use of DSS, which had once been reserved primarily for technical experts.

Listservs and chat rooms

Listservs and chat rooms are two means for engaging a wide range of people, particularly those working in disparate locations (Oostenbrink 2003). Listservs are similar to e-mail lists: when one person posts a message (usually by e-mail) to the listserv, it is sent to all the recipients on the listserv. Usually, only members of the listserv can post messages. Some listservs are "threaded", so that it is possible to visit a website where all the messages are archived and ordered by topic and responses. Listservs have the benefits of a targeted audience – namely, those interested in a particular topic – and temporal flexibility. Because the system basically relies on e-mail, people can respond when they are available – unlike with a chat room, they do not need to participate at the same time. This facilitates involvement of people who have other commitments – for example, those participating in their own time (after working hours). BECCnet has been a dynamic listserv for projects along the United States–Mexico Border (BECC 2003b). For example,

[BECCnet had] influenced decisionmaking about a half-dozen times [by early 1998]. When the [BECC] commission failed to adhere to self-imposed guidelines for a forthcoming meeting, for instance, e-mail protests were so numerous that the directors rescheduled the meeting. Similarly, at another meeting attended by about 200 people, the chairman gavelled the proceedings closed before allowing public comment; the cascade of protests on BECCnet led to a public apology and a binding modification of procedures for such comment. (Milich and Varady 1998)

These examples illustrate the ability of a listserv to channel public comment constructively and to engage the public in decision-making processes.

In contrast, a chat room allows for a simultaneous discussion by many people at once, even if the people are in many locations. This cuts down on travel expenses, and can be easier and more affordable than telephone conference calls. Although chat rooms are common for personal interests and hobbies, they have yet to be embraced fully in the professional context of watercourse management. However, one leading example is RioWeb's "virtual meeting rooms" to discuss matters relating to the Rio Grande and Rio Bravo Basin (Rio Grande/Rio Bravo Basin Coalition 2003).

Complaint mechanisms

A growing number of international institutions provide formal avenues for individuals to seek redress for harm (Bruch 2002). As Charles Di

Leva describes in chapter 10 of this volume, the World Bank Inspection Panel and a similar body for the Inter-American Development Bank have both considered applications alleging that projects on transboundary watercourses have not complied with the necessary institutional policies. Similarly, the North American Commission for Environmental Cooperation (NACEC) has received submissions alleging harm to international waters resulting from the failure to enforce environmental laws, described by Geoffrey Garver in this volume. Similarly, the International Court of Justice, in its Gabčíkovo–Nagymaros decision (analysed by Ruth Greenspan Bell and Libor Jansky in this volume) addressed the environmental implications of a proposed dam along the Danube.

These institutions have emerged gradually, and they continue to evolve. Internet-based tools increasingly bear on the operations of these institutions. Most have websites that provide information on how to file a complaint or make a submission, as well as a docket of pending and completed applications that frequently includes certain documents, such as the complaint and the final decision. Moreover, some of the institutions (such as the World Bank Inspection Panel) allow submissions by e-mail (Kuo 2004).

Evolution of Internet-based tools

The use of Internet-based tools to engage the public actively in decision-making processes is still evolving. Although a number of such tools have emerged, most of them tend to be in developed nations where a greater segment of the public at large has access to the Internet and uses the Internet to obtain information and to communicate with others. As discussed in the next section, there are many ways in which the lessons learned so far may be expected to be extended to other watercourses, particularly as Internet connectivity grows.

Future developments

The technological leaps made in computer and telecommunication technologies over the last few decades have revolutionized how people generate, disseminate, access, and use information. With such alterations have come changes in how decisions are made and who participates in the decision-making process. Considering how fast the changes have occurred to date, and their dramatic impact, it is hazardous to guess what types of Internet-based tools may emerge in the long run or how existing tools may evolve. In the next few years, though, certain developments are foreseeable – indeed, necessary.

More data on international watercourses and the factors affecting them are needed. Efforts to monitor and assess environmental conditions, emissions of effluent, and other related parameters should be structured to anticipate posting the data on the Internet in a user-friendly manner. Such data are already available for certain watercourses – largely in developed countries – and additional efforts should be made to support extension of such efforts in developing countries. This would entail financial and technical assistance from bilateral and multilateral sources, as discussed below.

The problem of generation and use of the data raises the matter of format. Different programs require varying formats for data inputs. If the raw data are to be made available, and not simply manipulated through database programs on the website, it may be advantageous for the leading data generators (namely researchers) to develop an informal but widely used methodological protocol for posting such data on the Internet.

At the same time, as more information becomes available on the Internet – whether raw data, reports, or legal or institutional documentation – it becomes increasingly important to organize the information. Some users may be sufficiently sophisticated to discern which sites provide rigorous and defensible information, but many are not. As Peter Steiner's prescient 1993 New Yorker cartoon reads, "On the Internet, nobody knows you're a dog." To some extent, organization and consolidation of information on international watercourses is already taking place. Nevertheless, there remains a compelling need for a limited number of meta-sites that do not necessarily generate information, but organize the vast body of existing relevant information.

There are many ways in which the Internet can be better used to improve public participation in decision-making regarding international watercourses. Although various Internet-based tools have been developed and tested, they have yet to be applied in many international watercourses. Some of the constraints have been related to access: many people still lack affordable access to the Internet; their low-bandwidth connection limits use of chat rooms or of other tools, or both. There are also unresolved legal questions – for instance, in filing formal complaints to compliance mechanisms, regarding electronic signatures on legal documents (Keiner 1999).

Nevertheless, these constraints are diminishing. Internet access and use has been growing rapidly, particularly in developing nations. As more people become familiar with – and come to rely on – the Internet, the more advanced uses become more feasible. At the same time, international, governmental, non-governmental, and academic institutions continue to expand opportunities for the public to participate in decision-making.

One of the areas that is likely to see growth in the coming years is the development and expansion of Internet-based DSSs (Great Lakes Commission 2003). The Web-based DSS for the Ganges River that Salewicz and colleagues are developing is at the cutting edge of such endeavours. As computational and communications technology increases, other approaches to Web-based DSS may be feasible. For example, the user may be able to define more parameters and run more scenarios. Moreover, Internet-based DSS needs to be expanded to other watercourses in different regions so that more people – decision makers and the public alike – become familiar with the potential of DSS to enhance informed decision-making.

Tools designed both to educate and engage need to account for linguistic differences. Translation can be expensive, but more translation is necessary. Increasingly, commercial translation is computer-assisted. Opportunities to utilize similar technologies for translating Web pages related to international watercourses should be explored. Currently, this might be more feasible for languages that are more widely used (such as the UN official languages), but many of the local languages in watersheds lack reliable computer-assisted translation programs, with few prospects of developing such capacity in the near future.

Ultimately, the further development, extension, and translation of various Internet-based tools will depend to a large extent on commitments of financial and technical assistance from international institutions, bilateral aid agencies, and charitable organizations. For example, the World Bank and the Global Environment Facility, which have placed a priority on improving water management, should examine ways to support the development and extension of Internet-based approaches for improving public involvement in international watercourse management. Such assistance need not be only for large projects: individuals or small groups of researchers can do a lot to introduce and extend the application of such tools to new water basins, and many small grants – perhaps administered through a small-grants project – could go a long way to improving the availability and use of Internet-based tools in many watersheds.

Conclusions

The rapid development and expansion, as well as widespread use, of various Internet-based tools attest to the critical role of such tools in involving a broad cross-section of the public in international watercourse management. Internet-based tools can be rapid, affordable, easy to use, flexible, and available 24 hours a day. Although this is not always the case, many of the limitations of Internet-based tools are being addressed.

Optimism about the role and potential of Internet-based tools does not mean, however, that other tools are outdated or unnecessary. Print dissemination of information remains valuable, especially for long-term recording. In-person consultations similarly have their unique values that cannot be fully accounted for by Internet technologies. Nevertheless, used appropriately, the Internet can disseminate information rapidly to a wide audience around a watershed or the world, and it can do so economically. Similarly, the Internet can help to engage people in decision-making processes who would not otherwise be able to participate – for instance, because they cannot afford to travel to the relevant meetings. Accordingly, the Internet offers a class of tools that constitutes "one of the tools in the toolbox" – a powerful tool, but not the only tool.

Acknowledgements

The author is grateful to the Carnegie Corporation of New York for financial assistance that made this research possible. Amber Benjamin and Steven Krieger assisted in background research, and Kazimierz Salewicz graciously reviewed and provided generous feedback on the text.

REFERENCES

Abrams, Len. 1999. "The African Water Page: An Experiment in Knowledge Transfer." Water International 24(2):140–146. Internet: ⟨http://www.iwra.siu.edu/win/pdf_file/abrams.pdf⟩ (visited 10 November 2003).

ACE (US Army Corps of Engineers). 2003. "Great Lakes Update." Internet: ⟨http://www.lre.usace.army.mil/⟩ (visited 10 November 2003).

Anderson, Faye. 1999. "The Challenge of Leveraging the Internet for a Sustainable Water Management Agenda: Enabling Global Cooperation and Local Initiatives." *Water International* 24(2):126–139. Internet: ⟨http://www.iwra.siu.edu/win/pdf_file/anderson.pdf⟩ (visited 10 November 2003).

AWRA (American Water Resources Association). 1999. "Conference Report: Water on the Web Workshop: Report of Recommendations, Fort Lauderdale, Florida, USA, October 23–24, 1998." *Water International* 24(2):172–175. Internet: ⟨http://www.iwra.siu.edu/win/pdf_file/conferencereport.pdf⟩ (visited 10 November 2003).

Baker, John C., Beth E. Lachman, David R. Frelinger, Kevin M. O'Connell, Alexander C. Hou, Michael S. Tseng, David Orletsky, and Charles Yost. 2004. *Mapping the Risks: Assessing the Homeland Security Implications of Publicly*

Available Geospatial Information. Santa Monica, CA: RAND Corporation. Internet: ⟨http://www.rand.org/publications/MG/MG142/MG142.pdf⟩ (visited 14 April 2004).

BECC (Border Environment Cooperation Commission). 2003a. Internet: ⟨http://www.cocef.org/⟩ (visited 10 November 2003).

BECC (Border Environment Cooperation Commission). 2003b. "BECCNET: Growing Source for Information." Internet: ⟨http://www.cocef.org/apartcom/beccneti.htm⟩ (visited 10 November 2003).

Bruch, Carl. (ed.). 2002. *The New "Public": The Globalization of Public Participation*. Washington, DC: Environmental Law Institute.

Bruch, Carl E. 2003. "Role of Public Participation and Access to Information in the Management of Transboundary Watercourses." In Mikiyasu Nakayama (ed.). *International Waters in Southern Africa*. Tokyo: United Nations University Press, 38–70.

CIA (US Central Intelligence Agency). 2003. *The World Factbook 2003*. Internet: ⟨http://www.cia.gov/cia/publications/factbook/⟩ (visited 10 November 2003).

Echeverria, John D., and Julie B. Kaplan. 2002. "Poisonous Procedural 'Reform': In Defense of Environmental Right to Know." Internet: ⟨http://www.law.georgetown.edu/gelpi/papers/poisonpaper.pdf⟩ (visited 14 April 2004).

ECOLEX. 2003. "ECOLEX: A Gateway to Environmental Law." Internet: ⟨http://www.ecolex.org⟩ (visited 10 November 2003).

E-LAW (Environmental Law Alliance Worldwide). 2003. "The E-LAW Network." Internet: ⟨http://www.elaw.org⟩ (visited 10 November 2003).

Fernández-Jáuregui, Carlos A. 1999. Hydrology and Water Resources on the Web in Latin America and the Caribbean." *Water International* 24(2):157–159. Internet: ⟨http://www.iwra.siu.edu/win/pdf_file/fernandez.pdf⟩ (visited 10 November 2003).

GEMS/Water (Global Environment Monitoring System). 2003. "The World of Water Quality." Internet: ⟨http://www.gemswater.org/⟩ (visited 10 November 2003).

GLERL (Great Lakes Environmental Research Laboratory). 2003. "GLERL Data." Internet: ⟨http://www.glerl.noaa.gov/data/⟩ (visited 10 November 2003).

GLIN (Great Lakes Information Network). 2003. "Great Lakes Level and Hydrology" Internet: ⟨http://www.great-lakes.net⟩ (visited 10 November 2003).

GLU (Great Lakes United). 2003. Internet: ⟨http://www.glu.org⟩ (visited 10 November 2003).

Great Lakes Commission. 2003. "Implementing Good Governance: Toward a Water Resources Management Decision Support System for the Great Lakes–St Lawrence River Basin." *Water Resources IMPACT* 5(4):13–16. Internet: ⟨http://www.awra.org/impact/0307impact.html⟩ (visited 10 November 2003).

GWP (Global Water Partnership). 2003. "GWP Toolbox for Integrated Water Resources Management." Internet: ⟨http://www.gwpforum.org/⟩ (visited 10 November 2003).

Hachenberg, Freider. n.d. "Rhine-River-Project-Our-Data." Internet: ⟨http://www.fh-koblenz.de/koblenz/remstecken/rhine/projektschulen/vergleich/frieder.html⟩ (visited 10 November 2003).

Halverson, Lynn, and Lisa Burton-Radzely. 1999. "Developing Consumer-

Friendly Water Websites." Internet: ⟨http://www.awra.org/proceedings/www 99/w08/index.htm⟩ (visited 10 November 2003).

IBWC (International Boundary and Water Commission). 2003a. "Colorado River." Internet: ⟨http://www.ibwc.state.gov/wad/colorado_river.htm⟩ (visited 10 November 2003).

IBWC (International Boundary and Water Commission). 2003b. "Rio Grande Water Flows." Internet: ⟨http://www.ibwc.state.gov/wad/rio_grande.htm⟩ (visited 10 November 2003).

IBWC (International Boundary and Water Commission). 2003c. Internet: ⟨http://www.ibwc.state.gov/⟩ (visited 10 November 2003).

ICPDR (International Commission for the Protection of the Danube River). 2003. "Home." Internet: ⟨http://www.icpdr.org/pls/danubis/danubis_db.dyn_navigator.show⟩ (visited 10 November 2003).

IFC (International Finance Corporation). 2003. "Project Documents Search." Internet: ⟨http://ifcln001.worldbank.org/IFCExt/spiwebsite1.nsf/d2354e2fa3c3610385256a5b006f1829?OpenForm⟩ (visited 10 November 2003).

IJC (International Joint Commission). 2003. Internet: ⟨http://www.ijc.org⟩ (visited 10 November 2003).

IRN (International Rivers Network). 2003. Internet: ⟨http://www.irn.org⟩ (visited 10 November 2003).

IRC (International Water and Sanitation Centre). 2003. Internet: ⟨http://www.irc.nl/⟩ (visited 10 November 2003).

IUCN – The World Conservation Union. 2002. "'Judicial Portal' Fact Sheet." Internet: ⟨http://www.iucn.org/themes/law/pdfdocuments/Judicial%20Portal FactSheet.pdf⟩ (visited 10 November 2003).

IUCN – The World Conservation Union. 2003. "Wetlands and Water Resources." Internet: ⟨http://www.iucn.org/themes/wetlands/⟩ (visited 10 November 2003).

Keiner, Suellen T. 1999. *From Pens to Bytes: Summaries of Court Decisions Related to Electronic Reporting*. Washington, DC: Environmental Law Institute.

Kemp, David. 2004. "Media Release: National Pollutant Inventory – Helping Cut Emissions." 30 January. Internet: ⟨http://www.deh.gov.au/minister/env/2004/mr30jan04.html⟩ (visited 14 April 2004).

Kuo, Jennifer. 2004. "Online Ombudsman." *Foreign Policy*, 93, January/February.

Levings, Gary W., Denis F. Healy, Steven F. Richey, and Lisa F. Carter. 1998. "Water Quality in the Rio Grande Valley, Colorado, New Mexico, and Texas, 1992–1995." *US Geological Survey Circular 1162*. Internet: ⟨http://water.usgs.gov/pubs/circ/circ1162/nawqa91.5.html⟩ (visited 10 November 2003).

MekongInfo. 2003. Internet: ⟨http://www.mekonginfo.org⟩ (visited 10 November 2003).

Milich, Lenard, and Robert G. Varady. 1998. "Managing Transboundary Resources: Lessons from River-Basin Accords." *Environment* 40:10.

MRC (Mekong River Commission). 2003. Internet: ⟨http://www.mrcmekong.org/⟩ (visited 10 November 2003).

NBI (Nile Basin Initiative). 2003. Internet: ⟨http://www.nilebasin.org⟩ (visited 10 November 2003).

OMB Watch. 2002. "Access to Government Information Post September 11th." Internet: ⟨http://www.ombwatch.org/article/articleview/213/1/104/⟩ (visited 14 April 2004).

Oostenbrink, Willem Tjebbe. 2003. "Electronic Tools Charge Activists." *The Bulletin* [of the Regional Environment Center for Central and Eastern Europe]. Autumn.

"Organization Summary: Environmental Law Alliance Worldwide (E-LAW)." 1997. *Colorado Journal of International Environmental Law and Policy* 8(1):129–136.

REC (Regional Environment Center) for Central and Eastern Europe. 2003. "The Danube River Protection Convention (DRPC)." Internet: ⟨http://www.rec.org/DanubePCU/drpc.html⟩ (visited 10 November 2003).

Ribot, J.C. 1999. "Decentralisation, Participation and Accountability in Sahelian Forestry: Legal Instruments of Political–Administrative Control." *Africa* 69(1):23–65.

Rio Grande/Rio Bravo Basin Coalition. 2003. "About RioWeb's Virtual Meeting Room." Internet: ⟨http://www.rioweb.org/messages.html⟩ (visited 10 November 2003).

RTKNet (Right-to-Know Network). 2003. "TRI Search." Internet: ⟨http://d1.rtknet.org/tri/⟩ (visited 10 November 2003).

SIL (Summer Institute of Linguistics). 2002. *Ethnologue: Languages of the World*, 14th edn. Dallas: Summer Institute of Linguistics. Internet: ⟨http://www.ethnologue.com⟩ (visited 10 November 2003).

UN Wire. 2004. "Internet Growth, Tourism Boosted Globalization in 2002." 25 February.

USEPA (US Environmental Protection Agency). 2003a. "Surf Your Watershed." Internet: ⟨http://www.epa.gov/surf/⟩ (visited 10 November 2003).

USEPA (US Environmental Protection Agency). 2003b. "Toxic Release Inventory (TRI) Program." Internet: ⟨http://www.epa.gov/tri/⟩ (visited 10 November 2003).

USGS (US Geological Survey). 2003. "Real-Time Data for the Nation." Internet: ⟨http://waterdata.usgs.gov/nwis/rt⟩ (visited 10 November 2003).

World Bank. 2003a. Internet: ⟨http://web.worldbank.org⟩, ⟨http://www.worldbank.org⟩ (visited 10 November 2003).

World Bank. 2003b. "The InfoShop." Internet: ⟨http://www.worldbank.org/html/pic/PIC.html⟩, ⟨http://www-wds.worldbank.org/⟩ (visited 10 November 2003).

Worldwatch Institute. 2002. *Vital Signs 2002*. New York: W.W. Norton & Company.

3WWF (Third World Water Forum). 2003. "Water Voice." Internet: ⟨http://ap.world.water-forum3.com/voice/en/⟩ (visited 10 November 2003).

19

Capabilities and limitations of decision support systems in facilitating access to information

Kazimierz A. Salewicz

Introduction

We make decisions all the time. The decisions range in difficulty from the very simple to the very complex, and in scope from the very narrow to the very broad. Simple decisions are made without much consideration of the factors affecting and affected by the decision. We normally give more complex decisions much more thought and consider more of the factors involved. Depending on the complexity and scope involved, the thought given may be a brief mental comparison of alternatives, or it may be a thorough analysis appropriate to a complex situation in which there are significant differences in the impacts of various factors considered and in impacts of various alternative courses of action.

Undoubtedly, decision-making processes associated with the utilization of natural resources, including water resource management, fall into the category of complex situations requiring thorough consideration and analysis. This complexity manifests itself not only through the sophistication of physical and chemical phenomena taking place in water resources systems, but primarily through rich and multidimensional interactions between various types of more or less thought-out human activities, their influence on natural systems, and consequent impacts resulting from the responses of these natural systems back to the human world. However, it is not the intention of this chapter to analyse the complexity of interactions between human activities and natural systems.

This chapter seeks to review the basic concepts and notions underlying development of decision support tools providing decision makers and other involved parties with various forms of information which can be then used during the decision-making process. The capabilities as well as limitations of these tools in securing access to information are analysed. This analysis is illustrated by examples of various solutions and applications, including presentation of the Internet-based prototype of the decision support system (DSS) that has been developed for the Ganges River. Finally, recommendations concerning future research needs and challenges are presented. All considerations contained in this chapter are made from the point of view of a technically minded professional, who has been involved for many years in the development of various tools and solutions supporting decision-making processes and managerial activities. Therefore, the psychological, social, political, and legal aspects of the decision-making processes are not considered here.

Decision problems, stakeholders, and basic concepts of systems analysis

The decision-making associated with the utilization of water resources is understood here as the process of selecting such actions affecting a given water resource system, which seeks to result in better fulfilling the goals and objectives by the system under consideration. The decision-making can be also understood as a process of seeking the "best acceptable" solution for a specific system.

The decision-making processes are taking place in a structure consisting of the following elements (see fig. 19.1):

- The system (in our case, a water-management system) under consideration representing material and physical reality.
- The problem that requires a decision. The term "problem" refers to the existence of a gap between the desired state and the existing state (Sabherwal and Grover 1989). Consequently, the decision-making process aims to fill (or at least to reduce) this gap and thus solve the problem.
- The decision maker (that is, the person or organization, who will decide upon an action or a set of actions to be undertaken in order to achieve certain objectives (fill or reduce the gap between the existing and desired state of the system). These objectives are provided by those to whom the decision maker is responsible. Most methodologies assume an individual decision maker; however, in a real-world situation, the decisions are usually made by a group (or even several

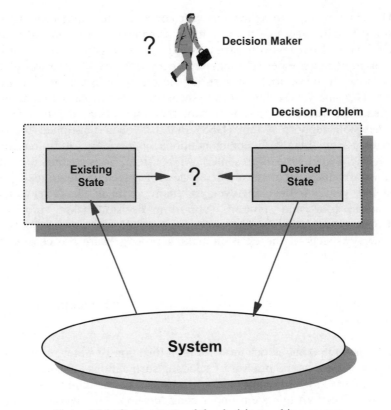

Figure 19.1 Components of the decision-making process

groups) of people representing different views, preferences, and expectations.

Political and social developments in various countries mean that the notion of a single decision maker has been losing its rationale. Complex economic, social, and political structures require decisions to be made in a framework of sophisticated processes involving many stakeholders, who more or less directly participate in the decision-making process. In a case of water management systems, the professional and institutional affiliation of decision makers has been changing over time. As Loucks (2003) points out, considering the United States as an example, originally civil (mainly structural) engineers dominated river basin development, and this led to a situation where engineers involved in managing river basins had to fit into multidisciplinary teams including ecologists, economists, environmental specialists, social scientists, water users, lawyers, and regulators. The same now applies to many countries all over the world.

What connects the elements of the structure underlying the decision-making process mentioned above is information, which is continuously gathered, exchanged, processed, enhanced, evaluated, and used during the decision-making processes. Decision-making processes associated with water resource management concern many areas – and decisions can be purely technical, technical with economic and social impacts, political, economic, social, and so on. There is no definitive scientific explanation of how decisions are made by individuals, why people make one and not another decision, or what information they use in making decisions. It is assumed that the decisions can be made faster and better when decision makers have access to the most up-to-date, complete, and correct information relevant to their particular decision problem.

The information used in the decision-making process may take different forms: these range from a collection of various historical data, literature, results of public opinion polls, or actual measurements of a physical system's parameters up to forecasts and simulation results of computations showing the consequences of considered decision alternatives. Depending upon the concrete (specific) decision situation, the information requirements and needs expressed or perceived by the stakeholders in the decision-making process can be very different. Experience shows that it is extremely difficult to specify beforehand what information is necessary and sufficient to make good decisions. Usually, the process of decision-making goes together with a learning process. In the framework of the learning process, stakeholders make decisions based on the information available; they learn about the impacts and consequences of those decisions; then they make further decisions influenced by the new knowledge and information that they have gathered. Consequently, in a repeatable process they enhance their knowledge and understanding of the decision problem and also identify needs for new types of information. Information needs and requirements therefore grow together with the growing understanding of the problem in hand.

An interesting discussion concerning this subject is provided by Simonovic (2000) in the context of a complexity paradigm relevant to water problems. Population growth, climate variability, and regulatory requirements are increasing the complexity of water resources problems. Water resource management schemes are planned for longer temporal scales in order to take into account and satisfy future needs. Planning over longer time horizons also extends the spatial scale. Extension of temporal and spatial scales leads to increases in the complexity of the decision-making processes and involves an increasing number of stakeholders. Consequently, together with the growing complexity of the decision-making problems, there are also growing demands and challenges concerning tools used to provide information and to support decision-making processes.

The methodological framework underlying the process of searching for solutions (decisions) of the decision problem is offered by the scientific discipline of systems analysis (Sage and Armstrong 2000), which evolved through parallel developments in mathematics, engineering, and economics. As system analysis has matured in recent decades, its applicability in water resource planning and management has steadily grown, and currently it is impossible to imagine water resource management without the use of methods and tools offered by systems analysis.

The notion of a system is a basic one for this scientific discipline (Nandalal and Simonovic 2002). We consider physical water resources systems as a collection of various elements interacting in response to natural and human-induced actions. The systems and related human actions are aimed at satisfying social and economic needs. Systems analysis enables the study not only of interactions between components of the system but also of the overall response of the whole system to various human actions associated with development and management alternatives.

The behaviour of a system as a whole, or the behaviour of some of its components, can be the subject of systems analysis only when the system or its elements can be modelled using mathematical representation (mathematical models). Models and their properties can vary greatly: the same physical phenomenon can be described using different types of models, depending on specific purposes which the model may serve. These different types of models may have different mathematical representation: for instance, a model of a water reservoir used for calculating water balance in a basin is represented by a very simple mass-balance equation, whereas the model of the same reservoir used to describe thermal or water-quality processes has a complex mathematical structure (partial differential equations) and data requirements. Therefore, the mathematical representations of the reality chosen by the model builder should be consistent with the overall accuracy required from the system. The mathematical representation should allow for a description of reality that is adequate to meet the purposes of the model. This model should provide decision makers with information relevant to the decision problem at hand and should address the information needs of the stakeholders.

A system may be understood as a part of physical reality and consisting of a finite number of interrelated, interacting elements. This system is identified through the functions that it fulfils and is influenced by uncontrollable (often, not exactly known) natural factors, as well as targeted, aim-oriented human actions. As shown in figure 19.2, both un-

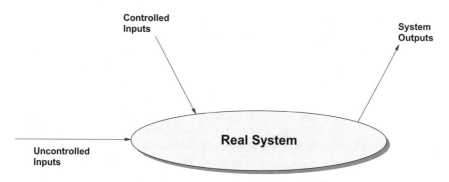

Figure 19.2 A system and its interactions with the surrounding world

controllable natural stimuli (uncontrolled inputs) and human-induced, targeted actions (controlled inputs) influence the behaviour of the system, which "responds" through physical values identified as outputs (system outputs).

Controlled inputs are equivalent to "decision variables," which must be selected in the framework of the decision-making process from the set of feasible alternatives. The transformation of the system due to the influence of both decision variables and uncontrolled inputs is described using a set of so-called "state variables," which are associated with mass and energy preservation. Internal properties of the system are described by system parameters.

Considering a storage reservoir as a sample system, the value of release from the reservoir represents the decision variable; the amount of water stored in a reservoir is equivalent to the state variable; and values such as storage capacity or storage–area relationship represent parameters of the reservoir.

Finally, physical values through which a given system acts on the surrounding context are known as output variables. The selection of the output variables often depends on the purpose of the system or its model. In the example of the reservoir considered here, in one situation we can select release from the reservoir as an output (when the reservoir is considered as a source of water supply); in another situation, when the reservoir operation serves hydropower-generation purposes and interacts with the energy system, the amount of energy generated by a power plant located at the reservoir site can be considered as an output variable.

With the functioning of every system, there are also certain associated goals that should be attained. The functional relationship between decision variables, state variables, and system parameters on one side and the quantitative description of the degree to which these goals are attained is termed objective function. Depending on the complexity of the system and specification of the goals, the objective function may have the form of a scalar (single value) function; however, it may also have a form of a vector function attaining multiple values. The process of selecting such values of decision variables, which allow achieving the best possible results (with respect to existing constraints on decision and state variables) is termed optimization (Rardin 1997). If the objective function is a scalar one, there is a single objective optimization; when the objective function has a vector representation, the notion of multiobjective (multicriteria) optimization applies (Rosenthal 1985; Miettinen 1998).

Practice shows that real-life decision-making problems rarely (if at all) boil down to solving clear-cut optimization problems. The search for a solution of the decision problem involves complex patterns of using optimization and simulation models of the system under consideration in order to find feasible and satisfactory values of decision variables (controlled inputs) in a framework of decision-making processes. The system model, consisting often of many sub-models and components, must also account for the presence of uncontrolled inputs influencing the system at hand. The information about these uncontrolled inputs is usually available in a form of forecasts or historical and/or generated time series representing the most significant uncontrollable inputs.

The decision-making process cannot take place in the absence of feedback information about results of previously applied (selected) controls. This feedback information is based on observations and measurements of the output system variables and state variables. Figure 19.3 shows schematically the major components of the decision-making process and the main directions of the information flow accompanying this process.

Intuitively perceived (and already mentioned), the complexity of the decision-making processes associated with the utilization and management of water resources calls for tools capable of mirroring the complexity of the problems under consideration. At the same time, these tools have to be able to cope efficiently with the multiplicity and amount of information to be processed during decision-making. The capability to process relevant information must be accompanied by capabilities to present this information to the user and consequently to the decision maker. Such capabilities are provided by DSSs.

History and basic concepts of decision support systems

Decision support systems can be defined as computer technology solutions that can be used to support complex decision-making and problem

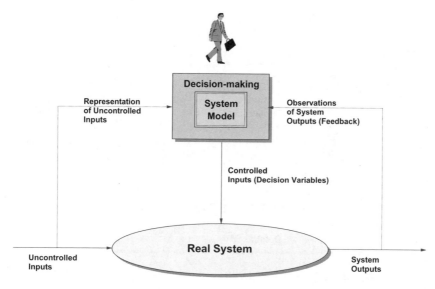

Figure 19.3 Schematic of the decision-making process

solving (Shim et al. 2002). Although this definition applies to decision-making in many purely technical areas, it does not reflect one, extremely important aspect of the decision-making process in water resource systems – the role of human factors.

Owing to the complex nature of water resource management problems, lack of consistent and complete data, uncertainties, and ill-structured decision problems, the process of finding decisions cannot be limited to solving mathematical optimization problems or performing complex simulations. Therefore, a DSS is understood to include a set of computer-based tools that provide decision makers with interactive capabilities to enhance understanding and the information basis for a decision problem through usage of models and data processing, which in turn allows decisions to be reached by combining personal judgement with information provided by these tools.

A simple Internet search performed by the author of this chapter on 29 January 2003 using the Yahoo search engine identified 3,450,000 websites thematically related to the subject search key "Decision Support Systems." This enormous number of "hits" demonstrates how widespread is the notion of DSS, as well as the broad scope of human activities related to this subject.

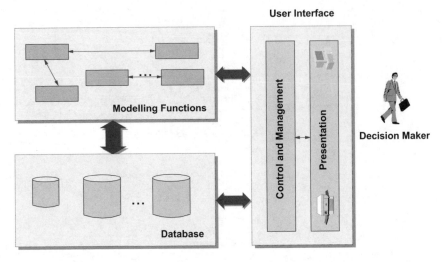

Figure 19.4 Main building blocks of the decision support system

The term DSS was born in the early 1970s. DSS has evolved from two main areas of research – the theoretical studies of organizational decision-making conducted at the Carnegie Institute of Technology during the late 1950s and the technical investigations carried out at Massachusetts Institute of Technology in the 1960s (Keen and Morton 1978). The classic DSS design, as shown in figure 19.4, comprises the components for the following:

- database-management capabilities with access to internal and external data, information, and knowledge;
- powerful modelling functions accessed by a model-management system;
- user interface designs that enable interactive queries, reporting, and graphic functions.

This view of DSSs concerns their technical architecture and building blocks, which have to be incorporated into the design and development of DSSs. Over the past three decades, the developers and users of DSSs have used a variety of constructs and definitions, while other solutions, not fully meeting the above-listed components, have emerged to assist specific types of decision makers facing specific kinds of problems. Nevertheless, the classic DSS architecture contains these three basic components.

Another, complementary way of looking at DSSs is associated with the role and functions that DSSs fulfil (Parker and Al-Utaibi 1986), as seen from their user's perspective:

- they assist managers in their decision processes in semi-structured tasks;
- they support and enhance, rather than replace, managerial judgement;
- they improve the effectiveness of decision-making, rather than its efficiency;
- they attempt to combine the use of models or analytical techniques with traditional data access and retrieval function;
- they specifically focus on features that make them easy to use interactively by people, even those who are not skilled computer users in an interactive mode;
- they emphasize the flexibility and adaptability to accommodate changes in the environment in which the decision maker acts, and the decision-making approach of the user.

The capabilities of DSSs to fulfil the functions listed above are particularly important for their practical usability and acceptance by a broad range of stakeholders involved in the decision-making processes. The degree to which a specific DSS meets these characteristics and capabilities has a direct impact on its abilities to satisfy the information needs of the decision makers as well as those of the stakeholders participating in a decision-making process.

The ability of a DSS to efficiently communicate with its users is an important aspect associated with its meaningful development. The communication is performed through a user interface (UI), as schematically shown in figure 19.4. From a functional perspective, the UI can be divided into two layers:

- an inwards-oriented "control and management" layer responsible for controlling and managing data flow and computational processes in the whole DSS;
- a user-directed "presentation" layer that organizes the process of communication between user(s) and internal structures of the DSS.

Functions of the UI are associated to a large extent with organizing processes of data input and data output. In this context, "data" includes any type of textual, numerical, graphical, or other information that can be exchanged between the DSS and user(s). For both data input and data output, the communication between the tool and the user must be designed and organized in such a way that:

- communication is consistent with the level of expertise of the user;
- the exchange of information between the user and the DSS must be efficient;

- there must be a clear and unmistakable distinction between data entered by the user and results produced by the system;
- communication with the system fulfils the information needs of the user.

Traditionally, mathematical models and various forms of decision support tools and systems incorporating these models have been developed by analysts and modellers for the same type of audience. Therefore, it was not necessary to pay any special attention to the design and implementation of user-friendly interfaces between the tool and its user. This state has continued for years, contributing to the creation and growth of a gap between modellers and analysts on the one hand and decision makers (not to mention the general public) on the other hand. As long as decisions were taken by a narrow circle of specialists, the gap was manageable and was not perceived as meaningful.

The situation became much more complicated when these tools began to be used not only by a limited range of modellers and analysts but also by emerging groups of other, less technically minded and less-experienced users seeking to use these tools to secure active and informed participation in decision-making processes. This caused developments in two areas:

1. substantive, concerning the phenomena and processes to be modelled (analysed);
2. communication, securing proper exchange of information between the model(s) and various types of users.

One of the biggest challenges of DSSs in facilitating access to information by a broad spectrum of stakeholders is associated with the fact that available information must directly address their concerns and information needs. Therefore, it is important to know how the information is obtained from and presented to non-specialists: what information is or should be presented; the form of the information; and how access is managed. The next challenge is associated with providing non-professionals involved in technical matters with the possibilities to obtain answers to relevant questions, especially when the questions and responses do not necessarily have to be expressed in technical terms. The information presented to non-specialists cannot substitute for, or conceal, real facts. This information must contain the same value as far as real consequences of considered decision alternatives are concerned, but the form of this information should allow for straightforward recognition of impacts, perils, and benefits. The only possible method of adequately responding to these challenges has been associated with the balanced and targeted use of technical and technological means combined with organizational forms of decision-making processes, when professionals in non-technical disciplines and various interest groups have the right to participate in the evaluation of alternatives and their respective impacts.

Technical and technological factors underlying capabilities of DSSs

The development of DSSs is closely connected with progress in computer technology. In fact, the advances in computer technology have facilitated access to information for broader and broader audiences.

Information technology is based on two complementary pillars – hardware (which includes all sorts of equipment used to process and store data) and software (which includes various types of programs that control hardware and allow it to perform desired computations and data processing). The technological and technical progress in hardware has continuously stimulated advances in software, while progress in software development has created a demand for new hardware capabilities.

Computing capabilities have been changing dramatically over the last 50 years. The first monolithic mainframe computers, created in the late 1940s and 1950s, performed computations using vacuum tubes. The user interface was limited to punch-card or punched-band readers for data entry, and primitive printers provided outputs to the users. The number of computer installations was very limited, and the circle of users comprised a narrow group of specialists with no possibility of providing direct access to information produced by computers to the wider audience.

The invention of the transistor in 1947 revolutionized communication and computing technology. The transistor and integrated circuit gave rise to the second generation of computers in the 1960s and 1970s. With the second and third generation of computers came major improvements in the user interface – namely, that the user could remotely communicate with a computer using a terminal and keyboard, which (together with the development of operating systems) opened up possibilities for time sharing and facilitated user interaction with the computer. Although this was a significant step to widen access by the broader public, it was not yet sufficient to allow wide circles of people to benefit from accessing information processed and produced by computers at that time. Another breakthrough brought the random access memory (RAM) chips, introduced by Intel in 1969. The biggest leap in computer technology was brought about by creation of the first microprocessor, again by Intel, in 1971. The first microprocessor had 2,300 transistors, but the number of transistors contained in consecutive versions of microprocessors has steadily grown, so that the Pentium 4 processors introduced in 2000 contained 42 million transistors. An empirical law formulated by Gordon Moore of Intel states that the computing power of a new chip doubles every 18 months (Honda and Martin 2002).

As the consequence of processor miniaturization, the computers became not only computationally more powerful but also smaller, less expensive, and more popular. The range of manufactured machines spread to include not only huge mainframes but also smaller mini- and microcomputers broadly installed in industry, military, government, and scientific and research organizations.

The first personal computer to enter the market was the Apple II computer released in 1977, but introduction of the PC by IBM in 1981 opened the way for a rapid proliferation of desktop computing, although not without its drawbacks. The early personal desktop computers consisted of a central processing unit (CPU) with small random access memory (RAM) (typically 64 kB, capable of reaching 640 kB at most), diskette drive, small hard disk (20 MB), keyboard, and monochrome monitor and had very limited interface capabilities. The user communicated with the computer using command line interface. The real revolution came in 1984, when the graphical user interface (GUI) was introduced by Apple Computer; this opened up the possibility of the use of computers by less technically minded and less well-educated people.

Not only were these advances in computer technology associated with the breaking down of a number of technical barriers but also widespread access to computer technology did away with several mental and social barriers. The critical mass was reached, and personal computers became an element of daily life, creating new possibilities for information processing and dissemination. Moreover, computers ceased to be perceived and treated as a special type of equipment reserved for particular pur-, poses and accessible only by privileged specialists.

Further advances in information technology, such as networking technology and client–server computing, enabled the creation of computer networks and data sharing between single computers or computer networks.

Creation of the transmission control protocol/internet protocol (TCP/IP) (Rodriguez et al. 2001), which was installed for the first time in 1980, opened the way for a revolution in the computing and communications areas – the Internet. The word "internet" itself is a contraction of the phrase "interconnected network." However, when written with a capital "I," the Internet refers to the worldwide set of interconnected networks. TCP/IP refers, in fact, to two network protocols or – in other words – methods of data transport used on the Internet; these are transmission control protocol and internet protocol, respectively. These two protocols work together to provide nearly all services available to today's "Net" surfer, including transmission of electronic mail, file transfers, and access to the World Wide Web.

The progress in computer technology underlying the development of hardware has been closely linked with advances in software. As with hardware, the software domain is not homogeneous and can be divided into three basic sub-domains. These are:

- *Operating systems* – that is, programs used to manage and control the use and operation of physical resources of the computer. Progress in this area allowed the creation of computers consisting of multiple processors performing parallel computations for multiple users that are also capable of communicating with other computers and computer networks.

- *Programming languages* used to secure communication between the user and machine and to provide means to write programs instructing a computer how to perform computations and operations. Primitive programming performed at the level of single registers has been replaced by procedural and then object-oriented languages and programming tools allowing for developing programs in a graphical mode and for use of code generators.

- *Databases* – that is, technology to store and manage huge amounts of data. The initially simple structures of data files have been replaced by hierarchical and relational databases allowing the storage of terabytes of data and its access within milliseconds.

As a result of progress in this domain, the computational capabilities have grown enormously, offering users the ability to solve mathematical problems to an extent that was hard to imagine a few years ago.

Examples of DSS implementation for water resource management

The developments in systems analysis and information technology have enabled significant progress in hydrology, water resource management, and environmental and decision sciences. Taking place over a number of decades, the evolutionary process of developing models and other tools for water resource management has closely reflected the progress in mathematical modelling, linear and non-linear optimization, stochastic modelling, programming languages, and data processing.

This dramatic progress is extensively documented in the rich literature on this subject. The multiplicity of works and publications means that even a superficial review of major publications exceeds the scope and space limitations of this chapter. The progress has witnessed development of various approaches and tools, sometimes reflecting certain "fashions"; nevertheless, some of the tools created even recently build upon still-valid concepts underlying water resource management and multiple-reservoir systems, such as storage zones and rule curves, which were developed many years ago (Loucks and Sigvaldason 1982). Much of the fundamental work has been done at the Hydrologic Engineering Center (HEC) of the US Army Corps of Engineers (USACE) at Davis, California, where a number of models and decision support tools have been developed over the past few decades, including:

- HEC-1 Flood hydrograph package;
- HEC-2 Water surface profiles model (USACE 1992);
- HEC-3 Reservoir systems analysis model (USACE 1985);
- HEC-5 Reservoir operation simulation model containing water-quality components (USACE 1982, 1986);
- HEC-RAS River analysis system containing graphical information systems extensions (USACE 1995);
- Decision Support Systems utility programs and components (USACE 1987).

Currently, these programs are widely used by specialists around the world. They have been adapted to new technological developments and can be purchased or downloaded from websites of various software and engineering services providers, such as http://www.hydroweb.com or http://www.bossintl.com.

The programs and decision support tools originally developed by HEC, like many other tools which have been developed for supporting decision-making processes, have been designed for use on powerful computers in a batch mode and did not allow (at least in their first years of development and operation) for interactive data input and operation. They were specifically designed for use by highly specialized professionals and did not provide any possibilities that would enable their use by less technically minded audiences.

With the advent and expansion of personal computers and powerful work stations, there are now capabilities for creating flexible and easily transferable tools suitable for users to work interactively. The following subsections consider three representative examples of DSSs for water resource management. All three are characterized by their common ability to interactively define the model of the water-management system under consideration. The main difference among these systems lies in the growing sophistication of the mathematical basis underlying their concept and implementation, and also in the gradually increasing difficulty of their usage. This aspect is particularly important as far as the use of decision support tools by the general public is concerned.

IRIS and IRAS modelling systems

The underlying idea of the work by Loucks and his collaborators has been to develop simple, interactive, graphics-based simulation models for estimating time series of flows, storage volumes, water qualities, and hydroelectric power produced in a particular water-management system. With the use of a simulation model, the impacts of alternative land-use and water-management policies and practices in a watershed could be evaluated and compared, even by inexperienced users. Models have been developed in such a manner that no experience or skills in programming and modelling have been necessary to apply and use them.

The first version of the system known as the Interactive River Simulation (IRIS) package was developed in the late 1980s (Loucks and Salewicz 1989; Loucks, Salewicz, and Taylor 1990). It was developed as a decision support and alternatives screening tool to assist decision makers and stakeholders in resolving conflicts associated with the management of international river basins (Salewicz and Loucks 1989; Venema and Schiller 1995; Salewicz 2003).

An extended and improved version is the Interactive River–Aquifer Simulation (IRAS) program (Loucks and Bain 2002). The simulation model has been developed primarily to assist those interested in evaluating the performance of watershed or regional water resource systems. The performance is associated with spatial and temporal distribution of flows, storage volumes, water quality, hydropower production, and energy consumption in water resource systems. Such systems can include rivers or streams, diversion canals, lakes, reservoirs, wetlands, and aquifers, together with various multiple water users. The model is data driven, and the user defines and has full control over the spatial and temporal resolution of the system being simulated.

The input data define the system configuration, the system components, their design parameters, and operational rules describing how each of those components operates. The system to be simulated is represented by a network of connected nodes (such as gauge sites, aquifers, consumption sites, and reservoirs) and links (such as river reaches, diversions, water transfers, and pipelines). The user must draw the network into the graphics terminal. The systems to be simulated using IRAS can include up to 400 links and up to 400 nodes. One-dimensional simulation is based on mass balances of quality and quantity constituents, taking into account flow routing, seepage, evaporation, and water consumption, as applicable. IRAS can simulate independent or interdependent water-quality constituents defined by the user, who must define not only the constituents to be simulated but also their growth, decay, and transformation rate constants together with other parameters necessary to perform water-quality simulation.

The results of any simulation run are initial or final storage-volume values, together with average flow, energy, and water-quality values for each within-year period expressed in the units defined by the user. These data can be plotted over time or space – for example, on digitized maps. Space plots can be dynamic, showing how values of selected variables change over time and space. User-defined functions of computed output variables, as well as statistical analyses based on these output variables, can also be calculated and displayed. These displays can include probability distributions of resilience and vulnerability criteria, based on either duration or failure and extent of failure.

The output data files, once created, then can be used for further display of the simulation results or they can be used as input data for utility programs to perform further analyses, evaluation, and display.

ModSim

ModSim is a general-purpose river and reservoir operation-simulation model. It was originally developed by Labadie in the mid-1970s to simulate large-scale, complex water resource systems (Labadie 1995; Fredericks, Labadie, and Altenhofen 1998; Department of Civil Engineering Colorado State University CSU/DOI 2000; US DOI 2000). It accounts for water rights, reservoir operation, and institutional and legal factors that affect river-basin planning processes. It is a water-rights planning model capable of assessing past, present, and future water-management policies in a river basin. From its initial development, the model has continually been upgraded and enhanced with various features and extended capabilities. A water resource system is represented as a connected network of nodes (such as diversion points, reservoirs, points of inflow/outflow, demand locations, gauge sites) and links that have a specified direction of flow and maximum capacities (such as canals, pipelines, and natural river reaches). This structure generally reflects the real system network that requires user knowledge and appropriate data. The tool allows for one-dimensional simulation of flows. In order to consider the demands, inflows, and desired reservoir-operating rules, ModSim creates internally (and on its own) a number of artificial "accounting" nodes and linkages that are intended to ensure mass balance throughout the system's network.

The graphical user interface provides a user with capabilities to construct a model of a river-basin network consisting of nodes and links and then to enter or import the necessary data and parameters. Geographic information system (GIS) tools can be used to prepare and attach necessary geographical data.

In ModSim, the network can be visualized as a resource-allocation system through which the available water resource can be moved from one point to another to meet various demands. Unlike the IRIS or IRAS systems, where the user defines the simulation sequence of nodes and links, the underlying principle of a network solver is based on the optimization principle minimizing the "cost" of water. The cost of water is based on water-right priorities serving to prioritize water allocation. ModSim employs an advanced optimization algorithm – the Lagrangian Relaxation Algorithm – that finds the minimum cost flow through the whole network within required limits (Bertsekas and Tseng 1994).

The form of the solution ensures that available flows in the system are allocated according to user-specified operational rules and demand priorities. ModSim simulates several types of water rights, (including direct-flow rights, instream-flow rights, reservoir-storage rights, and reservoir-system operation) and exchange and operational priorities. The model can also accommodate reservoir operations and accounting, hydropower, channel routing, and import and export of water from the network. Mod-Sim can also simulate the interaction between surface streams and groundwater aquifers.

The executable code of ModSim, together with documentation, tutorials, numerous examples, and supplementary routines, can be downloaded free of charge from the Internet at http://modsim.engr.colostate.edu.

RiverWare

RiverWare represents a new generation of tools for planning and managing river basin systems (Zagona et al. 1998, 2001). Many watershed models and decision support tools developed in the 1970s and 1980s were site specific and applicable to the particular watershed for which the model was developed. Although many decision support tools, such as IRAS and MODSIM, provide users with the capability to perform computations for a user-defined configuration and structure of a water-management system, their flexibility of accounting for various possible types of reservoir-operating policies is limited to rule curves and flow prioritization. These limitations result from the fact that those tools have been developed using algorithmic programming languages, such as Fortran. The algorithmic languages highlight the ordering of events in sequences of consecutive actions performed according to certain algorithms. New capabilities offered by object-oriented technology (Booch 1994) allow for the development of new software through the use of general modelling tools that are not specifically designed for river-basin systems by combining them in a single modelling framework.

RiverWare, developed at the Center for Advanced Decision Support for Water and Environmental Systems (CADSWES) of the University of Colorado, in cooperation with the US Bureau of Reclamation, utilizes object-oriented software to create a flexible modelling framework by combining building blocks that describe possible physical components of a water-management system with specific solvers capable of tackling operational problems through simulation and/or optimization. The RiverWare model construction kit allows a user to create a model of the system using graphical input and selecting appropriate objects to represent specific components of a water-management system, such as a storage reservoir, pumped storage reservoir, river reach, confluence, and many others (16 types in total). With every object, there is an associated mechanism for defining and entering data: (1) those concerning the physical parameters of the object (such as volume of the reservoir and the storage–area relationship); (2) time series data (such as flow sequences, evaporation, etc.).

The physical behaviour of each object is described in terms of so-called methods that are mathematical descriptions of certain properties of the object, such as mass preservation, water routing, and power generation. The user can select desired methods for each object. Currently, the following processes can be modelled: mass balance in level pool reservoirs; wedge storage in long reservoirs; river-reach routing; tailwater computations; hydropower generation; thermal system economics; diversions; water quality (temperature and salinity); evaporation; and bank storage.

The consequences of considered management alternatives can be evaluated using pure simulation, rule-based simulation, and optimization techniques. Pure simulation involves the solution of a precisely specified problem using various appropriate methods (functions) associated with objects constituting the system. Rule-base simulation is performed utilizing a verbal description of operating policies, which are defined using a specific rule language for RiverWare. This language is interpreted by a computer during the run time. The rule language is, in fact, a programming language intended to express policies formulated by the user (decision maker) in a form involving verbal formulations and if-then-else logic, as demonstrated by the following example referring to a simple flood-control rule for the reservoir:

If ReservoirElevation > ReservoirData.floodguide

Then ReservoirOutflow = ReservoirData.MaxRelease

RiverWare contains a built-in editor allowing the user to construct operating rules, which then govern the solution of the simulation process performed in accordance with the user-defined rules and methods defining the behaviour of the objects. The optimization is performed following the definition of the network and the construction of a model, which involves the selection of:

- policy variables for each object (for instance, in the case of a reservoir used for hydropower-generation purposes, the decision variables are turbine release, spill, outflow, and storage);
- linearization methods for the non-linear policy variables.

Using a policy editor, the decision maker can express the priorities of the policy objectives. The policy goals are entered into a graphical policy editor. Each objective can be given either as a simple linear programming objective, or as a set of constraints that is automatically converted to an objective by minimizing the deviations from the constraints.

A set of utilities facilitates the computational process as well as viewing and using the output. The data computed by RiverWare can be transferred to external sources for further processing. Output options include plots, data files, and spreadsheet files (such as Excel).

The efficient use of RiverWare requires advanced skills. Accordingly, in addition to purchasing a software licence, educational courses provided by developers of the system are recommended. Extensive information about the system and conditions of its availability and usage can be found on the RiverWare homepage: ⟨http://cadswes.colorado.edu/⟩. Further information concerning this system can be also found at: ⟨http://www.usbr.gov/rsmg/warsmp/riverware⟩.

More generally, the Internet offers a rich source of information regarding various models and decision support tools. The following addresses are particularly useful:

- "Decision Support Systems Resources" ⟨http://www.dssresources.com/⟩;
- Inventory of water resource management and environmental models ⟨http://www.wiz.uni-kassel.de/model_db/models.html⟩;
- The "USGS Surface-Water Quality and Flow Modeling Interest Group" ⟨http://smig.usgs.gov/SMIG/archives_commercial.html⟩;
- Independent "Water Page" also containing "The African Water Page" ⟨http://www.thewaterpage.com/⟩;
- Selected "World Wide Web Sites For The Water Resources Professional" containing numerous links to important water-related web-sites ⟨http://www.wrds.uwyo.edu/wrds/wwwsites.html⟩;
- The "Land and Water Management" site of the Delft University in Holland ⟨http://www.ct.tudelft.nl/wmg_land_water/⟩;

- "An Inventory of Decision Support Systems for River Management" 〈http://www.geocities.com/rajesh_rajs/inventary.html〉;
- "Environmental Organization Web Directory" (claiming to be the world's largest environmental search machine) 〈http://www.webdirectory.com/〉.

Internet implementation of DSSs

Unlike traditional DSS, which is implemented on a single computer or a network on which the user (decision maker or stakeholder) has an account, the development and use of Web-based DSSs faces many conceptual and technical challenges. In the case of a DSS implemented on a single machine or network, the user has access to all resources of the machine and the DSS, available either through the operating system or through the user interface to the DSS. The access to resources concerns not only physical resources of the computer (such as disk space, memory, and printers) but also software and data. The user working with the DSS in an interactive mode may also access and manipulate models built into the DSS and their parameters; activate or deactivate certain components of the system model; change preferences; and select display or printout alternatives. Data used by the DSS can be accessed and modified to allow the user to explore various situations and scenarios. Results obtained by the user can be stored for further use; working sessions can be suspended and then started again without losing information or data created during the session.

In the case of Internet-based DSSs, the situation is significantly different: the user accesses the Web through a special program called a browser, which does not offer the capabilities of an operating system. Moreover, capabilities of the DSS–user interface are not available to the browser. The Web user may access a certain Internet address and use resources offered to the user only in a range defined and controlled by the owner of a particular Web page. The user's computer, on which the browser is installed and which allows the user to communicate with the server hosting a particular website, frequently is connected with the Internet through low-end communication or telephone lines with quite often relatively low transmission rates (especially in developing countries). Thus, the time needed to load one page or to obtain a response to a choice made by the user can be relatively long (taking even minutes), not to mention the time necessary to perform computations on the server side.

Communications between the user and the server are in the form of messages – the user's request to the server and the server's response to the browser. The hypertext transfer protocol (HTTP) used in the Internet has no mechanism for keeping information about previous requests or storing information about the current request. Consequently, unless special and advanced Internet technologies are used on the server side, the Internet user has no direct possibilities to store on the server intermediate results for further use during future interactive sessions.

The distribution of computing power available for DSS in an Internet environment can be described by two conceptual models – namely: (i) a "thin client and thick server" concept, or (ii) a "thick client and thin server" option (Salewicz 2001).

The "thin client and thick server" concept for implementing DSS refers to the situation when the user is connected to the Internet and the user's PC acts as a communications terminal only. The user's PC, then, enters into an interactive mode for certain data (decisions and/or parameters chosen among available alternatives) and then displays results of the computations performed on a remote computer (server). All models and the database reside on the server, and all computations are performed on the server. Implementation of this concept means that the amount of data to be transferred back and forth between the server and the user's computer is relatively low, although the data have to be transmitted in small "portions" after each action initiated on the user's side.

There are a number of advantages associated with this concept. Relatively low amounts of data have to be transmitted, which is particularly important for users from countries where the telecommunications infrastructure is limited and transmission rates are relatively slow and unreliable. This approach also has the benefit of high security and consistency of data and models: since both the data and the models reside on the server, they are protected from manipulation and unauthorized modifications by users; such changes, in extreme cases, could lead to fraud. Another positive feature of this concept is that such a DSS can be built using already-existing simulation and optimization models that were developed in traditional programming languages such as Fortran and C, thereby limiting the programming effort associated with implementation.

The disadvantage of this concept, however, is that there are heavy computational burdens and data loads on the server side. This requires installation of powerful machines for servers.

The second option – namely, a "thick client and thin server" – means that the user's PC functions not only as a data entry and display terminal but also as a platform to perform all computations using programs and data downloaded from the server. The role of the server is therefore reduced to that of a repository for executable codes of all components of the DSS and, eventually, data sets that can be used with it. This approach is popular, and a number of solutions or DSSs can be downloaded (at least in a trial version) by anybody interested (Palomo, Rios-Insua, and

Salewicz 2002). The possibility of downloading and then using models or a DSS to address specific issues and decision problems is attractive, especially to professional and scientific communities in many countries (not only developing ones), because it gives easy and free access to tools already developed, or access to alternative solutions that may enhance capabilities of tools already available. However, the effort necessary to download these tools, to install them, and then to learn how to use them and to resolve the problem at hand, seems to exceed the interest and devotion of the average layperson.

If DSS tools can be freely downloaded, there is also a risk that the use of downloaded models or DSSs can be abused. Such development is plausible in a case of controversial problems or decisions – for instance, concerning an international dispute when one party, for unethical or politically motivated reasons, presents results supporting its position and obtained without using a particular and usually highly regarded tool downloaded from the Internet, but claiming at the same time that these results have been obtained with the help of the said tool. In such cases, it might be difficult to prove the wrongdoing, and the burden of proving that may fall on the authors of the model. Moreover, the reputation of the DSS or its authors, unintentionally involved in such abuse, may be significantly hurt.

In order to explore the technical possibilities and feasibility of developing DSSs using the Internet, the author of this chapter has initiated research to develop a prototype (pilot) installation of a DSS on the Web. This research was based on the technical concept of the "thin client and fat server" and assumes the following:

- the prospective user of the DSS is interested in assessing the potential consequences of a certain policy that is expressed in terms of clearly identified alternative actions;
- actions associated with the policy are formulated preferably in a qualitative manner, and not quantitatively;
- the user has no experience, and no desire to learn about the specifics, of any mathematical models and tools;
- the tool should allow for simple selection of available alternatives and should present the consequences of selected decisions in a meaningful way;
- the time interval between formulating a query and obtaining a response should be minimal.

Initial efforts were directed towards selection of an appropriate case-study system, with the following characteristics:

- it could attract a significant audience;
- it concerns a controversial issue (possibly international), involving conflicting objectives and interests;

- it has been described using sound, verified, and viable modelling techniques;
- it has been analysed and modelled by objective, unbiased, and independent specialists, who are not involved in the controversy.

An extensive search has led to selection of the Ganges River case study (see fig. 19.5), which has been the subject of extensive research performed at the Center for Spatial Information Science, University of Tokyo (Ministry of Land, Infrastructure and Transport 2001). The case study analyses the impacts of agricultural and urbanization policies applied in India. India's agricultural and urban development policies directly affect the amount of water in the Ganges River that flows into Bangladesh. Taking into account the mutual distrust and lack of cooperation between these two countries (Biswas and Uitto 2001), the availability of unbiased and independently developed models and DSSs capable of analysing the consequences of selected policy options could help both sides to establish a common basis for discussing and evaluating alternatives. The relevant policies that can be applied in India concern the following decision variables:

- the length of the stretch of river over which the agricultural and urbanization policies will be implemented;
- the intensity of the changes in land-use patterns;
- the intensity of the urbanization changes over the area considered.

These policies can be described in detail in quantitative terms, using precise values of the above-mentioned decision variables, and then the response of the system can be simulated for selected values. However, one run of the simulation to calculate the response of the system to selected policy alternative may require a few hours of computations (K. Rajan, Institute of Industrial Science, University of Tokyo, personal communication 2002). This property of the model could disqualify it, at least as far as the use of the model in an Internet-based, interactive DSS is concerned.

Taking into account the fact that the average user of the model does not have enough knowledge and experience to experiment with selecting precise numeric values for decision variables, we had to look for another approach. The approach that we found is based on the concept of a qualitative qualification of decision variables: the feasible range of each decision variable has been divided into a small number of sub-intervals. With all the values of the decision variable belonging to a certain sub-interval, there have been associated one single, qualitative, attribute characterizing this range in descriptive terms (i.e. low, medium, high). Such a process of qualitative categorization of decision variables can be performed only on the basis of a thorough sensitivity analysis and knowledge of models used to calculate the impact of policy parameters.

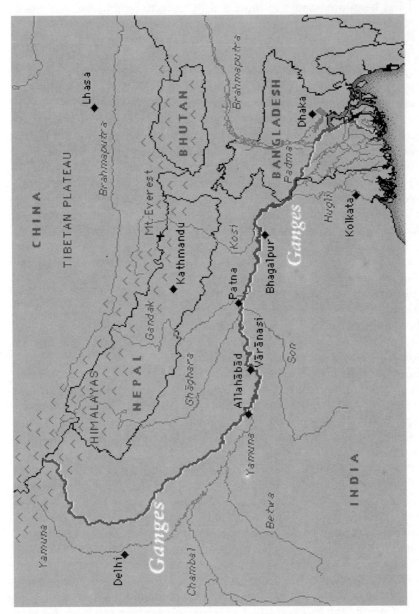

Figure 19.5 Map of the Ganges River Basin

Following this concept, the feasible decision variables expressed in descriptive terms are as follows (see fig. 19.6):

1. The length of the area upstream of the Harding Bridge, where the changes to land-use policies will be introduced, has been divided into three categories:
   - changes on the stretch shorter than 100 km;
   - changes on the stretch between 100 and 200 km;
   - changes on the stretch longer than 200 km.

2. The intensity of change in land-use patterns has been divided into four categories:
   - shift in the cropping pattern from the current one to one more intensive;
   - shift from the current pattern to one less intensive;
   - no change in the land-use pattern (retain current conditions);
   - increase in an irrigation command area, which is equivalent to the creation of bigger farms.

3. The intensity of the urbanization changes over the considered area has three alternatives:
   - no changes to the current population density;
   - increase of the population density by up to 50 per cent;
   - increase of the population density by up to 100 per cent.

Consequently, the user who wants to see the consequences of changes in Indian land-use policy selects the respective combination of policy parameters expressed in descriptive terms, as defined above.

The impacts of the policy alternatives may vary, depending upon natural climatic conditions which, in this region, are characterized by monsoons. Thus, in this case, a qualitative description of climatic conditions has been used: the impact of land-use policy is analysed using three alternative scenarios of climatic conditions extending over a one-year time horizon for (i) average, (ii) better than average (more rainfall), or (iii) worse than average (less rainfall) meteorological conditions. The impacts of a selected policy alternative are represented by the monthly time series of the following indicators:

- *normal water demand*, that is, the demand on water associated with currently used and unchanged conditions of the land use in the area of interest (upstream of the Harding Bridge);
- *expected water demand*, which is represented by the values of water demand calculated for the selected combination of decision variables;
- *normal water supply*, equal to flow rate at the Harding Bridge cross-section calculated for current (unchanged) land-use conditions;
- *expected water supply*, equal to the flow rate at the point of interest calculated for the user-selected land-use policy.

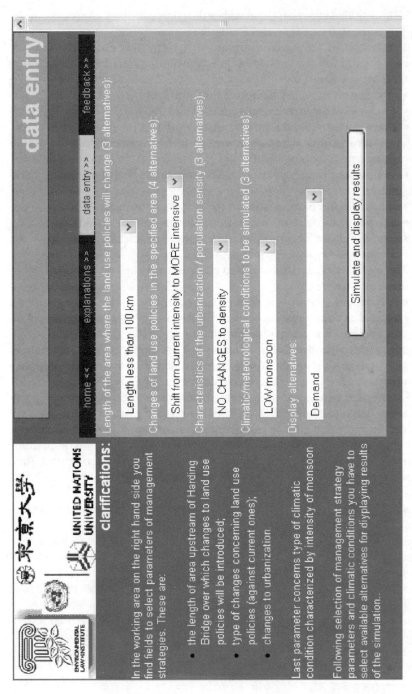

Figure 19.6 Example screen for selection of parameters for the strategic policy alternative

In addition, the user may select two other impact indicators, which are derived from values defined above, namely:
- the difference between water supply and water demand calculated by the simulation model for unchanged land-use conditions;
- the difference between water supply and water demand calculated for selected land-use policy options.

Time series with all impact indicators are presented to the user in the form of a graph, which can be printed out.

The system offers the user the ability to communicate with the developers of the DSS. The feedback is provided in a form of free text messaging, which can be composed and sent back. In order to obtain more specific feedback information from the users of the DSS, they are also asked to respond to a number of questions concerning:
- their country of origin;
- their professional background and affiliation;
- their opinions about the information that should be presented in visual form;
- their general opinion about the usability of the system.

Answers to these and, eventually, additional questions will serve as the basis for improving the system and for a better understanding of the reactions of the general public to tools such as this one. Consequently, materials and experiences collected in the framework of this study not only will allow the improving of this particular prototype system but also will provide the basis for improvements in the design and implementation of similar tools to be developed for other case-study systems and for formulating the future research agenda.

Summary: Capabilities, limitations, and challenges

Following the review of a broad scope of subjects related to basic concepts, technological foundations, development and example implementations of DSSs for water resource management, general conclusions may be drawn regarding the capabilities and limitations associated with creating and applying DSSs.

The ability of DSSs to describe real systems and calculate potential consequences of policy and operational alternatives results from the capabilities of mathematical models that are incorporated into the modelling base of a DSS. Advances in mathematical modelling and numerical methods, combined with progress in computer technology, that perform millions of arithmetic operations per second, have made it possible to build and implement models that can closely approximate physical reality. Even complex phenomena can be now modelled, using not only one-dimensional but also two- and three-dimensional models based on partial differential equations. The time-scales used by models can vary (depending on the phenomena modelled and types of models) from seconds to months and years; simulation horizons may extend even over hundreds and thousands of years.

Those complex, multidimensional models can use and process geographical and topological data available from various GISs. As a consequence, sophisticated multidimensional models solved using finite-element or finite-difference methods (Istok 1989; Wang and Anderson 1995) may be supplied with exact spatial data and parameters derived from GISs. The results of computations performed using those complex models can be presented in a graphical form by combining a display of numerical values with presentation on a map of the area under consideration.

These capabilities improve not only the viability of the models, their computational precision, and their ability to exactly describe physical phenomena, but have a significant impact on the ability to present results of computations in a meaningful and straightforward manner to a broad audience. For instance, data about the expected size of the area to be flooded that are presented as a map are much more informative and convincing than the same data presented as a table with numerical values only.

Graphical capabilities of contemporary models and DSSs relate not only to the display of computation results but also to data input. Using graphical user interfaces, it is possible to enter data and parameters by drawing in functions, shapes, and special configurations of the system. The data resulting from the simulation or optimization computations can be transferred easily to other models and tools (such as spreadsheets) for further analysis and processing.

The currently available vast technical and technological capabilities do not seem to constitute the main barrier for developing user-friendly and viable decision support tools; the difficulties and challenges are different. One of the main challenges is that of integrating the components to build a comprehensive and user-friendly DSS. Although many models and simulation and optimization algorithms exist, their integration into one consistent system efficiently addressing all issues important for stakeholders and decision makers often is close to impossible. This is due to differences in data requirements and data formats, inconsistencies in time steps used, lack of communication interfaces between various models, differences in programming languages in which those components have been developed, lack of standardization concerning output data, and display of the data. Together, these factors mean that often it is impossible to combine existing models into one system without significant (even uneconomic) effort. Accordingly, sometimes it is better to develop certain components anew instead of using existing ones.

This aspect is particularly important, if the development of Web-based DSSs is concerned. Because of the specific requirements and limitations imposed by Internet technology, many existing and proven models developed in procedural languages cannot be directly used for creating decision support tools in a Web environment: they have to be reprogrammed or adapted to specific requirements associated with Internet technology. The growing popularity and availability of object-oriented technology build around Java language (Flanagan 1999) can be seen as a basic mechanism to gradually overcome these problems. Currently, however, the number of specialists proficient in using and applying these technologies and in a water resource management sector is still very limited.

Together with technical problems associated with incorporating existing building blocks into a DSS, the availability of the right model at the right place and at the right time is often an issue. This issue is particularly difficult to overcome in developing countries, where the necessary information databases often are absent or inadequate (Turton, Earle, and Wessels 2002).

Despite great progress in recent years regarding the collection and storage of data (including the use of remote-sensing technology), the availability of reliable, credible, and consistent data has been – and will remain – a problem in the coming years. Collecting and storing data requires not only technical and technological infrastructure but also a great deal of investment in measurement networks and in processes of data validation and verification. Significant financial efforts, largely by governments and international agencies, will therefore be essential if significant process is to be achieved.

In addition to the technical challenges of DSS development, there are many areas for improvement in the "soft" side of developing and applying decision support tools. One of the most important difficulties concerning application and acceptance of these tools concerns their ability to communicate with a broad circle of users and stakeholders. In order to achieve progress in this area, the tools and models have to provide relevant, correct, and meaningful information to those involved in decision-making processes. The information presentation must be improved to allow users to grasp and quickly to understand important aspects and implications of considered policies and alternatives. As experience demonstrates, significant ways of improving the presentation and its relevance to the problem at hand can be addressed through joint development of decision support tools, when analysts, modellers, and users work together to build tools that are understandable and acceptable to all parties involved (Cuddy, Marston, and Farley 2000). Cultural and social aspects associated with developing and using DSS tools are also currently the subject of research efforts (Tung and Quaddus 2002).

The discussion concerning social, political, and organizational aspects relevant to developing and applying a DSS is beyond the scope of this chapter, but – as practical experience demonstrates – aspects such as mutual trust of the parties involved in a dispute, credibility of analysts and their models/tools, and the willingness to communicate and share information are difficult to handle and cannot be resolved by simple application of technical means.

Environmental conflicts concerning natural resource management can be solved in a framework of long and complex processes, in which formal tools and models can contribute to the growth of mutual understanding and objectification of the dispute by providing all parties with up-to-date, correct, and verifiable information. Internet technology and Web-based tools and information sources can play a particularly important role in this context. These efforts should be twofold: the first involves low-cost initiatives associated with creating and expanding traditional Internet sites providing free (and, possibly, unlimited) access to information, data, and literature and models to be downloaded; the second involves relatively expensive efforts to create Web-based DSSs. Such systems could be created by international organizations to provide independent, unbiased, and objective tools capable of addressing controversial issues arising between two or more countries in order to establish a communication and discussion basis to help to resolve the controversies.

Acknowledgements

The research reported here was supported by the United Nations University, Tokyo University of Agriculture and Technology, and the Environmental Law Institute. The author would like to express his deep and sincere thanks to all those who helped to perform the research reported in this paper. He is particularly grateful to Professor Mikiyasu Nakayama for his encouragement to undertake the research reported here and for very deep and useful discussions during the writing of this paper. The motivation provided by Carl Bruch appeared also very helpful and stimulating. Special thanks to Neven Burazor for his programming support.

REFERENCES

Bertsekas, D., and P. Tseng. 1994. *RELAX-IV: A Faster Version of the Relax Code for Solving Minimum Cost Flow Problems*. Completion Report under

NSF Grant CCR-9103804, Cambridge, MA: Department of Electrical Engineering and Computer Science, MIT.

Biswas, A.K., and J.I. Uitto. 2001. *Sustainable Development of the Ganges–Brahmaputra–Meghna Basins*. Tokyo: United Nations University Press.

Booch, G. 1994. *Object-Oriented Analysis and Design With Applications*, 2nd edn. Redwood City, CA: The Benjamin/Cummings Publishing Company, Inc.

CSU/DOI (Department of Civil Engineering, Colorado State University, and US Department of the Interior, Bureau of Reclamation, Pacific Northwest Region). 2000. *MODSIM: Decision Support System for River Basin Management, Documentation and User Manual*. May. Fort Collins, Colo.: Colorado State University.

Cuddy, S.M., F.M. Marston, and T.F.N. Farley. 2000. "Which Buttons and Bars? An Exercise In Community Participation In Decision Making Software Development." In R. Denzer, D.A. Swayne, M. Purvis, and G. Schimak (eds). *Environmental Software Systems: Environmental Information and Decision Support*, pp. 213–220. Boston, Mass.: Kluwer Academic Publishers.

Flanagan, D. 1999. *JAVA in a Nutshell*, 3rd edn. Sebastopol, CA: O'Reilly and Associates, Inc.

Fredericks, J., J. Labadie, and J. Altenhofen. 1998. "Decision Support System for Conjunctive Stream–Aquifer Management." *Journal of Water Resources Planning and Management*, March/April. 124(2):69–78.

Honda, G., and K. Martin. 2002. *The Essential Guide to Internet Business Technology*. Upper Saddle River, NJ: Prentice Hall PTR.

Istok, J.D. 1989. *Groundwater Modeling by the Finite Element Method*. Water Resources Monograph, 13. Washington, DC: American Geophysical Union.

Keen, P., and M. Scott Morton. 1978. *Decision Support Systems: An Organizational Perspective*. Reading, MA: Addison-Wesley Publishing.

Labadie, J. 1995. *River Basin Network Model for Water Rights Planning, MODSIM: Technical Manual*. Fort Collins, CO: Department of Civil Engineering, Colorado State University.

Loucks, D.P. 2003. "Managing America's Rivers: Who's Doing It?" *International Journal of River Basin Management* 1(1):21–31.

Loucks, D.P., and M.B. Bain. 2002. "Interactive River–Aquifer Simulation and Stochastic Analyses for Predicting and Evaluating the Ecologic Impacts of Alternative Land and Water Management Policies." In M.V. Bolgov (ed.) *Hydrological Models for Environmental Management*, 169–194. Dordrecht: Kluwer Academic Publishers.

Loucks, D.P., and K.A. Salewicz. 1989. *IRIS – An Interactive River System Simulation Program, General Introduction and Description*. Laxenburg, Austria: International Institute for Applied Systems Analysis.

Loucks, D.P., and O.T. Sigvaldason. 1982. "Multiple-Reservoirs Operation in North America." In Z. Kaczmerek and J. Kindler (eds). *The Operation of Multiple Reservoir Systems*, 1–104. IIASA Collaborative Proceedings Series CP-82-S3. Laxenburg, Austria: International Institute for Applied Systems Analysis.

Loucks, D.P., K.A. Salewicz, and M.R. Taylor. 1990. *IRIS – An Interactive River System Simulation Program, User's Manual Version 1.1*. Laxenburg, Austria: International Institute for Applied Systems Analysis.

Miettinen, K.M. 1998. *Nonlinear Multiobjective Optimization*. International Series in Operations Research & Management Science, 12, 1st edn. Boston, Mass.: Kluwer Academic Publishers.

Ministry of Land, Infrastructure and Transport, Infrastructure Development Institute. 2001. *Study of Advanced Technology for Use of Global Geographic Information System, Creation of the Global Map Data Base for the Ganges River Basin and Trial Construction of the Simulation Models for Water Resources Management and Disaster Prevention Using Global Map Data*. March. Tokyo: Infrastructure Development Institute.

Nandalal, K.D.W., and S.P. Simonovic. 2002. *State-of-the-Art Report on Systems Analysis Methods for Resolution of Conflicts in Water Resources Management*. Paris: Division of Water Sciences, UNESCO.

Palomo, J., D. Rios-Insua, and K.A. Salewicz. 2002. "Reservoir Management Decision Support." In *Proceedings from iEMSs 2002 Integrated Assessment and Decision Support, 24–27 June 2002, Lugano, Switzerland* 3:229–234. Manno, Switzerland: Elsevier. Internet: ⟨http://www.iemss.org/iemss2002/⟩ (visited 12 October 2003).

Parker, B.J., and G.A. Al-Utaibi. 1986. "Decision Support Systems: The Reality That Seems Hard to Accept." *OMEGA International Journal of Management Sciences* 14(2):135–143.

Rardin, R.L. 1997. *Optimization in Operations Research*. Upper Saddle River, NJ: Prentice Hall.

Rodriguez, A., J. Gatrell, J. Karas, and R. Peschke. 2001. *TCP/IP Tutorial and Technical Overview*. IBM Redbook Series. August. Research Triangle Park, NC: ibm.com/redbooks

Rosenthal, R.E. 1985. "Concepts, Theory, and Techniques – Principles of Multi-objective Optimization." *Decision Sciences* 16:133–152.

Sabherwal, R., and V. Grover. 1989. "Computer Support for Strategic Decision-Making Processes: Review and Analysis." *Decision Sciences* 20:54–76.

Sage, A.P., and J.E. Armstrong. 2000. *Introduction to Systems Engineering*. New York: Wiley-Interscience.

Salewicz, K.A. 2001. "Decision Support Systems – How They Could Be Developed in a Web Environment." Conceptual Paper, presented at ELI seminar: "Improving International Water Management through Transparency and Public Participation," 6 December 2001, Washington, DC.

Salewicz, K.A. 2003. "Building the Bridge Between Decision-Support Tools and Decision-Making." In M. Nakayama (ed.). *International Waters in Southern Africa*, 114–135. Tokyo: United Nations University Press.

Salewicz, K.A., and D.P. Loucks. 1989. "Interactive Simulation for Planning, Managing and Negotiating." In D.P. Loucks (ed.). *Closing the Gap Between Theory and Practice*, 263–268. IAHS Publication No. 180. Wallingford, Oxfordshire: International Association of Hydrological Sciences.

Shim, J.P., M. Warkentin, J.F. Courtney, D.J. Power, R. Shards, and Ch. Carlsson. 2002. "Past, Present and Future of Decision Support Technology." *Decision Support Systems* 33:111–126.

Simonovic, S.P. 2000. "Tools for Water Management – One View of the Future." *Water International* 25(1):76–88.

Tung, Lai Lai, and Mohammed A. Quaddus. 2002. "Cultural Differences Explaining the Differences in Results in GSS: Implications for the Next Decade." *Decision Support Systems* 33:177–199.

Turton, A., A. Earle, and K. Wessels. 2002. *Okavango Pilot Project*. Proceedings of Workshop One, Maun 9–12 September 2002. Report of African Water Issues Research Unit. University of Pretoria, South Africa: AWIRU.

USACE (US Army Corps of Engineers). 1982. *HEC-5 Simulation of Flood Control and Conservation Systems, User's Manual*. CPD-5A. Davis, CA: Hydrologic Engineering Center.

USACE (US Army Corps of Engineers). 1985. *Reservoir System Analysis for Conservation, HEC-3 User's Manual*. CPD-3A. Davis, CA: Hydrologic Engineering Center.

USACE (US Army Corps of Engineers). 1986. *HEC-5(Q) Simulation of Flood Control and Conservation Systems; Appendix, Water Quality Analysis*. Davis, CA: Hydrologic Engineering Center.

USACE (US Army Corps of Engineers). 1987. *HEC-DSS User's Guide and Utility Program Manual*. Davis, CA: Hydrologic Engineering Center.

USACE (US Army Corps of Engineers). 1992. *Water Profile of Open or Artificial Channels, HEC-2 User's Manual*. TD-26. Davis, CA: Hydrologic Engineering Center.

USACE (US Army Corps of Engineers). 1995. *HEC-RAS User's Manual*. Davis, CA: Hydrologic Engineering Center.

USDOI (US Department of the Interior, Bureau of Reclamation, Pacific Northwest Region). 2000. *River and Reservoir Operations Simulation of the Snake River – Application of MODSIM to the Snake River Basin*. May. Boise, Idaho: US Department of the Interior, Bureau of Reclamation, Pacific Northwest Region.

Venema, H.D., and E.J. Schiller. 1995. "Water Resources Planning for the Senegal River Basin." *Water International* 20(2):61–71.

Wang, H.F., and M.P. Anderson. 1995. *Introduction to Groundwater Modeling: Finite Difference and Finite Elements Methods*. New York: Academic Press.

Zagona, E.A., T.J. Fulp, H.M. Goranflo, and R. Shane. 1998. "RiverWare: A General River and Reservoir Modeling Environment." In *Proceedings of the First Federal Interagency Hydrologic Modeling Conference* 5–113. Las Vegas, NV: Subcommittee on Hydrology of the Interagency Advisory Committee on Water Data.

Zagona, E.A., T.J. Fulp, R. Shane, T. Magee, and H.M. Goranflo. 2001. "RiverWare." *Journal of the American Water Resources Association* 37(4):913–929.

20

Sketches from life: Adaptive ecosystem management and public learning

John Volkman

Introduction

In the United States, it is seldom hard to generate interest in environmental conflicts. Those concerned about the decline of species and other natural resources tend to be fervent and they are matched by landowners, developers, and others whose actions they challenge. Opportunities to engage these issues are not hard to find. Government agencies have legally mandated public-involvement requirements and hearings and opportunities for public comment abound. News media are quick to cover a good fight, particularly when endangered species and major developments are pitted against each other.

However, heat is one thing, and light is another. Notwithstanding public-involvement efforts, public opinion polls show that people feel cut off from decision-making and mistrust the government's ability to solve problems. Public agencies often feel under assault from people who take positions that seem to be scripted by organized lobbies. There is a felt need to engage these issues, not just as partisans but with a willingness to explore solutions that respect the risks implicit in choice.

The problem has several aspects. One is the difficulty in seeing resource problems in a large enough ecological context to enable them to be dealt with effectively. Ecosystems have been likened to chains of favourable environmental conditions: different habitats and ecological processes form links in the chain (Lichatowich 1999). Whether a particular habitat is an important link and whether a proposed action will break that link is often unclear. In restoration programmes, it may make little sense to restore one or two links if other links in the chain cannot be fixed; a chain with one failed link is still broken. However, although these concepts may be clear enough, information about ecosystem dynamics usually is too sketchy to enable reliable conclusions to be drawn. One can develop models of these systems, but they are at best incomplete sketches of obscured landscapes. A second aspect is institutional. Because ecosystems are not limited by institutional boundaries, addressing ecosystem problems requires the involvement of an array of government and non-government interests. Even superficially bridging these interests so that ecological problems can be dealt with requires a major effort. Both problems – the lack of ecological understanding and the mismatch between institutions and ecosystems – are likely to grow more difficult as population growth and global climate change put more pressure on natural systems and humans.

This chapter discusses a technique termed adaptive ecosystem management, which has been tested in the Columbia River and elsewhere, that seeks to help manage ecological uncertainty in decision-making (Walters 1997; Doremus 2001; Neuman 2001; Ladson and Argent 2002). The chapter describes the background of the idea, problems in its implementation – particularly difficulties in using models not just to sketch what is known but to prod experimentation to deepen understanding of these systems – and how these problems might be addressed. Recognizing that important elements of these problems are institutional, the chapter discusses how local adaptive management initiatives appear to be managing some of these institutional issues, and how these efforts might be integrated on an ecosystem scale.

Theory and practice of adaptive management

Ecosystem management requires learning

Trying to understand how ecosystems work in a scientific sense is difficult enough – it is an open question whether ecosystems can remain still long enough to be understood. Ecosystem management suggests something much harder: that humans not only can understand these systems but also can restrain human activities in order to protect them. The mixture of human and ecological challenges in a single proposition poses an enormous challenge, and little in our history suggests we are equal to it. No one in nineteenth-century America imagined that hunting sea otters to extinction would lead to the collapse of the Pacific sardine fishery, but such appears to have been the case (McEvoy 1986). To have avoided it would have required the development of new laws or institutions to protect species whose ecological function was not understood at the time, at a price that people probably would not have paid. It is debatable whether we can do better today. We have more information, but it is rarely definitive, raising alarm in some quarters and suggesting a need for measured compromise in others. Middle-ground solutions may be good politics, but if they are not based on good information they may be ecologically irrelevant. A species that cannot live in water at 62°F (17°C) takes little solace from a riparian restoration programme that lowers stream temperatures to 68°F. We might or might not do better now.

Managing in the midst of complexity, uncertainty, and politics is nothing new to human experience, of course, and we should not overstate the problem. Medical science is said to be only about half right, and we are not sure which half (Sanders 2003). Although we are not always happy with medical judgements, medicine is viewed as an important endeavour, imperfect but workable. Ecosystem management needs to find its own brand of imperfect workability.

That said, it is also true that ecosystem management faces unprecedented problems. Growing populations and climate change promise to put natural systems under greater pressure with unforeseeable consequences. Distinguishing problems that occur locally from those that are caused by regional, national, or global phenomena, and finding solutions that account for effects on multiple scales, pose novel scientific problems. Because solutions will require broader political support to be effective, government and non-government actors will have to find ways to interact across political boundaries. It is no comfort that the same things are likely to put pressure on other areas of life, not just on natural systems. As many things become less stable at once, coming to grips with causes and effects and constructing workable remedies will only become harder and more essential.

Ecosystem management blurs lines between government and non-government interests

Federal, state, and tribal governments in the United States have limited capacity to deal with ecological problems. In this volume and elsewhere, Bradley Karkkainen uses the term "post-sovereign governance" to describe the problem – not a withering away of the state, but the emergence of problems that government can influence only in alliance with others – tribes, non-government organizations (NGOs), and private parties (Karkkainen 2003).

The list of mismatches between governments and ecosystems is well known. There are mismatches between government and ecosystem boundaries: governments operate on election and budgetary cycles, while ecological problems evolve over decades and centuries. In constitutional democracies, governments have limited authority to control human activities that affect ecosystems. Ecosystem problems implicate so many entities equipped with lawyers and political champions, that action can be frozen at many points. In some parts of the United States, subjects such as water law policy can hardly be discussed without invoking the metaphor of gridlock. It is not clear that other forms of government fare better. Although totalitarian regimes may dictate some aspects of human behaviour, they are not insulated from economic or environmental consequences.

Government institutions will be increasingly strained as the pressures of population growth and climate change mount. Changing natural systems with unforeseeable consequences require institutions that can adapt more quickly and flexibly. Government institutions limited by these mismatches badly need management strategies that allow them to address problems in collaboration with others. The question is, if ecosystem problems can be dealt with only in coalitions with other government and non-government organizations, how can these broad and shifting coalitions engage broad and shifting ecological problems?

Adaptive management

"Adaptive management" is a broad rubric that refers to learning from decisions and applying this new knowledge in future rounds of decision-making. This broad concept has been operationalized in various ways, one of which (termed "adaptive ecosystem management") was developed by C.S. Holling (1978) and Carl Walters (1986). Adaptive ecosystem management proposes a specific procedure, as follows:
- Develop computer models to organize scientific information about how the landscape at issue (watershed, river, species, etc.) works. This is done collaboratively, so that people involved in the issue can understand and debate the data and uncertainties.
- Use models to see what may be needed to protect the landscape, explore ways to avoid conflicts, and shape the outcome into management programmes.
- Implement management programmes in experimental frameworks to test them; revise the assumptions in the computer model according to what is learned; and use this new information in further decisions.

The approach has many attractions as a tool of ecosystem management. It suggests a device for keeping in view the ecosystem context for resource problems. By drawing a picture of the system in a computer model, questions about ecological connections can be, if not answered, at least asked. Adaptive ecosystem management also suggests a potential cure for institutional fragmentation. If disparate government and non-government actors buy into the same computer model, use the same data, and test the same assumptions, they will begin to develop common understandings and knit their actions together.

Adaptive ecosystem management has been implemented in a variety of settings and, although there is still great interest in the idea (Doremus 2001; Neuman 2001; Ladson and Argent 2002), implementation problems have emerged. Carl Walters, one of the progenitors of adaptive ecosystem management, describes the experience thus:

Unfortunately, adaptive-management planning has seldom proceeded beyond the initial stage of model development, to actual field experimentation. I have participated in 25 planning exercises for adaptive management of riparian and coastal ecosystems over the last 20 years; only seven of these have resulted in relatively large-scale management experiments, and only two of these experiments would be considered well planned in terms of statistical design (adequate controls and replication). In two other cases, we were unable to identify experimental policies that might be practical to implement. The rest have either vanished with no visible product, or are trapped in an apparently endless process of model development and refinement. Various reasons have been offered for low success rates in implementing adaptive management, mainly having to do with cost and institutional barriers. (Walters 1997)

The Columbia River is one of the places where adaptive management has hit rough water, if not foundered. Adaptive management has been a subject of interest in the Columbia River Basin since the 1980s, when Kai Lee, a member of the Northwest Power Planning Council, advanced the idea to guide conflicts between hydropower dams and salmon (Lee and Lawrence 1986; Volkman and McConnaha 1993; McConnaha and Paquet 1996). By the early 1990s, when the first salmon runs in the Columbia River Basin were listed under the Endangered Species Act (ESA), the idea was still being discussed but implementation proved difficult. After the ESA listings, decision makers initiated a more structured process for implementing adaptive management in a project known as the Process for Analyzing and Testing Hypotheses (PATH) (Marmorek and Peters 2002).

PATH proposed to use computer models to test hypotheses explaining the decline of Snake River salmon, which were the first ESA-listed Northwest salmon populations. The PATH project, endorsed by the National Marine Fisheries Service and funded by the Bonneville Power Administration, drew a broad range of interested parties. Diverse technical analysts were convened to review data and identify alternative explanations and potential solutions for the declines in the salmon population. The explanations were treated as hypotheses in computer simulations. If a hypothesis produced results that approximated historic data, the hypothesis would be strengthened; if not, it would be weakened. PATH thus proposed to sift hypotheses, narrow the range of explanations, and draw inferences about potential solutions. The process was exhaustive. Almost any hypothesis that PATH's large and diverse group of analysts offered was surveyed. Where there were disagreements over analytical models, PATH incorporated alternative analyses using multiple models and used the results to test the models.

In the autumn of 1998, after several years of work, PATH prepared a "weight-of-evidence" report that summarized the analyses, complete with alternative views and explanations. This material was presented to a Scientific Review Panel composed of four respected scientists from outside the basin. The Review Panel evaluated the evidence and issued its own report ascribing weights and probabilities to competing hypotheses and evaluating models and potential solutions. Regarding solutions, the Review Panel said that breaching four Snake River hydropower dams would make it more likely that Snake River spring chinook would reach recovery thresholds, but this conclusion was hedged by uncertainty (Carpenter et al. 1998).

Although its data analysis had been exhaustive, PATH spent little time in identifying management experiments that would verify assumptions used in modelling. The Review Panel report commented only generally on the subject of experimentation, describing:

three possible experimental manipulations: dam removal, elimination or substantial reduction of hatchery releases, and transportation turn-off. Implementing these actions in a well-designed experimental fashion can provide stronger evidence of each factor. (Carpenter et al. 1998)

The Panel outlined two ways to approach such a programme: (1) take the cheapest action first, monitor the effects, and then take progressively more costly steps; or (2) take all the actions at once and then restore each one and evaluate the results (Carpenter et al. 1998).

The Scientific Review Panel's report qualified endorsement of dam removal produced an electric reaction: some people portrayed the Review Panel's report as a definitive resolution of the scientific issues; some argued with the data; others criticized the "dysfunctionality" of the process (Anderson 1999).

Growing controversy over the PATH report coincided with other events that tended to shift attention from the Snake River dams. Several Columbia River salmon populations were added to the endangered species list in 1998, some of which were in worse shape than the Snake River populations. In April, the National Marine Fisheries Service issued a report that noted the PATH results, agreed that removing four dams would probably help Snake River fish, but observed that it was not clear that dam removal was either necessary or sufficient for recovery of Snake River fish and would do nothing for Columbia River populations (Northwest Fisheries Science Center 1999). The Service subsequently issued a biological opinion on the effects of operation of the Columbia River dams. The opinion did not recommend Snake River dam removal, deferring that question pending implementation of a variety of habitat-rehabilitation programmes. A federal court later found the opinion to be "arbitrary and capricious" and sent it back to the agencies for reconsideration. In September 2004 the agencies responded with a new opinion proposing various measures, but reaffirming the agencies' view that their approach would not jeopardize salmon. The opinion virtually ensures further litigation.

The PATH process reflects many of the same problems that Walters identified: on the large ecosystem scales, adaptive ecosystem management can get stuck in modelling, struggle to identify practical experimental policies, and dissolve in institutional conflict over underlying substantive issues.

Could it work better?

Modelling and other ways to learn

Modelling has become a key tool in resource management, and the fact that models have trouble in capturing the shifting mass of fuzzy variables known as ecosystems is no reason to ignore their use. Models help to structure the central concept of ecological thought – that ecosystems are chains and that changes in one link affect all the others. Models also can be a group discipline, forcing people to explicitly identify their data and assumptions. Although models may oversimplify problems, simplification can be a virtue, allowing one to reduce complex problems to more tractable forms. Nevertheless, this virtue can also be a vice. Given the complexity of ecosystems and how poorly we understand them, how confident can we be that a model has not simplified vital processes out of the picture? Models are sketches of a reality we only partly see. They can help one to detect blind spots (Peterson et al. 1997), but they also may deepen one's blindness if they are mistaken for reality.

A case in point is as follows. In the Columbia River, salmon life-cycle models were first developed at a time when fisheries management was in a pre-ecosystem management stage of thinking. Managers thought of the system they were managing, in this way: if more fish were produced from hatchery A and fewer fish were harvested at point B, will enough fish return to hatchery A to supply eggs for the next generation? Remedial efforts were built on similar assumptions. Hatcheries would compensate for lost spawning and rearing habitat (Wilkinson and Conner 1983; Northwest Fisheries Science Center 1999); dam turbines could be screened to keep fish out of them; and fish could be transported around dams in barges. Managers assumed that if enough of these actions were taken, salmon populations would persist in harvestable numbers.

The initial salmon life-cycle models were suited to this kind of thinking. They were essentially calculators, counting how many salmon survived at each point in the life cycle – as salmon emerged from eggs, when juvenile salmon reached the river's mainstem, at each of the dams and reservoirs, in the estuary, the ocean, etc. The models were used to play "what if?" games – what if we killed fewer fish at this point or that, produced more fish from hatcheries, improved bypass systems at dams, added barges, etc? The first generation of computer models pushed managers to connect these parts explicitly and justify their assumptions about what they would produce, and in this sense these were "system" models. However, the variables in the system were more technological than ecological, reflecting the assumptions of the time. Refined versions of these models were used in the PATH process.

A richer and more comprehensive sketch of the salmon ecosystem emerged from two narrative reports in the mid-1990s. In the early 1990s, when Snake River salmon were put on the list of endangered and threatened species, Congress asked the National Academy of Sciences to review the science underlying the salmon declines. The National Research Council's Committee on Protection and Management of Pacific Northwest Anadromous Salmonids published a final report in 1996 (National Research Council 1996). The report emphasized that the salmon problem is an ecosystem problem for which there is no single solution (Magnuson 1995). It pointed out the limitations of technological mitigation strategies and concluded that no combination of them would lead to recovery. Rather, the report contended that the Columbia River ecosystem needed a higher degree of ecological function if salmon were going to recover to self-sustaining levels. These findings, along with a long list of difficulties and dilemmas were apparent in the report's title: *Upstream: Salmon and Society in the Pacific Northwest*. Subsequently, a science panel commissioned by the Northwest Power Planning Council filed a similar report with a similar verdict. *Return to the River: Restoration of Salmonid Fishes in the Columbia River Ecosystem* (Independent Scientific Group for the Northwest Power Planning Council 2000) concluded that unless more ecological function could be restored to the Columbia River ecosystem, wild salmon face extinction over the next 50–100 years. No combination of technologies could do anything more than delay the inevitable.

The authors of *Return to the River* illustrated their point with a "working river" metaphor. The contemporary working river does the work of power generation, irrigation, flood control, and navigation, the report observed. This river "works" because dams and other technologies have simplified the river's complexity. The dams enable us to manage flow releases to respond to human demand for electric energy, protect against floods, and float over the cataracts that once made the river so difficult to navigate. Hatcheries, barges, turbine screens, and other mitigation programmes aim for a simplified salmon population, one that is released from hatcheries and migrates in time periods that fit harvest plans and minimize conflicts with hydropower generation. In Richard White's phrase, the river is an organic machine (White 1995), although a vastly simplified version of a natural river.

The biological problem, according to *Return to the River*, is that biologically productive rivers are complicated. They have braided channels, intricate hydrological processes, and huge populations of insects. They have rapids and falls. They may flood and recede, change channels, and push sediment and gravel around. These complex rivers are also "working rivers" because their natural functions work to transform energy into nutrients and support diverse species. If the Columbia were this kind of working river, there would be a resilient salmon population with many populations migrating at different times, returning to different habitats, and interacting in obscure and unpredictable ways. Salmon recovery, *Return to the River* contended, would require a shift towards ecological complexity.

These two narrative reports did a great deal to widen awareness of factors that the Columbia River models left out. Everyone knew that the models had limitations, including their failure to account for unquantifiable ecological relationships, genetic diversity, and other factors. *Upstream* and *Return to the River* are not just literature reviews or model critiques, but works of scientific imagination that fill gaps left by the models with well-grounded hypotheses. Their conclusion that technological solutions cannot compensate for a basic lack of ecological function tended to refocus the political debate and reshape the models. After the reports were published, a new generation of models was developed that pay much closer attention to ecological connections. A similar phenomenon has been observed in ecological science generally, through genetic architecture, structured populations, and spatial processes (Kareiva 2003).

However, although these reports added richness to the pictures that had been painted by the models, much of this richness was hypothesis.

Without experimental data to validate their ideas, these reports were still a mixture of imagination and reality. The lesson of Columbia River model development and the scientific reflection in these reports is that no single technique is enough. One's field of vision is always constrained in some manner – by the pressures of the moment, the weather conditions of the season, or the characteristics of the model (Gaddis 2002). One needs models, a broad and imaginative scientific context, and a feel for recurring patterns that a mathematician or historian may see and a biologist may not (Holling 2002), but one also needs data based on observation.

Management experiments and the problem of scale

The difficulty that PATH encountered in developing a systematic programme of management experiments with which to generate better data in the Columbia had two causes. One was that modelling seemed to become the entire focus of the scientific debate. Rather than accepting the models as quick sketches, decision makers took them as producers, if not of answers then at least of consensus on which decisions could be based. As a result, the modelling process became more and more complex, expensive, and exhausting. The complexity of the process screened out non-technical parties. The technical experts who remained, squeezed the models so hard and so long that they had little time, money, or patience to address the need for management experiments.

The second cause was the intrinsic difficulty of experimenting with large-scale management programmes. The Scientific Review Panel's report in the PATH process, with its discussion of turning large human systems on and off, reflects the dilemma. In big ecosystems, experiments must be large enough to produce measurable change; however, the risk and expense of large-scale change in important human systems are so large as to be virtually untouchable. Gauging how far experiments have to go to be informative is largely a scientific question, but judging how far they can go before they trigger legal and political land-mines is not. The experimental manipulations suggested by the Scientific Review Panel set off alarm bells in many corners and contributed to the perception that the adaptive management process had failed (Volkman 1999).

The emerging wisdom is that, in so far as management experiments have real consequences for choices, they are less viable as the size of the ecosystem and the stakes increase (Ladson and Argent 2002). The Columbia River Basin is roughly the size of France. The stakes involved in experimental manipulation of the hydropower system are large, implicating half the Northwest's electric energy supply (Volkman and McConnaha 1993); therefore, experimental management is unrealistic.

In contrast, there seems to be growing evidence of adaptive management working at lower scales – an upward percolation of initiatives from decentralized regulatory programmes. Federal Energy Regulatory Commission (FERC) licensing negotiations, for example, are using adaptive management to manage conflicts over fish and wildlife licence terms (EPRI 2000). The Fish and Wildlife Service and National Oceanic and Atmospheric Administration (NOAA) Fisheries encourage the use of adaptive management in the ESA granting of permits (US Departments of Interior and Commerce 1996; *Federal Register* 2000). At Glen Canyon Dam on the Colorado River, a promising adaptive management programme emerged from a legislatively mandated National Environmental Policy Act process. The Grand Canyon Protection Act of 1992 mandated an environmental impact statement and long-term monitoring of dam-operation effects on the river (Committee on Grand Canyon Monitoring and Research 1999). The impact statement adopted a flow programme for the dam, recognized uncertainties about how those benefits could be achieved, and called for flows to be managed experimentally in an adaptive management programme. Even in the Columbia River Basin, a sub-basin planning process is under way to identify salmon recovery measures at the local level, although the extent to which this will fit into an adaptive management framework is unclear (Volkman 1999). The jury is still out on whether these initiatives will produce meaningful experimentation: they may be examples of adaptive management as a negotiation tool rather than as a framework for experimental management; if so, they may merely postpone conflict. Nevertheless, at this point, they may be taken as positive signs that adaptive management is finding a more hospitable environment at smaller scales.

If large-scale experimental manipulations are infeasible, the next question is whether one can learn greater lessons by aggregating small-scale, low-stakes experiments across large landscapes? Some commentators think so:

> [W]e are likely to find that the slow rate at which we can gain experience with complex systems can be greatly accelerated if we can pool and compare observations of subsystems. Experiences in similar, proximate subsystems can be aggregated into a relevant collective experience applicable to the scale of those subsystems. This is probably as close as we can come to controlled experiments in these systems (Walters 1986), but it is possible to learn a lot this way. (Wilson 2002)

Cross-scale learning seems to make sense conceptually. Whether it is feasible depends partly on who would do it – an institutional question.

Institutional learning

Many of the problems of adaptive management are institutional. Adaptive management needs to find a home in agency missions in order not

to be merely a free-floating idea. Someone needs to systematically identify propositions that need to be tested to answer important questions,
manage modelling, and ensure that a suitable degree of effort is devoted
to management experiments. When experiments are designed, someone
needs to ensure that they are carried out and the results fed back to analytical and policy bodies. One of the chief problems in implementing
adaptive management is the lack of someone to do these things: no one
is charged with the day-after-day, year-after-year, care and feeding of
such a process. This takes us back to a question posed at the beginning
of this chapter: if broad, shifting, ecological problems can be dealt with
only by broad, shifting, institutional coalitions, how can long-term learning occur? Learning requires on-going management, long-term data sets,
and experimental designs across large landscapes. "Post-sovereign" coalitions that coalesce locally and fall apart every few years cannot fulfil
the need.

Broken into pieces, the problem may not be as dire as it sounds. Some
things can be done on smaller scales. In the FERC licensing, ESA granting of permits, and other processes mentioned earlier, government and
non-government interests were able to institutionalize adaptive management programmes using existing regulatory mechanisms. Existing regulations limit the geographic scale of these processes, establish time-frames
for learning (30–50-year permit terms), and check points to accommodate new information and systems of accountability, including future opportunities for wholesale reconsideration of issues.

What parties in smaller-scale settings cannot be expected to do is to
plan and coordinate learning across entire ecosystems. The missing link
is an entity to identify questions that should be addressed by experiences
in subsystems, to develop experimental designs by which comparative
studies could generate larger insight, to coordinate with smaller-scale
processes, and to aggregate and report the resulting information.

Notwithstanding the interest that some Columbia River Basin parties
have had in adaptive management, they have been unable to implement
a system-wide monitoring, evaluation, and adaptive management system.
There are probably many reasons why this link is missing in the Columbia River. Determining which questions should be addressed, and how, is
a great deal more difficult than it sounds. Given the urgency of environmental problems, a seemingly abstract learning enterprise can find it difficult to compete for funds. There may also be political reasons – some of
the same problems that plague large-scale management experiments. Because adaptive management works from a scientific model, it challenges
the underpinnings of policy. One of the tools of science is the null hypothesis, an assertion that contradicts the experimenter's belief. "Truth"
emerges only as alternative explanations (null hypotheses) are rejected.

This process serves science and policy by helping to avoid false explanations. However, null hypotheses are not likely to be welcomed by programme managers fighting budget battles or politicians looking for more straightforward answers than are given by programmes whose underpinnings are in question. In the short term, proving a thing true may be less effective than acting as though it were true.

For these and other reasons, the institutional support for large-scale experimental frameworks with which to develop more powerful depictions of ecological processes has not materialized. However, if the null hypothesis is that it cannot be done, the experience to date suggests only that it has not been done yet.

Conclusions

Experience with adaptive ecosystem management has demonstrated the difficulty of the process and the limitations of the tools. Nevertheless, the objectives of adaptive management remain compelling and the tools are evolving. As people gain experience with computer models, they are improving such models and understanding their limitations better. Large-scale ecosystem-management experiments may be intrinsically rare, but it is demonstrably possible to experiment on smaller scales. Dealing with institutional problems to achieve cross-scale learning is a further challenge, but it does not seem intrinsically unmanageable.

If one could make adaptive management work, larger problems would remain, of course. Whether people can respond effectively to environmental change is not just a matter of environmental policy, but a test of the idea that humans can adapt intentionally. Intentional, cultural, adaptation may be better addressed by methods other than ecological analysis: the techniques of history, economics, neuropsychology, and literature may be better adapted to study this messy subject. Virginia Woolf once asked, "Do you think it is possible to write the life of anyone? I doubt it, because people are all over the place" (Lee 1999). Capturing the essence of human interactions with ecosystems is an even more dubious prospect, but it may be a tool of adaptation.

REFERENCES

Anderson, J. 1999. Quoted in: *Clearing Up* 862:10 (Jan. 25).
Carpenter, S., J. Collie, S. Saila, and C. Walters. 1998. "Conclusions and Recommendations from the PATH Weight of Evidence Workshop." 8–10 September.
Committee on Grand Canyon Monitoring and Research, National Research

Council. 1999. *Downstream: Adaptive Management of Glen Canyon Dam and the Colorado River Ecosystem.* Washington, DC: National Academy Press.

Doremus, H. 2001. "Adaptive Management, the Endangered Species Act, and the Institutional Challenges of 'New Age' Environmental Protection." *Washburn Law Journal* 41:54.

EPRI. 2000. *Hydro Relicensing Forum: Relicensing Strategies.* National Review Group Interim Publication, December.

Federal Register. 2000. 65:35242. Addendum.

Gaddis, J. 2002. *The Landscape of History: How Historians Map the Past.* Oxford: Oxford University Press.

Holling, Buzz. 2002. "Adaptive Management for Water Resources." Conference Proceedings of Simon Fraser University's Workshop, Water and the Future of Life on Earth (22–23 May). Internet: ⟨http://www.sfu.ca/cstudies/science/water/pdf/Water-Ch13.pdf⟩ (visited 14 October 2003).

Holling, C.S. (ed.). 1978. *Adaptive Environmental Assessment and Management.* New Jersey: Wiley.

Independent Scientific Group for the Northwest Power Planning Council. 2000. *Return to the River: Restoration of Salmonid Fishes in the Columbia River Ecosystem.* Portland, Oreg.: Northwest Power and Conservation Council.

Kareiva, Peter. 2003. "Why Worry About the Maturing of a Science?" National Center for Ecological Analysis and Synthesis. Internet: ⟨http://www2.nceas.ucsb.edu/fmt/doc?/search/⟩ (visited 14 April 2003).

Karkkainen, Brad. 2003. "Adaptive Ecosystem Management and Regulatory Penalty Defaults: Toward a Bounded Pragmatism." *Minnesota Law Review* 87:943.

Ladson, A., and R. Argent. 2002. "Adaptive Management of Environmental Flows: Lessons for the Murray–Darling Basin from Three Large North American Rivers." *Australian Journal of Water Resources* 5:89.

Lee, H. 1999. *Virginia Woolf.* New York: Vintage.

Lee, K., and J. Lawrence. 1986. "Adaptive Management: Learning From the Columbia River Basin Fish and Wildlife Program." *Environmental Law* 16:431.

Lichatowich, Jim. 1999. *Salmon Without Rivers: A History of the Pacific Salmon Crisis.* Washington, DC: Island Press.

Magnuson, John. 1995. Quoted in National Research Council November 8 Press Release. (On file with the author).

Marmorek, D., and C. Peters. 2002. "Finding a PATH toward Scientific Collaboration: Insights from the Columbia River Basin." *Conservation Ecology* 5(2):8. Internet: ⟨http://www.consecol.org/vol5/iss2/art8/⟩ (visited 14 October 2003).

McConnaha, W., and P. Paquet. 1996. "Adaptive Strategies for the Management of Ecosystems: The Columbia River Experience." *American Fisheries Society Symposium* 16:410. Bethesda, Md.: American Fisheries Society.

McEvoy, A. 1986. *The Fisherman's Problem: Ecology and Law in the California Fisheries, 1850–1980.* Cambridge, UK: Cambridge University Press.

National Research Council. 1996. *Upstream: Salmon and Society in the Pacific Northwest.* Washington, DC: National Academy Press.

Neuman, J. 2001. "Adaptive Management: How Water Law Needs to Change." *Environmental Law Reporter* 31:11432.

Northwest Fisheries Science Center, National Marine Fisheries Service. 1999. "An Assessment of Lower Snake River Hydrosystem Alternatives on Survival and Recovery of Snake River Salmonids." Appendix to the US Army Corps of Engineers' "Lower Snake River Juvenile Salmon Migration Feasibility Study," pp. 6–7. 14 April.

Peterson, G., G.A. De Leo, J.J. Hellmann, M.A. Janssen, A. Kinzig, J.R. Malcolm, K.L. O'Brien, S.E. Pope, D.S. Rothman, E. Shevliakova, and R.R.T. Tinch. 1997. "Uncertainty, Climate Change and Adaptive Management." *Conservation Ecology* 1(2): article 4. Internet: ⟨http://www.ecologyandsociety.org/vol1/iss2/art4/⟩ (visited 14 October 2003).

Sanders, Lisa. 2003. "Medicine's Progress: One Setback at a Time." *New York Times Sunday Magazine*. 16 March.

US Departments of the Interior and Commerce. 1996. *Habitat Conservation Planning and Incidental Take Permit Processing Handbook*. Washington, DC: US Departments of the Interior and Commerce.

Volkman, J. 1999. "How Do You Learn From a River? Managing Uncertainty in Species Conservation Policy." *Washington Law Review* 74:719.

Volkman, J., and W. McConnaha. 1993. "Through a Glass, Darkly: Columbia River Salmon, the Endangered Species Act and Adaptive Management." *Environmental Law* 23:1249.

Walters, C. 1986. *Adaptive Management of Renewable Resources*. New York: Macmillan.

Walters, C. 1997. "Challenges in Adaptive Management of Riparian and Coastal Ecosystems." *Conservation Ecology* 1(2):1. Internet: ⟨http://www.consecol.org/vol1/iss2/art1⟩ (visited 14 October 2003).

Wilkinson, C.F., and D.K. Conner. 1983. "The Law of the Pacific Salmon Fishery." *Kansas Law Review* 32:17.

Wilson, J. 2002. "Scientific Uncertainty, Complex Systems, and the Design of Common-Pool Institutions." In Elinor Ostrom, Thomas Dietz, Nives Dolšak, Paul C. Stern, Susan Stonich, and Elke U. Weber (eds). *The Drama of the Commons*, 346. Washington, DC: National Academy Press.

White, Richard. 1995. *The Organic Machine*. New York: Hill and Wang.

21

The Colorado River through the Grand Canyon: Applying alternative dispute resolution methods to public participation

Mary Orton

Introduction

Environmental disputes are among the most difficult conflicts to resolve. Frequently, they reflect fundamental differences in values and include highly complex scientific and technical issues that are not easily understood by members of the public or by stakeholder groups (Daniels and Walker 2001). Allocation of scarce resources among competing stakeholders with legitimate claims can be a challenging part of the conflict. The disputes are often characterized by a significant amount of scientific uncertainty and they resist simple, unilateral solutions. The involvement of multiple parties, issues, and political jurisdictions compounds these difficulties (Dukes 1996; Kriesberg 1997).

These dilemmas present themselves in environmental disputes at both national and international levels. This chapter presents a domestic example, the lessons from which may be useful for tackling other national and international disputes.

When environmental disputes involve a US government agency, the National Environmental Policy Act of 1969 (NEPA) is often a factor. NEPA prescribes a process by which federal agencies must produce environmental analyses – environmental assessments or environmental impact statements (EISs) – when they undertake "major federal actions" (NEPA 1969). NEPA regulations also require the agencies to solicit and use public input at specified stages in the process (CEQ 2002a).

NEPA processes can involve a high degree of controversy and contentiousness between agency decision makers and their constituents, particularly if the issues are complex and highly contested and the stakeholder groups are polarized.

Applying alternative dispute resolution (ADR) techniques, such as those used in mediation, to NEPA requirements for public participation can reduce contentiousness and aid in resolving conflicts. The use of these techniques can improve communication between agencies and the public, thereby enhancing trust among the parties. It can also increase an agency's ability and willingness to include public values in its public-policy decisions. The result is an enrichment of the decision-making process, increasing the likelihood of producing durable management decisions and reducing the probability of litigation (Bingham and Langstaff n.d.).

This chapter illustrates the value of applying ADR methods to public participation during a NEPA process by examining the Grand Canyon National Park's revision of its 1989 Colorado River Management Plan. The Park, illustrated in figure 21.1, was required by NEPA to produce an environmental analysis and ensure public participation before finalizing the plan. Managing the process internally, the Park held several public meetings in 1997 and 1998 to gather input from stakeholders and constituents (Jalbert 2003a). [Unless otherwise cited, information about the 1997 public-participation process and events before that time is from personal communication with Linda Jalbert.] Primarily because the process became so contentious, the Superintendent of the Park halted the planning process in 2000 (Arnberger 2000b).

When the Park recommenced the planning process in 2002, its management chose a different approach. They retained the author's company, one that specializes in ADR approaches to environmental and public-policy disputes, to assist them with the public-participation processes. It is from that perspective that this chapter reviews the history of the process that began in 1997, and compares it with the process that the Park used when planning resumed in 2002.

This case description begins with an account of the Park's 1997 NEPA process to revise the Colorado River Management Plan, starting with identification of the stakeholder groups involved and the major issues that concerned them, followed by a description of the process used, including outcomes. After a brief account of the termination of the process in 2000 and its resumption in 2002, there follows an explanation of the process developed by the author and used by the Park in 2002 and a comparison of outcomes from the 1997 and 2002 processes.

The comparison demonstrates that the use of ADR techniques had a positive effect on the process of revising the Colorado River Manage-

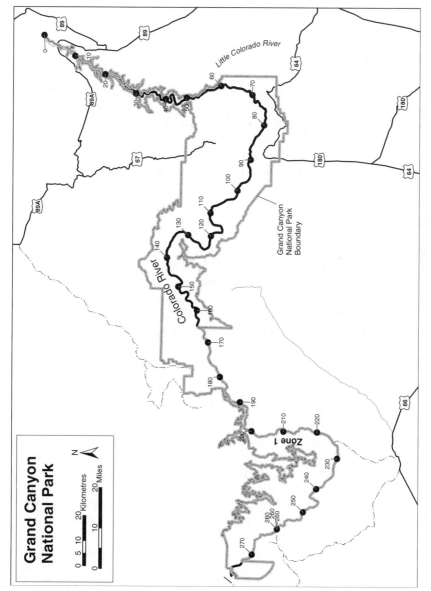

Figure 21.1 Map of the Colorado River through the Grand Canyon National Park (Source: Grand Canyon National Park)

ment Plan, and that use of ADR in these processes should be considered by public agencies involved in difficult and contentious NEPA processes.

1997: Colorado River Management Plan revision begins

The Colorado River

The Colorado River is the largest river in south-western North America, stretching for 1,450 miles from Wyoming to the Sea of Cortez in Mexico. On its route, it drains 246,000 square miles in parts of seven states in the United States and two in Mexico (McHenry 1993). Perhaps best known is the reach through Grand Canyon National Park, which provides the longest stretch of navigable white water in the continental United States (Jalbert 2003b).

The stakeholders and the issues

This chapter focuses on non-commercial, or private, boaters and the commercial river outfitters as two principal stakeholder groups represented during the Colorado River Management Plan revision process in 1997. Non-commercial boaters are also sometimes called "private boaters," although some in this category prefer the term "public boaters." In this chapter, the terms "private" and "non-commercial" are used interchangeably.

The two most contentious issues in the Colorado River Management Plan between these two groups were (1) how recreational river use should be allocated between commercial and non-commercial river users, and (2) whether motorized boats should be allowed on the river.

The allocation of recreational use between commercial and non-commercial boaters had been controversial for years. Limits on usage, established by the Park to protect natural and cultural resources and the visitor experience, made recreational use a scarce commodity. The allocation between the sectors in the 1989 Colorado River Management Plan was 68 per cent commercial and 32 per cent non-commercial. This was the status quo when the Park began to revise the plan in 1997 (US Department of the Interior 1989).

The commercial and non-commercial sectors used different systems to distribute their allocation to the end-user. By 1989, a waiting list to lead a non-commercial river trip had been established. It contained 6,800 names, resulting in a wait estimated by one group to be eight to ten years (Aronson 1997). By contrast, a reservation system distributed the commercial allocation. If a commercial customer were not able to reserve space on a river trip in a particular season, that customer would be required to contact the outfitter again for the next season (Grisham 2003).

In January 1997, the inaugural newsletter of the new Grand Canyon Private Boaters Association (GCPBA) described the obtaining of a permit to lead a non-commercial river trip on the Colorado River through Grand Canyon as "nearly impossible" (Martin 1997a). GCPBA's primary objective was to increase the allocation of user days to non-commercial boaters. Conversely, the Grand Canyon River Outfitters Association, comprising the 16 outfitters with contracts from the Park to provide the public with river trips through the Grand Canyon, favoured retaining the current allocation (Grisham 2003).

The issue of whether motorized boats should be allowed on the Colorado River through Grand Canyon, like the allocation question, had been controversial for years. This complex dispute stems from different ways of interpreting three public documents – namely (1) the Wilderness Act, (2) a proposal for wilderness designation submitted by the Park, and (3) National Park Service policy *vis-à-vis* proposed wilderness.

The Wilderness Act describes requirements for designation and management of wilderness areas (Wilderness Act 1964). Section 3(c) of the Act prohibits "motor vehicles, motorized equipment [and] motorboats" in designated wilderness areas. The Act also specifies in Sections 4(c) and 4(d) that "the use of aircraft or motorboats, where these uses have already become established, may be permitted to continue subject to such restrictions as the Secretary ... deems advisable."

In 1980, after a NEPA process that included public participation and an environmental analysis, the Park transmitted a wilderness proposal in accordance with the Wilderness Act. This wilderness proposal recommended that the Colorado River be designated as potential wilderness, and was accompanied by a formal plan to gradually eliminate motorized boats on the river. This recommendation was never conveyed to Congress, and Congress – which has the sole authority to designate wilderness on federal land – has never acted on a wilderness bill for the Grand Canyon.

According to some parties' interpretation of the Wilderness Act and National Park Service policy, the Park was required to remove motorized rafts from the river because the wilderness proposal included this provision. Other parties – including the National Park Service – did not share that interpretation, and believed that motorized craft could remain unless and until Congress acted to prohibit them. As the 1997 process for the revision of the Colorado River Management Plan began, many non-commercial boaters were in favour of removing motorized craft from the Grand Canyon. The outfitters strongly favoured maintaining the status quo, believing that their position reflected the public interest (Grisham 2003).

The 1997 scoping process

The Park began its 1997 process to revise the Colorado River Management Plan with detailed plans for significant public participation. The public was invited to attend several meetings and to submit comments by regular mail or e-mail through December 1997. Although there was a high level of public involvement, the process proved to be contentious.

The first step in a NEPA process for which public participation is required is termed "scoping" and is outlined in Section 1508.25 of the Council on Environmental Quality (CEQ) regulations (CEQ 2002a). Scoping is an issues-surfacing process that is designed to identify the "range of actions, alternatives, and impacts to be considered in an environmental impact statement" (CEQ 2002a). This process is important to stakeholders because scoping gives the public a chance to suggest to the agency which issues should be addressed and which alternatives should be analysed. More importantly, alternatives that are not analysed in the EIS cannot be included in the final plan. During the development of the EIS, federal agencies are required to develop "all reasonable alternatives, which must be rigorously explored and objectively evaluated." These alternatives reflect different ways to accomplish the major federal action that the agency is contemplating. When making the decision on how to proceed, "a decision maker must ... consider all the alternatives discussed in an EIS," and "must not consider alternatives beyond the range of alternatives discussed in the relevant environmental documents" (CEQ 2002b).

According to a Park press release, the focus of the scoping meetings was "identifying the full range of river management issues and solutions that are important to the public" (Oltrogge 1997). The goal of the meetings was to encourage communication between stakeholders and Park staff and among stakeholders, and for the Park thereby to obtain their ideas for the Colorado River Management Plan revision. They did not attempt to produce consensus among stakeholders; rather, they wanted to develop an understanding of the stakeholder interests and positions. They were also interested in the overlap between those interests, where trade-offs and solutions may lie that could help satisfy the needs of all the stakeholders (Chesher 2003).

Process details

The meetings were widely publicized, and there was a high level of participation. Attendance was encouraged through a semi-annual Park newsletter, *The Canyon Constituent*, a press release, and a mailing to individuals who had expressed an interest and to all the stakeholder and boating groups that were known to the Park staff. The press release suggested that participants come with the "issues they wanted to see addressed, as well as proposed solutions to those issues" (Oltrogge 1997). Because of the publicity and the level of interest, the turnout at the three public meetings was higher than expected. Park staff anticipated (and would have been satisfied with) a turnout of 50, and were overwhelmed by more than 100 participants at each meeting.

The meetings were carefully planned with the intent to maximize public participation and input. They took place in Portland, Oregon; Salt Lake City, Utah; and Phoenix, Arizona, on three consecutive weekends in September 1997. The meetings spanned Friday night and most of Saturday, for a total of 11 hours each. Participants were asked to attend both days of the meeting. One or two volunteers from the River Management Society, a national non-profit professional society dedicated to the protection and management of North America's river resources (River Management Society 2003), assisted Park staff with the meetings, with an intent to have a neutral facilitator of the meetings.

Friday night was designed to be an introductory session. The room was arranged in theatre style, with rows of chairs facing the front and a table and chairs for three Park staff members at the front. Other planners and Park staff were also at the front of the room, off to the side. Distributed throughout the room were pads of flipchart paper on stands, labelled with a question: "What issues would you like the Park to consider?" When participants arrived, they signed in and were invited to write their concerns and issues on the pads of paper. A River Management Society representative opened the meeting by welcoming the attendees and describing the agenda for the evening and the next day. Next, Grand Canyon Science Center Chief David Haskell reviewed the overall process and timeline in some detail, noting that the Park anticipated that the process would be completed in approximately two years. After the introductory comments, the attendees were invited to form small groups to generate a list of issues they wanted the Park to consider. A Park employee or River Management Society member facilitated these small group meetings, while a second person served as recorder, using pads of flipchart paper. Ground rules included the following: listen to others; there is not just one correct answer; consensus is not the goal; allow everyone to speak in turn; and participants may pass if they do not wish to speak (Martin 1997b; Jalbert 2003a).

For Saturday, the second day of the meeting, the Park designed another series of small group meetings focused on the issues that had been expressed on the previous day. The issues in all three cities were similar:

use of motorized boats on the river; allocation of recreational use; natural and cultural resource protection; the non-commercial river trip permit-distribution system; helicopter use; recreational trip attributes (size, length, etc.); and range of visitor services (which referred primarily to the desire on the part of educational institutions to have more access to the Colorado River for educational purposes). The purpose of the small group meetings was for the Park to obtain in-depth information on issues that had been raised on the previous night. These were concurrent sessions, and attendees were able to attend more than one of them in the course of the day.

Public discussion

The discussions on both days were difficult and argumentative. Stakeholders contradicted and challenged each other's facts and opinions. Some stakeholders chose a particular small group because their perceived adversaries joined that group, augmenting the friction. While some participants dominated the discussions, others did not participate much at all. Because the staff had not been trained on how to facilitate difficult discussions, they found the contentiousness hard to control. At times, Park staff members facilitating the sessions expressed their opinions about the issues under discussion, causing some attendees to feel that the issues had already been decided (Grisham 2003; Jalbert 2003a). Some stakeholders even felt that there was animosity directed towards them from Park staff (Grisham 2003). The perception of bias on the part of Park staff and the rising tensions among the participants added to the feelings of mistrust and suspicion towards the Park and among stakeholders (Anon. 2002). [Unless otherwise cited, information about stakeholder views is from interviews with 36 stakeholders by Mary Orton, June 2002. Anonymity was promised to the stakeholders who were interviewed.] A reporter characterized one of the meetings as follows:

Crowded into an airport hotel conference room, participants were watched by uniformed Park Service law enforcement officers wearing side arms. Grand Canyon National Park's top management, charged with making the decisions, did not even attend. The hearing erupted into near chaos and some people walked out when park staffers announced they would not allow verbatim comments at a microphone but merely wanted focus group discussions. (Smith 2002)

Scoping process outcomes

Despite the difficulties, Park staff reported several positive outcomes from the scoping period in general and the public meetings in particular. The Park, for the first time, had made a strong effort to keep their constituents informed and involved. For example, they issued at least 10 newsletters over three years on the subject of the management-plan revision. For many members of the planning team, the scoping meetings had been their first face-to-face encounter with constituents. Stakeholders now knew whom to call at the Park when they had a question or suggestion. Through the process, the public was educated about the issues and the constraints under which the Park operates, and was able to offer constructive suggestions for change. As a result, the Park received a considerable amount of useful information about issues of concern to the public (Chesher 2003).

However, most of the Park staff and stakeholders felt that contentiousness, disagreement, misunderstandings, and polarized people and issues were the primary results of the process. Relationships between agency staff and stakeholders, and among stakeholders, were characterized by mistrust and acrimony. Descriptions of the process from both Park staff and stakeholders included the terms "not constructive," "contentious," "conflict," "grandstanding," "painful," "ranting," and "screaming" (Anon. 2002; Jalbert 2003a).

The situation would worsen before it was mitigated.

Post-scoping public participation

Although public participation is not required by NEPA in the time between the end of scoping and the issuance of a draft EIS, the Park sponsored several public meetings and workshops to keep stakeholders informed and involved after the scoping period ended in December 1997.

The first workshop focused on the private-permit distribution system. Although this workshop was contentious, argumentative, and difficult, Park staff again gleaned useful information from the participants. In fact, this workshop resulted in administrative changes to the waiting-list system, using ideas suggested by stakeholders (Chesher 2003).

In the summer of 1999, the Park sponsored two more workshops on the subject of a new river-trip simulation model. This computer-based model allowed testing of different combinations of various types of river-trip launches – commercial and non-commercial, oar-powered, and motorized – to evaluate the impact on two variables – (1) crowding at attraction sites and (2) encounters between trips. (Crowding and number of encounters are two important indicators of the quality of the visitor experience on a river trip.) The principal investigator for the model had interviewed outfitters, guides, and non-commercial boaters so that the model could mimic their decision-making logic on river trips. The workshops were viewed by many stakeholders as one of the more significant exercises in public involvement, because of the potential for significant impact of the model on the management plan and the substantial level of stake-holder involvement that was needed for its development.

Other planned and publicized workshops never occurred. As described in the next section, the release of a different, unrelated, draft manage-ment plan caused enough controversy not only to prevent the scheduling of the remaining workshops but also to derail the entire Colorado River Management Plan planning process.

2000: Colorado River Management planning interrupted

Wilderness Management Plan released

In order to understand why Grand Canyon National Park Super-intendent Robert Arnberger halted the Colorado River Management Plan planning process in 2000, awareness of another, concurrent park-planning process would be helpful. The Park had begun developing its Wilderness Management Plan in 1995. This management plan addressed issues relating to undeveloped parts of the Park, excluding the river cor-ridor. The draft Wilderness Management Plan and environmental assess-ment were released on 1 June 1998, two weeks after a public meeting in Flagstaff at which the scoping comments from the Colorado River Man-agement Plan were released.

This provoked confusion among stakeholders and members of Con-gress on two counts. The first source of confusion was the release of a dif-ferent management plan in the midst of the revision of the Colorado River Management Plan. Many found it difficult to differentiate between the two plans (Arnberger 2000a). The second source of confusion was in the name of the plan. The Wilderness Management Plan was an update of the Park's Backcountry Management Plan. Park staff used the word "wilderness" in the title, instead of "backcountry," to be consistent with National Park Service wilderness policies and the Park's wilderness pro-posal. Some stakeholders and members of Congress thought this was an-other proposal to designate wilderness in the Grand Canyon. Grand Can-yon river outfitters were concerned that this entailed another threat to the use of motorized rafts on the Colorado River (Grisham 2003).

Controversy soon erupted on several fronts. Some stakeholders, ad-vocating a management approach that included the entire ecosystem, criticized the Park for planning backcountry management in a process that was separate from the river-planning process. Additionally, many of the issues that were important to stakeholders on all sides of the issues could not be resolved because Congress had not yet acted on the Park's wilderness proposal, adding to the discord and uncertainty (Arnberger

2000a). In September 1998, Arnberger was called to testify before the House Subcommittee on National Parks and Public Lands Oversight Hearings. As a measure of how significant and controversial these issues had become, members of Congress questioned Arnberger at length on the issues of wilderness and motorized rafts on the Colorado River.

In 1999, under pressure from Congress and stakeholder groups, Park staff and management engaged in several months of analysis of their options for combining the two planning processes. This effectively prevented the convening of the rest of the Colorado River Management Plan workshops. Arnberger ultimately decided that, if they combined their efforts for the two management plans, a much larger planning process would be required (Arnberger 2000a).

Management planning discontinued

On 23 February 2000, Arnberger announced his decision to "halt work on any combined planning process and on the Colorado River Management Plan" (Oltrogge 2000). To explain his decision, he noted that "polarization among the backcountry and river user groups and interests have intensified to the point of reducing the park's ability to bring together divergent perspectives toward collaborating and reaching acceptable resolution." He also cited "the inability to resolve many of these issues prior to the resolution of the park's wilderness recommendation, and to the lack of available fiscal and human resources to complete a comprehensive planning effort."

This announcement was disappointing and disheartening to the Park employees who had been involved in the planning process (Chesher 2003; Cross 2003b; Jalbert 2003a) and shocking to stakeholders (Martin 2000; Grisham 2003). One stakeholder group expressed "extreme disappointment" and characterized the action as having "done nothing less than strip the American people of their greatest opportunity to have a voice in the controversial issues that have created an uncertain future for our nation's most famous natural wonder" (The Wilderness Society 2000). Park staff felt that years of building relationships and trust had been squandered and that the Park had betrayed their stakeholders (Chesher 2003).

Lawsuit and settlement

Nineteen weeks later, on 7 July 2000, the Park was sued over the cessation of the planning efforts. The plaintiffs were four organizations – Grand Canyon Private Boaters Association, American Whitewater,

National Parks and Conservation Association, and American Canoe Association – and four individuals (United States District Court 2002).

In September 2000, Arnberger left his position of Superintendent of Grand Canyon National Park, and Joseph Alston became the new Park Superintendent in November 2000.

On 17 January 2002, the parties filed a settlement of the lawsuit, in which the Park agreed to recommence the planning process for the Colorado River Management Plan within 120 days after the dismissal of the suit (United States District Court 2002). The settlement specified no admission of wrongdoing or liability on the part of any of the parties. The settlement also required that the Park issue the final EIS and Record of Decision on the Colorado River Management Plan by 31 December 2004.

2002: Colorado River Management Plan planning begins anew

A different approach to public participation

Jeffrey Cross became Chief of the Grand Canyon National Park Science Center in April 2000, shortly after the Superintendent decided to discontinue the planning process. When it became clear that the Park would resume the development of the Colorado River Management Plan in 2002, he elected to contract with an outside neutral party for the public-participation component. "Neutral party," in this instance, meant an expert on process who had no stake in the substantive outcome of the process.

He decided on this course of action for reasons that involved both stakeholders and his staff. First, he knew from his interactions with stakeholders that they were dissatisfied with the previous scoping efforts and there was little trust of the Park. He wanted to rebuild stakeholder confidence by sending the message that the Park valued their input, and he felt that a neutral party could accomplish this more easily than could Park staff. Second, Park employees had told him that the previous scoping effort had been a difficult and painful experience. He wanted to support his staff by enabling them to focus on their substantive expertise, not expecting them to design and implement a public process. Finally, he wanted both employees and stakeholders to understand that the Colorado River Management Plan development was a high priority for the Park management, and that the Park would expend the resources to have a professionally designed and implemented process (Cross 2003b).

Cross contracted with the author's company because he had experienced her work with other groups that had struggled with complex and contentious environmental issues. Although she specializes in environmental and public-policy dispute resolution – that is, mediation – it is not unusual for environmental mediators to work in the public-participation field (Dukes 1996). As the next section describes, the nexus between the two fields is large and growing.

Alternative Dispute Resolution techniques

Many conflict-resolution researchers and theorists have described ADR techniques and strategies as applied to public processes, such as those mandated by NEPA (Bingham and Langstaff n.d.; Dukes 1996; Kolb et al. 1994; Susskind, McKearnan, and Thomas-Larmer 1999). Typical components include the following:

- Inclusion of all identifiable stakeholders: those who are involved and affected by the issues; those who will implement any agreement that is reached; and those who could potentially block implementation of an agreement.
- Direct communication among stakeholders, through which they are able to "exchange information, understand one another's interests and concerns, and develop options that address those concerns" (Bingham and Langstaff n.d.).
- Flexible design of the process, tailored to the needs of the situation and modified as necessary as the process unfolds.
- Stakeholder involvement in process design.
- Transparency in process implementation.
- Use of a neutral party (mediator or facilitator) who has skill and experience in assisting groups in decision-making and who has no stake in the outcome of the process.

The need for a neutral party is increased when "the issues are complex or contentious, when many parties are involved, when there is a history of distrust between the [agency] and other parties,... or when past efforts to resolve differences have failed" (Carlson 1999). All of these criteria were present in this case.

A NEPA process does not necessarily lead to the formation of consensus. E. Franklin Dukes describes typical goals of processes that address public conflict but do not necessarily result in consensus:

- Educating disputants, stakeholders, and/or the general public about the issues under consideration;
- Discovering public interest in, concern with, and ideas about particular issues;
- Raising the level of awareness among a particular audience about an issue;
- Demonstrating to adversaries that even on the most divisive issue there are

items which can be discussed and people on the other side(s) worth talking to;... [and]
• Building public support for consequential decisions. (Dukes 1996)

Cross echoed many of Dukes' goals when he indicated that his expectations of this new design were as follows:

[to] have a public process that would allow park staff to talk about the issues, particularly the resource issues that seemed to get lost in the controversies over motors and allocation ..., and to give the public the opportunity to tell us what's important to them so we would have a firm basis for developing the Colorado River Management Plan. I also wanted the park staff to have a better experience during the public part of the scoping process. (Cross 2003b)

When asked to develop a public-participation plan for the Colorado River Management Plan planning process, the author (hereinafter "the mediator") felt that the best way to proceed – given the history, the current situation, and the Park's goals – was to fuse ADR techniques with the requirements of NEPA. The next sections detail the approach.

Stakeholder interviews

In accordance with the settlement of the lawsuit, the resumption of the Colorado River Management Plan process began with a reopening of scoping; thus, the first task was to design a scoping process. The meetings were held in August, September, and October 2002. The Park was required by the settlement to hold scoping meetings in four western cities (United States District Court 2002), and the Park added meetings in four additional cities including one on the Hualapai Indian Reservation. The Hualapai Tribe had indicated an interest in working closely with the Park on the development of the management plan; in addition to the scoping meeting on their reservation, tribal members interviewed elders during the scoping period so that their ideas would be included.

Cross understood that, by hiring a neutral party, the scoping process would be designed to address not only the needs and interests of Park management but also those of stakeholders. To this end, from June 9 to 29, 2002, the mediator interviewed 36 stakeholders, including the Superintendent and Deputy Superintendent of the Park. All but two east-coast stakeholders were interviewed in person. The interviewees represented non-commercial boaters, wilderness advocates, researchers, Grand Canyon river guides, commercial outfitters, Park management, educators with an interest in educational river trips through the Grand Canyon, and Native American tribal members. Park staff members who had participated in the 1997 process generated the original list of interviewees. In the course of the interviews, the mediator invited stakeholders to suggest additional people to interview, and added several of those to the list. The objectives of the interviews included the following:

- Introduce the mediator to stakeholders, including her background and potential conflicts of interest.
- Gather information about stakeholders' experience with the 1997 planning effort, and their views on that effort.
- Gain an understanding of stakeholders' needs and concerns, the issues they were interested in, and the main points of agreement and disagreement.
- Educate stakeholders on the Colorado River Management Plan planning process, including the mediator's role.
- Begin to establish relationships with stakeholders, and set the tone for a cooperative, collegial process.
- Give stakeholders the information they needed in order to be engaged, constructive, participants in the process.
- Secure stakeholders' assistance in planning the scoping meetings so that their needs would be addressed in the meeting design.
- Obtain their suggestions for publicizing the meetings, as well as for securing input from those who could not attend the meetings.

Appendix A to this chapter includes the instrument used for the interviews.

During the interviews, with few exceptions, the stakeholders requested that the meetings be less contentious than those in 1997. For the majority of the respondents, this meant finding an alternative to the traditional concept – often used by government agencies but mostly avoided by the Park in 1997 – of an open microphone from behind which attendees address Park planners in front of a roomful of people. Two stakeholders who preferred the open-microphone format felt this was a valid way of expressing their strong negative feelings about Park management.

When asked for their ideas, stakeholders made many constructive and useful suggestions for the scoping process. One stakeholder mentioned an open-house format that a federal agency had sponsored for another planning process. Several stakeholders said it was important for everyone to understand what would happen at the meetings before they arrived, so that they could be prepared. Although one person commented that the facilitators did a good job in 1997, others suggested that having facilitators who were not Park employees would produce a process that was (or appeared to be) less biased. Several requested that the meetings be carefully facilitated so that one person could not dominate the meeting, quiet attendees would be comfortable to comment, community members would not be overwhelmed by professionals, an agenda would be followed, comments would be recorded, and the group would not focus on only one topic. A few stakeholders suggested that the meetings should be an opportunity to learn about other points of view. The mediator and her team were responsive to all of these suggestions from stakeholders in their process design.

At this point in the interview, the mediator suggested some meeting-format ideas in order to test their acceptability with the stakeholders. These included the following:

- an open-house format with no formal presentation and no open microphone;
- a variety of ways for people to provide input, both anonymously and publicly;
- focus on a vision of what the river corridor could become by asking the question, "What do you want to see on the Colorado River through the Grand Canyon in 20 years?"

The stakeholders almost universally approved of these concepts. They applauded the open-house format, saying that it would eliminate the "grandstanding" and public arguments that accompany a traditional public hearing, and could help educate stakeholders. In view of this response from stakeholders, the open house was used in the design. One stakeholder had some concern about focusing on a time horizon of only 20 years, and suggested that the attendees should also be asked what they value about the river corridor today. The mediator incorporated this suggestion, as well.

The idea of facilitated small group meetings as a part of the scoping meetings was also tested with stakeholders. This concept drew mixed responses. Some stakeholders thought that they would work well, but more were concerned that the contentiousness of the 1997 process would recur. One stakeholder was concerned that the people who belonged to organized groups would intimidate attendees who were new to the process. Several stakeholders requested that, if small group meetings were held, they should be identical, so that stakeholders would not have to attend all of them. The mediator decided to include small group meetings in the scoping meetings while addressing stakeholder concerns in their design, as described in the next section.

The 2002 scoping process

Meeting design

The open-house format was the most significant change from the 1997 scoping meetings. Members of the public were invited to drop in, rather than being asked to stay for the entire meeting. They were much shorter than the 1997 scoping meetings – only four hours in one weekday evening, compared with eleven hours over two days. The open houses featured a dozen stations positioned around a large room, with a poster on an easel at each station that described a subject that the Park would consider during the planning process. Subjects included management framework, NEPA, wilderness, administrative use, adjacent lands, Hualapai tribal concerns, concessions management, the issuing of permits, cultural resources, natural resources, visitor experience and values, and range of visitor services. A Park staff member conversant in the station's subject stood beside the station and engaged stakeholders in discussion as they walked by.

Before the meetings, Park employees were provided with strategies for interacting comfortably with their constituents. For example, they were given suggestions on how to communicate with someone who was angry, using role-plays to help them to practise answering difficult questions. They were encouraged to differentiate between their personal opinions and Park policy, and to express only the latter.

Because people have different preferences and comfort levels for modes of communication, there were six avenues for attendees to give their comments to the Park:

1. There were easels with large pads of paper and felt-tipped pens next to each station on which attendees were encouraged to write their comments. Each blank pad was posted with the two questions that constituted the theme of the meetings: "What do you value about the Colorado River through the Grand Canyon today?" and "What would be desirable on the River in 20 years?"
2. In an area set with tables and chairs, and close to a table spread with cookies and bottled water, comment forms were scattered on the tables. These forms had space to write the answers to the two questions. Attendees could deposit the forms in a box on one of the tables, or mail them to the Park by the end of the scoping period.
3. A stenographer recorded verbatim comments at each meeting.
4. For stakeholders who were most comfortable communicating electronically, computers allowed stakeholders to send e-mail directly to the Park.
5. A large map (eight feet in length) of the Colorado River corridor through the Grand Canyon was spread out on a table, with coloured markers available and a sign that read: "Draw what you want to see on the River."
6. Small group meetings began every 30 minutes to give stakeholders a sixth way to give their comments to the Park. These meetings had identical agendas, and the same two questions were asked: "What do

you value about the Colorado River through the Grand Canyon today?" and "What would be desirable on the River in 20 years?" The
purpose of these meetings was not to achieve consensus but to give
participants the opportunity to hear others' comments and to learn
from them or react to them. Professional facilitators (who were not
Park employees and who were experienced in contentious public processes) led these meetings and recorded comments from the attendees.
They were designed so that most of the communication was between
attendees and the facilitator to prevent the contentious interactions
that had been experienced in 1997. However, the facilitators were also
encouraged to promote conversations among the participants if they
were able to do so in a mutually respectful manner. This turned out
to be the case in many of the small group meetings. In fact, in one of
the meetings, representatives of the non-commercial boating community and the commercial outfitters successfully collaborated to persuade the rest of the group to adopt consensus language about the
Colorado River Management Plan.

Publicity and other opportunities to participate

During the interviews early in the process, the mediator asked stakeholders how best to publicize the meetings to their constituents and
what would best serve those who could not attend. Using their suggestions, Park staff established a "virtual tour" of the open house on the
Park's Colorado River Management Plan website. Stakeholders who
could not attend a scoping meeting could view the posters from the stations, fill out the comment forms, and electronically transmit them to the
Park. They could also send comments through regular mail.

The mediator's team and Park staff publicized the meetings and
the other opportunities for commenting, including extensive outreach to
news media. Electronic and street addresses for stakeholder organizations were gathered, and these organizations were asked to publicize the
process to their members. The Park sent messages via e-mail and regular
mail to their constituents. Information repositories were established at
local libraries, where members of the public could obtain background
documents and comment forms. These background documents, such as
the current management plan and the comments from the 1997 process,
were also available on the website.

Outcomes: Comparison between 1997 and 2002

Park staff and stakeholders reported increased satisfaction with the scoping process in 2002 compared with the 1997 process. This section details

the outcomes, compares them with those from 1997, and describes the reasons for the differences.

More-numerous comments

The "product" of scoping in a NEPA process is comments from the public. The 2002 scoping process produced about 50,000 comments, compared with about 3,000 comments from the 1997 process (Cross 2003b). Linda Jalbert, recreation resources planner at Grand Canyon National Park, credited the process for this significant increase thus:

I think we have received more [comments] because people are able to talk to us, they are able to talk to each other, they can go to the stenographer, put their ideas down in different ways, they can feel like they are being heard.... [I]n the small group discussions ... [in 1997], people couldn't really say what they felt without being picked on. (Grand Canyon National Park 2002)

Education of Park staff and stakeholders

Park staff and stakeholders both felt that the 2002 process educated Park staff about the concerns of the stakeholders more thoroughly than had the 1997 process. The open-house format allowed the conversations between Park staff and their constituents to be low-key and personal (Grand Canyon National Park 2002; Jalbert 2003a). Park employees were therefore more likely to understand the values and opinions that they were hearing than they had been during the fractious 1997 process and (it was hoped) would be more likely to include them when drafting the Colorado River Management Plan.

The open-house and stations format also better served to educate the stakeholders about the complexities of the planning process. Many of the stakeholders were already well informed about the issues they cared about, but few (if any) had a comprehensive understanding of the totality of the Colorado River Management Plan or the planning process that was needed for its revision. The stations in the open house gave the attendees an overview of all the issues that the Park would address during the planning process. By increasing their level of understanding, stakeholders were able to understand how the issues most important to them related to the overall planning process, and what constraints the Park faced as it revised the plan.

Both Park staff and stakeholders felt that the stakeholders' comments were more thoughtful, informed, and useful than those of 1997, because of increased understanding of the Colorado River Management Plan and of the myriad issues that the Park would need to address during the planning process (Grand Canyon National Park 2002; Jalbert 2003a).

A reduced level of contentiousness

The lack of contentiousness was surprising to many who had attended the 1997 meetings: the 1997 process was painfully difficult, whereas in 2002 the meetings were enjoyable. One senior member of the Park staff said: "I've been to a lot of public meetings in my NPS career, but this was the first one at which I had fun" (Pergiel 2002). The purpose of the meetings was more fully achieved because of the absence of divisiveness. This is not to say that attendees were in agreement on the issues; on the contrary, stakeholders fervently held their positions and many were passionately opposed to others. However, the personal attacks and loud, vocal, arguments among stakeholders, and between stakeholders and Park staff, were almost completely absent in 2002.

The open-house format contributed to the lack of discord. Most people are not comfortable speaking in front of a group. Those who do manage to present their comments in that stressful situation tend to be the most committed to their point of view, and thus may represent more extreme viewpoints than most stakeholders (Daniels and Walker 2001). Other attendees, with less-polarized views, may feel that they have no place at the meeting. Dukes describes in detail the reasons why citizens feel traditional public hearings are "often inflexible, stilted, adversarial, episodic, and generally intimidating for non-professionals":

Consider a typical public hearing. Speakers stand with their backs to the audience. They face an array of microphones on an unfamiliar podium. Speaking time is restricted and carefully monitored. The authorities hearing comments, seated behind their own desks and their own microphones, look down on the speaker from an elevated stage. Little or no response by these authorities is offered to the comments. If there is any negative response by following speakers there is no further opportunity for rebuttal, much less engagement in dialogue. (Dukes 1996)

The open-house format addressed this criticism by encouraging dialogue and providing a comfortable venue for personal interaction.

There were other factors contributing to the reduced conflict. One was the training for Park staff on how to handle difficult questions and how to present their particular area of expertise. Another was the use of professional facilitators in the small group meetings, along with the careful planning of those meetings. No one expressed suspicion of bias on the part of the facilitators (Grand Canyon National Park 2002; Jalbert 2003a). Finally, the use of a neutral party to design the process enabled the Park employees to listen to stakeholders, concentrate on their substantive expertise, and answer questions from the public, instead of having to focus on the process.

The presence of top Park management also contributed. Unlike the 1997 process, the Superintendent, Deputy Superintendent, and Science Center Chief attended every scoping meeting, along with other members of the Park management team. They spent the entire four hours engaging stakeholders in conversation and addressing, on a one-on-one basis, the controversial issues facing the Park. This gave the message that they cared about the concerns of their constituents and reinforced the importance of the process to the Park.

Better relationships

Closely connected to the reduced level of contentiousness, and perhaps more important in the long term, was the positive impact of the 2002 process on relationships between Park employees and stakeholders. In contrast to the painful interactions of the 1997 process, Park staff – from the resource specialists to the Superintendent – consistently reported enjoying the interaction with the public at the meetings (Grand Canyon National Park 2002).

Relationships among stakeholders were also positively affected. Stakeholders known to have strong adversarial positions had lengthy, friendly discussions with each other during the course of the meetings (Grand Canyon National Park 2002). One stakeholder noted that "you realize the other person is a human being," after talking face to face (Grand Canyon National Park 2002).

The process design allowed for the development and enhancement of relationships. The relaxed tenor of the open house was a major factor, as was staff training and orientation toward the public. Another significant contribution came from the stakeholder interviews and extensive personal outreach during the process-design stage. In effect, these meetings belonged to the stakeholders, as they had helped to plan them.

The map of the river corridor provided an unexpected benefit to relationships. At every meeting, stakeholders and Park staff, regardless of differences in their vision for the future, huddled over the map, sharing river-trip stories, and discussing this place to which all felt a strong and personal connection.

Another unanticipated relationship-building feature came as a result of asking stakeholders for suggestions for improvements after each of the scoping meetings. A suggestion was made after the first meeting to provide nametags for attendees; these were provided for the remaining seven meetings. Many stakeholders knew each other by name but not by face, from communicating via electronic mail. The provision of nametags enabled stakeholders to build community among themselves.

Reactions from stakeholders

Park management received praise and gratitude from stakeholders and staff for the efforts they had made to change the tone of the scoping process. As a stakeholder said,

I would recommend this format. The proof is in the pudding. You want a format not just to reduce conflict, or reduce contention. You want a format that actually draws in or encourages more constructive input. Fresh thinking, if you will. Some new ways of looking at the issue ... I think this format supports that. (Grand Canyon National Park 2002)

Another stakeholder observed that, "I heard several comments from participants to the effect of 'I really think my voice counts' and 'I think I'm being heard at this meeting'" (Jalbert 2002).

A newspaper article written after one of the meetings (Smith 2002) quoted the reactions of several stakeholders:

"We all remember how bad those 1997 meetings were, and this new format is so much better," Richard Martin, president of the Grand Canyon Private Boaters Association, said during last week's open house in [a Salt Lake City suburb]. "What we learn through this sort of public process is that these issues are not black and white. There is a gray area, a middle, where most of us live. That's the change I see."

Added Mark Grisham, director of the Grand Canyon River Outfitters Association: "There is a sense of shared frustration and camaraderie on all sides that is helping this new format succeed. The situation is not as intractable as it has seemed and a reasonable outcome seems possible."

Tom Martin, founder of River Runners for Wilderness and one of the most outspoken critics of the park's current management of the river corridor, said he was dismayed when he learned of the new format, fearing it would "dumb down" the public input and diffuse deserved criticism of the park's policies. After attending Thursday's open house in Flagstaff, he changed his mind.

"When you sit down with that court stenographer, you get a lot more than just five minutes behind a mike to explain your ideas," he said. "In essence, the park is casting a wide net to try to capture one or two groundbreaking ideas that may lead to the solutions they are gleaning for. I don't know if the public is seeing all the park's dirty laundry on those displays, but as someone who has been in the trenches and taken my own lumps, it's a much better start." (Smith 2002)

This reporter described the meeting as follows:

With subdued lighting and sugary treats, professional mediators and cafe seating, federal land managers are tweaking the format of the typical public input meeting to cool tempers and warm hearts. The trend eventually may mean the demise of a cherished Western tradition: standing in front of a rowdy crowd and behind a microphone to vent your spleen at a government bureaucrat ... The result was a public airing of divergent opinions in an atmosphere as laid-back as a Starbucks coffee house. (Smith 2002)

Post-scoping public participation

The success of the scoping meetings encouraged the Park again to provide public participation opportunities not required by NEPA.

In the period between the end of scoping and the issuance of the draft EIS, the Park asked the mediator to sponsor two stakeholder workshops. Their purpose was to build on the information contained in the scoping comments and to clarify values and preferences of stakeholders with regard to two important and controversial issues in the Colorado River Management Plan – the non-commercial river trip permit-distribution system and the range of visitor services to be offered to the public. As noted above, in 1997 "range of visitor services" primarily reflected the desire on the part of educational institutions to have more access to the Colorado River for educational purposes. In 2002, in addition to the educational purposes, this phrase reflected at least two other desires on the part of some stakeholders. One was additional access for special populations, including the disabled, disadvantaged youth, and low-income people. The other was a blurring of the distinction between commercial and non-commercial river trips, so that a non-commercial river-trip leader might, in future, be able to hire assistance (such as a guide, a medical officer, or a cook). This is currently prohibited.

The workshops were, in effect, focus groups. The mediator identified 10 stakeholder groups and invited each to send a fixed number of participants. She used an interactive decision support technology to enhance the effectiveness of the stakeholder workshops. This computer-based technology provided the ability to collect and document real-time opinions, and instantly and graphically to present them to the group in an anonymous manner for the participants to explore. The mediator and her team were able to isolate and compare data across stakeholder groups, while enhancing the results through a rich discussion by the participants. For the workshop on the permit system, participants were given a hand-held keypad that was connected via radio signals to a laptop computer. Participants used the keypad to rate the importance of various attributes of a permit system – such as fairness, predictability, ease of use, length of wait, and flexibility. The results of their ratings were projected instantly onto a large screen to generate discussion and clarifications. The stakeholders discussed values and trade-offs, with the Colorado River Management Plan planning team in the audience to hear the discussion. The planning team was able to learn about the interests behind the positions that the stakeholders had taken in the scoping comments, and, as they gained a better understanding of the interests of the other groups, stakeholders acknowledged how difficult the Park's decision-making would be. The stakeholder workshop on range of visitor services was conducted in a similar manner.

During this period, the Park also asked the mediator to sponsor two expert panels on controversial issues about which the Colorado River Management Plan planning team needed more information: these were (1) allocation of recreational use among user groups and (2) carrying capacity of the river corridor. These panels could have been conducted in private for the sole benefit of the planning team; instead, they were open to the public and held on days adjacent to the stakeholder workshops as a benefit to stakeholders. Stakeholders were able to hear the opinions of the experts at the same time as the planning team, and had the opportunity to ask questions of the panellists.

Park staff and stakeholders were pleased at the additional opportunities for interaction and participation. Cross felt that the activities had achieved their purpose:

The experts gave us some limits for alternative analyses in the draft EIS. The stakeholders ... heard some creative ideas that will appear in the alternatives. And [from the stakeholder workshops,] the park got a mandate to change the existing private permit system. (Cross 2003a)

Although there was frustration expressed by some stakeholders that the workshops did not go further in exploring solutions, all stakeholders who responded to a request for feedback, and Park staff, found the exercise to be beneficial (Ekker 2003; Ghilieri 2003; Grisham 2003; Johnson 2003; Martin 2003; Odem 2003).

Conclusions

This description of two different public-participation approaches illustrates how processes developed by those with the best of intentions can produce unfortunate and unanticipated results, and how ADR methods can be applied to improve the design of participatory processes under NEPA. Insights gained from comparing the two processes include the following:

1. The presence and support of upper management of the Park in 2002 was crucial in making possible the public perception that the Park understood that this was an important process, and that they truly wanted to listen to their stakeholders.
2. Use of a neutral party probably increased trust on the part of the

stakeholders, and certainly increased the enjoyment of the process on the part of Park staff. Although many public processes are conducted well without the aid of a neutral, as conflict-resolution practitioner Melinda Smith writes: "Professional assistance can help groups achieve sound process practice" (Smith 1999).

3. Involving the stakeholders in the design of the process, through pre-process interviews and feedback requests, both ensured that their needs would be met and increased their confidence in the process.
4. Thorough staff training, and a meeting format that allowed for personal, one-on-one conversations between staff and constituents, increased the opportunities for mutual education and relationship building.
5. Providing multiple avenues for stakeholders to submit comments enabled them to feel that their voices were heard and increased the number of comments submitted.
6. Trust of the process and the Park was enhanced through the use of professional, neutral facilitators of the small group meetings.

All of these factors served to improve relationships and to reduce the negative impacts of conflict.

Dukes suggests that public processes can "inspire, nurture, and sustain ... an engaged community, invigorate the institutions and practices of governance,... and enhance society's ability to solve problems and resolve conflicts" (Dukes 1996). In the case of the Colorado River Management Plan, with highly contentious issues and polarized stakeholders, the use of ADR strategies helped achieve at least part of that potential. It increased trust and mutual understanding, engendered positive communication, improved relationships, and generated creative ideas for solutions to problems.

Whether the full promise will be fulfilled remains to be seen. At the time of writing, the EIS is not yet completed. However, the results to date suggest that, when faced with similar difficult NEPA processes, agencies would be well advised to consider applying ADR principles and techniques.

Appendix: Colorado River Management Plan Stakeholder Interview Questions

Introduction of mediator

Introduce myself:
• I'm a mediator, I took this job as a neutral (explain neutrality, examples, working for client vs group, code of ethics, conversation with Park and their agreement to this approach).

- I want to develop a process that works for everyone, not just the Park. Thus, I wanted to interview you to get your input as to the process. I won't be writing the EIS; my focus is on the process, not the substance.
- Potential conflicts of interest: explain; ask for, and discuss, any concerns.
- I will not attribute anything you tell me without explicit permission, plus I will keep confidential whatever you ask me to.

Review of the process

- There will be public meetings in five cities, plus other ways of giving input.
  - I need your input on both.
  - Notice of Intent indicates that the scoping process will last 60 days, but will last until 30 days after last scoping meeting. The public scoping meetings will probably be August this year.
- As you know, this is primarily a river use management plan. The Park is concerned about their primary mandate to protect resources.
- What issues will not be addressed in the plan.
- Scoping: the Park will use 1997 scoping information as well as comments generated by this process in 2002.
- We anticipate that the focus will be on what stakeholders want the river to be like in the future.
- What are the conditions and qualities that make a river trip special to you? What do you think about that approach?
- What would be the components of such a vision?

Scoping meetings

- There will be meetings in five cities: Phoenix, Flagstaff, Salt Lake City, and Denver (required by the settlement) plus Las Vegas
- Any ideas for how to conduct the meetings?
- Some of the things we're thinking about (no decisions made):
  - Open house plus small groups
  - A variety of ways to give input, both anonymously and publicly
  - Laptops available at the meetings for people to give comments directly, right there
  - Stenographers present to take down comments
  - Paper available for people to draw pictures of what they want to see on the river.
  - Reactions?
- Ideas for stations at open house?
- Folks who can't attend the meetings – any ideas about how to reach out to them?
  - We are considering Web-based opportunities, including, for exam-

ple, posting any exhibits we may have at the meetings on the Web for folks to react to. We may have a documents repository at public libraries. (All Arizona libraries have free Internet access for the public, not sure of the other states.) Certainly e-mail and letters.
○ Any other ways that have worked for you before, or that you would like us to consider?
○ Are there ways to contact your constituency that we should be aware of? Ways to contact others?
• Are there any conferences or meetings in the next 18–24 months, to which someone from the Park could come, give an update, and solicit comment?

Draft Environmental Impact Statement

• The Park may have a preferred alternative in the draft EIS because of the time constraints in the settlement. However, the preferred alternative may be conceptual, without all the details. Then they are free to incorporate input from the public. We will have another round of public meetings for the DEIS – I will probably talk to you again about that when the time comes. Any reaction to this?

Interim period

• After the scoping period is over, the Park will issue a summary of scoping comments, and perhaps offer a regular update (e.g. newsletter, website updates) from the planning people.
• What else do you want to see in the period between scoping and presentation of DEIS? Would you be interested in working on developing some options for addressing the private boater wait list issue?
○ Are there other issues we should focus on?
○ What would be helpful to you?
○ Work on alternatives, private permit allocation plan?

Other

• Do you have any other thoughts you'd like to share?

REFERENCES

Anonymous. 2002. Interviews with 36 stakeholders by Mary Orton, June. Anonymity was promised to the stakeholders who were interviewed.
Arnberger, Robert. 2000a. Memorandum to Director, Intermountain Region, Na-

tional Park Service, from Superintendent, Grand Canyon National Park, Sub-ject: "Cessation of Planning on Undeveloped Areas Plan." February 10. *Grand Canyon National Park Colorado River Management Plan Administrative Record*, #644-13.

Arnberger, Robert. 2000b. Letter to "Grand Canyon Constituent." 23 February. *Grand Canyon National Park Colorado River Management Plan Administrative Record*, #645-1.

Aronson, Jeffe. 1997. "The Process and Outfitters." *The Waiting List: Newsletter of the Grand Canyon Private Boaters Association.* January. Internet: ⟨http://gcpba.org/pubs/waitinglist/97_01.html⟩ (visited 26 March 2003).

Bingham, Gail, and Lee M. Langstaff. n.d. *Alternative Dispute Resolution in the NEPA Process.* Internet: ⟨http://www.resolv.org/pub_articles/NEPA.htm⟩ (visited 26 March 2003).

Carlson, Chris. 1999. "Convening." In Lawrence Susskind, Sarah McKearnan, and Jennifer Thomas-Larmer (eds), *The Consensus Building Handbook: A Comprehensive Guide to Reaching Agreement*, 169–197. Thousand Oaks: Sage Publications.

CEQ (Council on Environmental Quality). 2002a. *Code of Federal Regulations. Title 40*, Secs 1500–1517. Internet: ⟨http://ceq.eh.doe.gov/nepa/regs/ceq/toc_ceq.htm⟩ (visited 21 February).

CEQ (Council on Environmental Quality). 2002b. *40 Frequently Asked Questions: Answers to 1–10.* Internet: ⟨http://ceq.eh.doe.gov/nepa/regs/40/1-10.htm⟩ (visited 20 February).

Chesher, Greer. 2003. Former Resource Specialist, Grand Canyon National Park, personal communication to author, April 7.

Cross, Jeffrey. 2003a. Grand Canyon National Park Science Center Chief, personal communication to author, 2 February.

Cross, Jeffrey. 2003b. Grand Canyon National Park Science Center Chief, personal communication to author, 7 April.

Daniels, Steven E., and Gregg B. Walker. 2001. *Working through Environmental Conflict: The Collaborative Learning Approach.* Westport, Connecticut: Praeger.

Dukes, E. Franklin. 1996. *Resolving Public Conflict: Transforming Community and Governance.* Manchester: Manchester University Press.

Ekker, TinaMarie. 2003. Policy Coordinator, Wilderness Watch, personal communication to author, 25 March.

Ghilieri, Michael. 2003. President, Grand Canyon River Guides, personal communication to author, 27 March.

Grand Canyon National Park. 2002. Videotape of Colorado River Management Plan scoping meeting in Mesa, Arizona, on file at the Grand Canyon National Park public information office. 15 August.

Grisham, Mark. 2003. Executive Director, Grand Canyon River Outfitters Association, personal communication to author, 1 April.

Jalbert, Linda. 2002. Recreation Resources Planner, Grand Canyon National Park, personal communication to author, 22 August.

Jalbert, Linda. 2003a. Recreation Resources Planner, Grand Canyon National Park, personal communication to author, 19 March.

Jalbert, Linda. 2003b. Recreation Resources Planner, Grand Canyon National Park, personal communication to author, 9 July.

Johnson, Jo. 2003. Co-director, River Runners for Wilderness, personal communication to author, 4 April.

Kolb, Deborah M., and Associates. 1994. *When Talk Works: Profiles of Mediators*. San Francisco: Jossey-Bass Publishers.

Kriesberg, Louis. 1997. "The Development of the Conflict Resolution Field." In I. William Zartman and J. Lewis Rasmussen (eds). *Peacemaking in International Conflict: Methods and Techniques*. Washington, DC: United States Institute of Peace Press.

Martin, Richard. 1997a. "HEY, Let's Go Boating!" *The Waiting List: Newsletter of the Grand Canyon Private Boaters Association*. January. Internet: ⟨http://gcpba.org/pubs/waitinglist/97_01.html⟩ (visited 26 March 2003).

Martin, Richard. 1997b. "1997 Colorado River Management Plan Public Meeting Report – Round One ... Done." *The Waiting List: Newsletter of the Grand Canyon Private Boaters Association*. November. Internet: ⟨http://gcpba.org/pubs/waitinglist/97_11.html⟩ (visited 26 March 2003).

Martin, Richard. 2000. "Plannus Interruptus." *The Waiting List: The Grand Canyon Private Boaters Association Quarterly*, March. Internet: ⟨http://gcpba.org/pubs/waitinglist/2000_04/March_2000_1_to_8.pdf⟩ (visited 8 April 2003).

Martin, Richard. 2003. President, Grand Canyon Private Boaters Association, personal communication to author, 18 March.

McHenry, Robert (ed.). 1993. *The New Encyclopaedia Britannica* (15th edn, Vol. 3). Chicago: Encyclopaedia Britannica, Inc.

NEPA (National Environmental Policy Act of 1969). *United States Code, Title 42*, Secs 4321–4347.

Odem, Willie. 2003. Former President, Grand Canyon Private Boaters Association, personal communication to author, 25 February.

Oltrogge, Maureen. 1997. Press release from Grand Canyon National Park: "Colorado River Plan Meetings Set to Begin." 21 August.

Oltrogge, Maureen. 2000. Press release from Grand Canyon National Park: "Grand Canyon National Park Moves in Different Direction with Planning Efforts for River and Backcountry." 23 February. *Grand Canyon National Park Colorado River Management Plan Administrative Record*, #643-1.

Pergiel, Chris. 2002. Chief Ranger, Grand Canyon National Park, personal communication to author, 8 August.

River Management Society. 2003. Internet: ⟨http://www.river-management.org/⟩ (visited 28 March 2003).

Smith, Christopher. 2002. "Land Debates Lose Nastiness." *The Salt Lake Tribune*, 12 August, p. D1.

Smith, Melinda. 1999. "Catron County Citizens Group." In Lawrence Susskind, Sarah McKearnan, and Jennifer Thomas-Larmer (eds), *The Consensus Building Handbook: A Comprehensive Guide to Reaching Agreement* 985–1009. Thousand Oaks: Sage Publications.

Susskind, Lawrence, Sarah McKearnan, and Jennifer Thomas-Larmer (eds). 1999. *The Consensus Building Handbook: A Comprehensive Guide to Reaching Agreement*. Thousand Oaks: Sage Publications.

United States Department of the Interior. 1989. *Colorado River Management Plan for Grand Canyon National Park.* September. Grand Canyon National Park, Ariz.: US Department of the Interior.

United States District Court. 2002. *Grand Canyon Private Boaters Ass'n v. Alston* Case No. CV-00-1277-PCT-PGR-TSZ (D. Ariz. 2000), Settlement. Filed 17 January.

Wilderness Act. 1964. 16 U.S.C. Secs. 1121, 1131–1136.

The Wilderness Society. 2000. Press release: "Statement by William H. Meadows, President, The Wilderness Society, Regarding the Decision to Halt the Planning Process for Grand Canyon National Park." 28 February. *Grand Canyon National Park Colorado River Management Plan Administrative Record,* #646-1.

22

Public participation in the development of guidelines for regional environmental impact assessment of transboundary aquatic ecosystems of East Africa

George Michael Sikoyo

Introduction

The Treaty for the Establishment of the East African Community was signed in Arusha on 30 November 1999, forming the East African Community (EAC) – the regional intergovernmental organization of the Republics of Kenya and Uganda, and the United Republic of Tanzania. These three East African countries cover a total area of 1.8 million square kilometres and have a population of over 83 million people who share a common history, language, culture, infrastructure, and livelihood strategies.

The EAC seeks to enhance livelihoods by widening and deepening cooperation among the partner states in, among other areas, political, economic, social, cultural, health, environment, education, science and technology, defence, security, and legal and judicial affairs for their mutual benefit (EAC 2000). The EAC also aims to achieve its goals and objectives through the promotion of a sustainable growth and equitable development of the region, including rational utilization of the region's natural resources and protection of the environment, as well as enhancement and strengthening of participation of the private sector and civil society and promotion of good governance. In this context, good governance includes adherence to the principles of democracy, rule of law, accountability, transparency, social justice, and equal opportunities.

Economically, the East African countries are among the poorest in the world. Their economies are largely dependent on exploitation of the natural resource base, mainly through agriculture and tourism. The live-lihoods of the peoples of East Africa are wholly tied to the environment, both to meet basic household requirements and as a basis for production and employment. The environment supports agriculture and other sectors of the economy, such as minerals, fisheries, arts and crafts, forestry and tourism (Bisanda 2003; Kairu 2003). Thus, biodiversity is important because it has provided (and continues to provide) people with diverse choices that have helped humankind to exist over the years.

The major transboundary aquatic ecosystems of East Africa include Lake Victoria shared by Kenya, Tanzania, and Uganda; Lake Jipe shared by Kenya and Tanzania; the Minziro–Sango Bay swamp forests located in south-western Uganda and north-western Tanzania beside Lake Victoria; and the coastal strip of Kenya and Tanzania along the western Indian Ocean (Twongo 2003) (see fig. 22.1). The destruction and degradation of these ecosystems may undermine both national and regional economies (ACTS 2000). Indeed, these phenomena potentially undermine prospects of achieving economic recovery and environmental sustainability.

The causes of environmental degradation and the destruction of trans-boundary aquatic ecosystems are numerous, complex, and interrelated (ACTS 2000; Bisanda 2003; Kairu 2003; Twongo 2003). They include the following:

- Human population densities in and around aquatic ecosystems are growing rapidly. Many rural households living around, and sometimes in, these ecosystems lack appropriate technologies to practise environmentally sound economic activities.
- National environmental policies and programmes have not explicitly provided for regional management of aquatic ecosystems. Rather, they are focused on promoting resource management within national territorial boundaries. The absence of common regional policies for transboundary aquatic ecosystems management is, perhaps, one of the major factors accounting for the lack of a coordinated approach to some of the environmental problems associated with lakes Victoria, Jipe, and Natron and Minziro–Sango Bay. For example, each of the three East African countries tends to apply its own, often isolated, environmental impact assessment (EIA) procedures.
- There is also a limited knowledge of structures and productive potentials of many of the aquatic ecosystems. Indeed, our knowledge of the content of ecosystems such as those of lakes Victoria, Jipe, and Natron is fairly limited. Moreover, our understanding of their potential to regenerate and to provide new goods and services is also inadequate.

In the light of these shortcomings, there is a growing urgency for the EAC to institute policies, guidelines, laws, and programmes to promote cooperation in the conservation and sustainable use of transboundary ecosystems in the subregion. The member states need to develop and

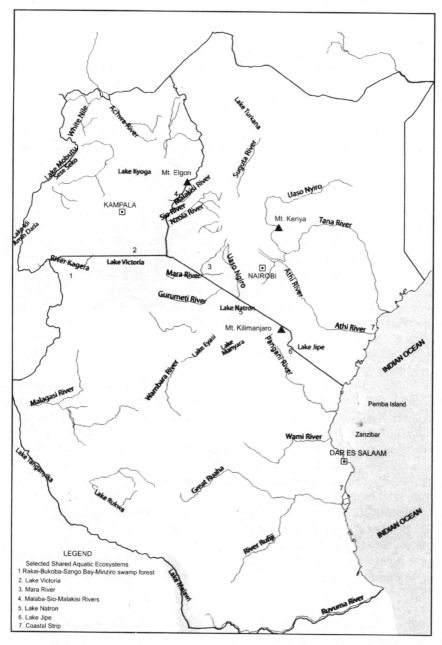

Figure 22.1 Selected shared or transboundary aquatic ecosystems of East Africa

435

adopt long-term guidelines and policies to assess the impacts of economic and development activities on these ecosystems (ACTS 2000; EAC 2000). EIA enables the identification and application of environmentally sound approaches to manage and ensure the sustainability and biophysical integrity of transboundary ecosystems. The choice and application of such approaches must be informed and guided by a properly conducted EIA in order to ascertain the economic, social, and environmental costs and benefits.

The African Centre for Technology Studies (ACTS) with the Regional Economic Development Support Office (REDSO) of the United States Agency for International Development (USAID) has entered into a co-operative agreement to support the EAC's efforts to develop guidelines for regional EIA (REIA) of shared ecosystems of East Africa. Transboundary aquatic ecosystems are but a part of the wider plan of action for the development of guidelines for REIA of such shared ecosystems of East Africa. A set of activities, including agenda setting and constituency building, were undertaken to achieve this goal. The aim is to involve the public through consultations to create a constituency for the entire process. This would ensure that diverse stakeholders – ranging from non-governmental organizations (NGOs), donors, and government departments and ministries in the partner states to local communities living around these transboundary aquatic ecosystems – support and contribute to the development of guidelines for REIA of shared ecosystems of East Africa.

This chapter is divided into four sections including this one (Introduction). The next section (pp. 436–437) analyses the region's policy and legislative framework, including provisions relating to the rights of citizens to be consulted in their administrative process as well as their access to environmental and other types of information. Also included in this section is a description of the regional legal framework for public involvement in the EAC. The subsequent main section (pp. 447–455) addresses the importance of, and nature of, public participation in the development of guidelines for REIA in East African transboundary aquatic ecosystems. It also reports on public participation and coordination with other institutions in the development of these guidelines. This chapter ends with some conclusions regarding the overall issue of including public participation in the development of guidelines for REIA in this context.

The policy and legal environment for public consultation in development of REIA

In keeping with Principle 10 – the participatory principle set forth in the 1992 Rio Declaration on Environment and Development (UNCED

1992) – environmental procedural rights at both the international and domestic levels include the following: the right to information, including the right to be informed in advance of environmental rights; the right to participate in decision-making in environmental issues; the right to legal redress, including *locus standi*; and the right to effective remedies for environmental damage. It also includes freedom from discrimination regarding decisions and actions that affect the environment. In this context, Principle 17 of the Rio Declaration, articulating the importance of EIA, is also relevant.

In effect, a procedural or participatory approach seeks to promote environmental protection by way of democracy and informed debate, the rationale being that democratic decision-making will lead to environmentally friendly policies. It is argued in favour of this approach that, in creating legal avenues for participation, it is possible to redress the unequal distribution of environmental costs and benefits. In this way, marginalized groups who currently suffer the most deleterious effects of environmental degradation – groups including women, the dispossessed, and communities closely dependent upon natural resources for their livelihood – can be included in the social determination of environmental management and change. If the people who make the decisions are the same as those who pay for (and live with the consequences of) the decisions, they will be more inclined to act to protect the environment. Further, the environment is not a unified concept and is very difficult to codify in legal language across cultures, communities, and ecosystems. A flexible and honest approach that takes into account unique environmental contexts has the potential to better guarantee people's right to a healthy environment.

This section reviews the policy and legal environment for public participation among the EAC partner states at the national level. It also explores how public consultation is viewed in the wider society, particularly with respect to the rights of citizens to be consulted in the administrative processes as well as to their access to environmental and other types of information. The regional framework for public involvement in the EAC is also analysed.

The policy and legal environment for public participation in Kenya

The policy environment

The Government of Kenya has recognized the role of procedural rights in decision-making. The 1994–1996 National Development Plan showed this awareness when the government categorically recognized that "[N]ational plans can only be implemented successfully with the support and cooperation of an informed public. To sustain [Kenya's] transition to sustainable development there is need to increase public information, awareness and participation" (Republic of Kenya 1994–1996). The government further promised to promote the provisions of Agenda 21 for improving public information, awareness, and participation for sustainable development in Kenya. This was to be achieved through policy, legal, and institutional measures to create and strengthen advisory bodies on public information for sustainable development, including sustainable development issues in the curricula and training programmes for journalists and other members of the media. The government also promised to assist and ensure that local people are fully involved in decision-making in all activities that impact on the environment, with a focus on involving women and young people in the age range 18–35 years.

Notable attempts have been made to actualize these pronouncements, one such attempt being the District Focus for Rural Development (DFRD) initiative. This initiative has decentralized the decision-making process by stressing the need to initiate development at the sub-location level, by the community. To this end, the government posts District Environmental Officers (DEOs) and District Environment Protection Officers (DEPOs) in order to provide technical advice and support for the District Development Committees and community projects.

Another example of public participation in Kenya is the National Environment Action Plan (NEAP) process (NES 1994). The process was strikingly participatory. The NEAP was prepared by nine task forces, whose membership included a broad representation of institutions and sectors including the public and private sectors, NGOs, and local communities. The preparation went through several drafts in an active participatory process, and the drafts were presented and debated in an interactive process by stakeholders. These took place at five regional workshops throughout the Republic of Kenya; the comments, criticisms, and recommendations from these workshops were incorporated into the final report. Consultations were also carried out at the district (sub-national) level. This process established a holistic approach to management of natural resources through interdisciplinary working groups at all levels. To steer and guide the NEAP process, the following institutional structure was adopted: the Ministerial Level Policy Steering Committee; the NEAP Coordinating Committee; the Secretariat, headed by the Coordinator; task forces that addressed the key environmental issues; and the NEAP Advisory Committee comprising donors, government representatives, private sector, NGOs, and international organizations.

In line with this increased government interest in public input to policy planning, many sectoral policies have incorporated provisions for community participation in environmental management. The new Water Policy, stated in the *Sessional Paper No. 1* of 1999 (Republic of Kenya 1999), acknowledges that communities have been marginalized in water management. It underscores the need to involve communities at all stages of developing water projects. It notes the need to strengthen water management institutions through the establishment of water support units for water supply and sanitation at the district level, training community workers in low-cost water supply and sanitation technology, hygiene promotion, and community participation. The government further undertakes to recognize gender aspects of water management. The draft Forest Policy and Forest Bill is also premised on community management principles (Republic of Kenya 2000). Section 57 of the draft Forest Bill provides that the Chief Conservator of Forests may allow communities living adjacent to a forest to assist in managing the forest. In return for this participation, the Conservator may allow the community access to forest products. This set of mutually beneficial incentives provides a possible model for systems of participatory management of other natural resources in Kenya.

Recognizing the integral role of the right to information, the Government of Kenya has since expressed its intention to improve information resources and management in order to institutionalize the systematic flow of (and access to) information. According to the *National Development Plan* 1997–2001, the government undertook to give priority to "the development of information centres and other documentary sources, of both domestic and foreign materials relevant to all sectors of the economy during the plan period" (Republic of Kenya 1997–2001). It also undertook to facilitate networking among libraries and information stores, including universities, schools, research centres, archives, and herbaria during the plan period. Among the issues that it promised to consider were the following: information-sharing principles; modalities; required tools; cost sharing (within sectors and at national intersectoral levels); and the development of appropriate models for geographic and intellectual organization of the information resources and services, ensuring comprehensiveness of coverage, operational feasibility, compatibility, and systems interconnection.

The legal environment

Kenya's current constitution does not contain explicit environmental provisions; however, it does provide for fundamental rights and freedoms, which has been construed to include environmental rights. Similarly, although there is no express provision on the right to information in Kenya, this right is implied in the provision for the protection of fundamental rights of freedom of the individual. The right to freedom of assembly is enshrined in the constitution, and so are the rights to freedom of speech and expression, including the freedom of the press. These fundamental rights and freedoms provide a legal basis for the citizens to exercise their right to public involvement in environmental decision-making processes. The right to sue (i.e. *locus standi*) is inextricably linked to the constitutional provisions for the protection of fundamental rights and freedoms of the individual, and the Kenya Constitution includes the right of access to the High Court for redress in respect of enforcement of those rights and freedoms.

The Kenyan government's commitment to public participation in the draft Constitution (Constitution of Kenya Review Commission 2002) and the Environmental Management and Coordination Act, 1999 (EMCA), which is currently in force, are particularly relevant to the present discussion. The draft Constitution states that every Kenyan has a right to a healthy environment and also places duties on Kenyan citizens to protect the natural environment. The draft Constitution also includes the basic principles of sustainable development – notably, public participation, but also equity within generations, equity between generations, the precautionary principle (which has to deal with situations where science and technology cannot provide a full response to issues, leaving a degree of uncertainty in terms of the effects of certain activities, technologies, products, etc.), the "polluter pays" principle, respect for traditional environmentally friendly ways, and international cooperation.

The draft Constitution recommends that the government should set up and ensure the effective functioning of environmental-management tools including EIA, environmental audits, and monitoring, and should ensure that environmental standards enforced in Kenya keep up with standards developing internationally. The proposed Bill of Rights guarantees the rights to life and liberty, which many courts around the world have interpreted as necessarily including the right to a healthy environment in which to live that life (Bruch, Coker, and VanArsdale 2001). The draft Constitution is emphatic that the administration of natural resources must involve the participation of local communities, while not losing sight of the need for natural resources to be protected and developed for the benefit of the nation as a whole. Thus, the draft Constitution recognizes the contribution that citizens can make to protecting and managing natural resources; indeed, this is required of citizens as their public duty.

In addition to the Constitution, several other laws provide for the procedural right to information. This includes the Environmental Management and Coordination Act, 1999. Part VI of the Act deals comprehensively with EIA. As stated, Section 58 of the Act enjoins every proponent of a project to conduct an EIA study before commencing any project. The report of the EIA study must then be submitted to the National Environmental Management Authority (NEMA). Upon receipt of the report, NEMA shall cause the following to be published for two successive weeks in the *Kenya Gazette* and in a newspaper circulating in the area (or proposed area) of the project: a summary description of the project; the place where the project is to be carried out; the place where the EIA study may be inspected, and a time limit of not more than 60 days for the submission comments. For the first time in Kenya, the Act provides a clear avenue by which citizens can express their views on specific development projects.

Other sectoral legislation such as the Wildlife (Conservation and Management) and Water acts have provisions for public participation. The Wildlife (Conservation and Management) Act (Republic of Kenya 1977) has devolved the management of game reserves to local authorities. The local authorities hold land in trust for the people. In effect, the game reserves can be said to be under the control of the local people through the county councils. Likewise, the Water Act, 2002 (Republic of Kenya 2002) also provides for the delegation of management powers to local authorities, boards, committees, or people. The Water Act also provides that the Minister may appoint any person or any number of persons to a local water authority for the management of water or for the management, drainage, or reclamation of lands. The person(s) so appointed may investigate, operate, construct, or maintain any community project or any other project for the provision of water.

Policy and legal environment for public participation in Uganda

Uganda has made significant strides through policy and legislation in enshrining public participation as an essential tenet of public planning, development, and implementation of policies, plans, programmes, and projects.

Policy environment

Uganda's policy on environmental management provides for the rational and sustainable use of natural resources in order to conserve the resources for the present and future generations (Republic of Uganda 1994). Among the many objectives, the policy seeks to enhance public participation in activities related to environmental management. Specific objectives emphasizing public participation include (a) integrating environmental concerns in all development-oriented policies; in planning; and in activities at national, district, and local levels; with participation of the people; (b) ensuring individual and community participation in activities to improve the environment.

In pursuance of these policy objectives, the EIA process is integrated in the overall environment policy. It is the general policy of the Government of Uganda that EIAs should be conducted for all planned policies and projects that will have (or are likely to have) significant impacts on the environment, so that adverse impacts can be foreseen, eliminated, or mitigated. It is also the policy that the EIA process be conducted by an interdisciplinary team that is fully transparent, using a participatory approach so that stakeholders have access to information.

Other sectoral policies that have embraced public participation include the Water Policy, 1995 (Republic of Uganda 1995d); Draft Forestry Policy, 2000 (Republic of Uganda 2000); the Fisheries Policy, 2001 (Republic of Uganda 2001); National Policy for the Conservation and Management of Wetlands, 1995 (Republic of Uganda 1995c); and Wildlife Policy, 1995 (Republic of Uganda 1995e). These policies for the first time explicitly have provisions for community participation in resource management and benefit sharing. In addition, these policies define the roles of civil society and the private sector in the management of the natural resources and environment in general.

The legal environment

Uganda's 1995 Constitution and the National Environment Statute (NES), 1995 provide for public participation (Republic of Uganda 1995a,b), as do a number of sectoral laws that have been reviewed to conform to the framework environmental law.

Uganda's Constitution enshrines basic principles and rights upon which public involvement in environmental decision-making is premised. It is contained in the section on "National Objectives and Directive Principles of State Policy." While this is contested as not necessarily creating enforceable rights and obligations, at a minimum it forms the basis for future policy development and juridical interpretation of the more substantive (referring to rights and duties) constitutional and legislative commitments to public involvement. The objectives and principles provide this juridical basis for public involvement in environmental decision-making in a number of ways, as follows:

1. By imposing an obligation on the state to protect the environment, the Constitution establishes a basis for citizens to provide oversight and to hold the state accountable for development decisions that impinge on environmental integrity.
2. By enshrining the principle of respect for international law and international treaty obligations, the state commits itself to fulfil international obligations and standards for public involvement. This would include, for example, those enshrined in the Rio Declaration (UNCED 1992), the Convention on Biological Diversity (UNEP 1992), the Memorandum of Understanding (MOU) on Co-operation in Environment Management between Kenya, Uganda, and Tanzania (EAC 1998), and other international instruments to which Uganda is a party or has subscribed.

3. By imposing an obligation on citizens to uphold and defend the Constitution, it provides standing for citizens to demand to be involved in decision-making processes that impinge on the environment, and to hold the state accountable in court when its conduct undermines its constitutional obligations. In addition to these declaratory principles, the constitution also enshrines substantive provisions that provide a legal basis for public involvement in environmental decision-making. The rights include the following:

- A clean and healthy environment: Article 39 of the Constitution provides that "Every Ugandan has the right to a clean and healthy environment."
- Freedom of association: Article 29(e) provides for the right to freedom of association, stating:

Every person shall have the right to freedom of association, which shall include the freedom to form and join associations or unions including trade unions and political and other civic organizations. The importance of the provision guaranteeing freedom of association is that in practice, public involvement tends to be more effective when it is pursued through organized groups.

The framework environment law in Uganda is the National Environment Statute (NES), 1995. The objective of the framework environmental law is to "provide for sustainable management of the environment; to establish an authority as a co-coordinating, monitoring and supervisory body for that purpose; and for other incidental to or connected with the foregoing" (Republic of Uganda 1995b). This law provides for a right to healthy environment [Sections 3(2)(a) and 4], public participation [Sections 3(2)(b) and 20(8)(c)] and access to information [Section 86(1)] as key principles in the development of policies, plans, and processes for the management of the environment.

All the sectoral laws that have been enacted after the NES contain provisions on public participation in promoting sustainable management of the environment. The NES imposes an obligation on developers to conduct an EIA if it is determined that a project (a) *may* have an impact on the environment; (b) *is likely to* have a significant impact on the environment; or (c) *will* have significant impacts on the environment. Detailed guidelines for conducting an EIA were developed and adopted by the National Environment Management Authority (NEMA), and EIA regulations were enacted in 1998 (Republic of Uganda 1998). The regulations detail how the public can be involved in the EIA processes and the type of information that should be made available.

Enforcing EIA requirements is primarily a responsibility of NEMA, which determines whether a proposed project should be subjected to an EIA, approves consultants to undertake the EIA study, invites public comments, and has the statutory authority to issue the certificate of approval. Lead agencies play a secondary role in the EIA process, mainly limited to recommending that an EIA be undertaken for particular projects in their relevant sectors and commenting on the EIA reports. Lead agencies are required to provide technical advice to NEMA as to whether an EIA should be approved, including the viability of mitigation measures. The NES was the first legislation to provide for the representation of civil society in environmental management: according to Section 9 and the second schedule, establishing the Board of the Authority, two representatives of local NGOs are to be appointed to the NEMA board.

Public-interest law and advocacy organizations are increasingly taking on an oversight role to ensure environmental accountability and compliance. However, the efforts of these oversight independent organizations are still hampered by a lack of clarity of procedural issues related to environmental litigation. Despite remarkable achievements in legislating environmental procedural rights in the Constitution and national legislation, effective public involvement has yet to be adopted as a guiding ethic for decision-making and project implementation.

Policy and legal environment for public participation in Tanzania

Policy environment

Just like the other EAC partner states, a number of Tanzania's strategies, policies and laws have provisions for public participation in the formulation of development programmes, policies and plans. These include the *Tanzania Development Vision* 2025 (Planning Commission 1999), *Tanzania Poverty Reduction Strategy Paper* (PRSP) 2000 (VPO 2000), *Tanzania Assistance Strategy* (TAS) 2000 (Planning Commission 2000), *National Environmental Policy* (VPO 1997), *National Forest Policy* (1998) (Ministry of Natural Resources and Tourism 1998a), *National Beekeeping Policy* (1998) (Republic of Tanzania 1998a), *Wildlife Policy* of 1998 (Ministry of Natural Resources and Tourism 1998b), and *National Tourism Policy* (1998) (Republic of Tanzania 1998b).

The *National Environmental Policy* (VPO 1997) specifically calls for public participation in environmental matters, asserting that environmental issues are best handled with the participation of all citizens at the relevant level. The policy also recognizes that the fundamental principle for achieving sustainable development is broad public participation in decision-making – including participation of individuals, groups, and organizations in EIA decisions, particularly those which potentially affect the communities [DoE (Division of Environment) 2002].

The legal environment

Despite the policy being in place, Tanzania lacks a coherent code of supporting legislation that ensures public participation in sustainable environmental management. The National Environment Management Council Act No. 19 of 1983 (Republic of Tanzania 1983) established Tanzania's National Environment Management Council (NEMC). The main function of the NEMC is to advise the government on all matters relating to the environment: in particular, it has a mandate to formulate policy on environmental management, coordinate the activities of all institutions concerned with environmental matters, evaluate existing and proposed policies and activities on pollution control and enhancement of environmental quality, and recommend measures to ensure that government policies take adequate account of environmental impacts. The Act gives the NEMC the power to formulate proposals for legislation on environmental matters and to recommend their implementation by the government.

The Act does not provide for EIA; however, NEMC has, over the years, promoted the use of EIA as one way to control pollution in Tanzania. In 1997, the NEMC developed a set of "Environmental Impact Assessment Guidelines and Procedures" which, although not legally binding, were primarily designed to guide developers in the initiation and implementation of development projects that do not degrade the environment. The guidelines and procedures outline three stages of preparing an environment impact statement (EIS) – namely, scoping, preparation of terms of reference, and preparation of the final EIS report (NEMC 2002).

Scoping is required to be performed by the project proponent or by the proponent's consultants to discern the main issues of concern. This is undertaken in consultation with the NEMC, relevant sectoral authorities, and affected and interested persons. The project proponent is required to ensure that all interested parties are fully involved by giving them sufficient opportunity to participate in the exercise. According to Paragraph 2.3.1 of the guidelines, the overall purpose of involving the affected persons is to see how their views could be taken into account in the terms of reference (ToR) and EIA study. To ensure that members of the public are fully involved in the process, the project proponent is required to initiate a public-information campaign in the area likely to be affected by the proposed project and to record any concerns raised by the members of the public and address such concerns in the EIA. Volume two of the guidelines deals with "Screening and Scoping Guidelines," and elaborates on the requirement for a public-information campaign and the developers' responsibility for the scoping process, which the guidelines require, to include the following information:

- which authorities and members of the public are likely to be affected by the proposed project;
- how stakeholders will be notified;
- what methods will be used to inform the stakeholders of the project proposal and solicit their comments;
- at what stage of the assessment process opportunities will be provided for public participation and input.

Thus, the guidelines provide for public consultation whenever an EIA is being carried out. To this end, the guidelines require that the project proponent should, at least, consult with the principal stakeholders, inform them about the proposed activity, and ask their views about the project. When potential problematic activities are identified, more extensive consultation is warranted. The findings of the entire process are supposed to be explicitly stated and shown in the EIA.

The guidelines require the project proponent to comply with the public-participation requirements listed above. The project proponent must provide the background information on the nature of the proposed project (purpose, proposed actions, location, timing, method of operation, and probable effects) to assist interested and affected parties in commenting constructively and from an informed position during the scoping process. Paragraph 2.3.3 of the guidelines requires the project proponent to establish a list of interested and affected parties, as well as to develop methods of notifying them about the proposed project. The guidelines also require that public concerns, interests, and aspirations feature in the record.

Regional legal framework for public involvement in the EAC

The most authoritative regional legal instrument providing a basis for public involvement in environmental decision-making in East Africa to date is the MOU on Co-operation in Environment Management (EAC 1998) between the three East African countries. The MOU was signed in Nairobi on 22 October 1998. Among its objectives is the development and implementation of environmentally sound principles, international agreements, instruments, and strategies for environmental and natural resource management.

The MOU sets out elaborate provisions on environmental procedural rights. The partner states commit themselves to promoting public-awareness programmes and access to information, as well as measures to enhance public participation on environmental management issues. Article 16(2)(d) provides that partner states agree to develop measures, policies, and laws to grant access, due process, and equal treatment in administrative and juridical proceedings to all persons who are (or may be) affected by environmentally harmful activities in the territory of any of the partner states.

Although it was concluded and signed approximately four months after the Aarhus Convention, there is no record to suggest that the MOU was influenced by the Convention's procedural rights provisions. However, a number of observations can be made with respect to the MOU. First, it is a pioneer legal instrument for promoting public involvement in a regional context. Second, the legal status of the MOU still needs to be clarified: Article 142(1)(i) of the Treaty Establishing the East African Community states that the MOU "shall not be affected by the coming into force of the treaty, but shall be construed with such modifications, adaptations, qualifications and exceptions as may be necessary to bring [it] into conformity with the Treaty...". Article 27 of the treaty, relating to the jurisdiction of the East African Court of Justice, provides that "the Court shall initially have jurisdiction over the interpretation and application of this Treaty." Article 30 provides that:

Subject to the provisions of Article 27 of this Treaty, any person who is resident in a Partner State may refer for determination by the Court, the legality of any act, regulation, decision or action of a partner state or institution of the community on the grounds that such as an act, regulation, directive, decision or action is unlawful or is an infringement of the provisions of this Treaty.

These are wide-reaching provisions that can guarantee enforcement of regional commitments to public involvement in environmental decision-making; but they are yet to be tested. The status of regional instruments in national legal jurisdictions is still contested. The national constitutions of the three East African countries contain highly nationalistic sovereignty provisions. Consequently, constitutional reforms might be necessary to make some of the MOU's provisions operational.

Public participation in the development of REIA guidelines

The review of national policies and laws and the regional framework on public participation in the previous section have a number of implications for the development of guidelines for REIA of transboundary aquatic ecosystems. These include the following:
• Governments often listen to organized civic groups more than to individuals.
• When the issue at hand is a controversial development project or policy, organizations insulate their members against individual attacks and isolation by the state.

- By targeting civic groups, it is often possible to significantly reduce the transaction costs associated with public involvement, compared with those of individual consultations with citizens; this also encourages the use of EIA and public participation.
- Securing the right to free association as part of fundamental human rights guarantees the legitimate existence of environmental and other associations. Because of this guarantee, civic organizations are able to confront governments regarding environmentally controversial projects without fearing threats of loss of legal status.

This section therefore examines (a) why public participation was considered important in the development of the REIA guidelines and (b) the actual procedures taken to involve the diverse stakeholders.

Why public participation?

For a long time, decision-making has remained the domain of governments and resource managers, particularly with regard to public investments, natural resource management and access, and the distribution of benefits derived from ecosystems. This excluded local communities who, from time immemorial, had the *de facto* rights over the resources. More recently, though, local communities were denied access rights and were not given the opportunity to contribute to the management of the resources that affected their daily lives. With the change in societal attitudes, and the requirements for attaining sustainable development (which call for public involvement in the management of natural and environmental resources), such management practices and approaches are no longer acceptable. This explains why many policies and legislation that the EAC partner states (described in the previous section) have put in place in recent years have provisions for active involvement of stakeholders affected by land-use approaches and management decisions, development programmes, and projects.

In the implementation of these programmes and projects, stakeholder participation process is based on the needs to inform those people who will be affected by any likely development or management within any given aquatic ecosystem, to give room for discussions around the subject, and to create opportunities not only for recommendations to be made in the development of the guidelines but also to solicit the stakeholders' proposals to realize the objectives of the programme or project. The ultimate result is an optimum trade-off, with stakeholders understanding the overall project objectives and suggesting issues that need to be considered in the development of the guidelines. (This would also improve the acceptance of the final outcome of the project.) Full stakeholder participation from the earliest planning stages was, therefore, perceived as the best process to ensure the long-term success of the project and smooth integration into the national EIA frameworks of the EAC partner states.

The implementation of the ACTS–REDSO Plan of Action, mentioned above, seeks to stay clear of the "DAD" principle ("Decide, Announce, and Defend"; Braack and Greyling 2003), which is a traditional approach where the public is informed only of the outcome of a project. The ACTS project, thus, intends to promote good stakeholder-participation processes by engaging with interested and affected parties from the earliest stages of planning, before significant commitments have been made. The aim is to provide information to the diverse stakeholders, by promoting an "open-day" policy, thereby creating understanding and trust among stakeholders.

Indeed, over the last five years, the EAC partner states have witnessed a situation in which policy-focused environmental NGOs have taken on more challenging and controversial projects. This is in addition to the advocacy-oriented NGOs.

It is in line with these developments that the development of guidelines for REIA cannot be undertaken in isolation of the opinions of the diverse stakeholders (including local communities) or of the valuable site-specific knowledge that can be obtained from local groups, NGOs, and other institutions with a stake in transboundary aquatic ecosystems of Lakes Victoria, Jipe, and Natron; the Minziro–Sango swamp forest; and the coastal and marine ecosystems of East Africa. Public participation in the development of REIA guidelines serves two purposes – first, the need to communicate to various stakeholders the aims of the project and, second, the recognition that the public is an important source of technical, economic, and social information that is relevant to developing guidelines for REIA. Because the public is knowledgeable about the local conditions and issues necessary for good management of transboundary aquatic ecosystems, public involvement is expected to give the stakeholders a sense of local ownership, social acceptability, and commitment to both the development and the implementation of the guidelines when fully promulgated by the EAC and adopted by the partner states.

Procedures for public consultation in the development of REIA guidelines

The procedures for public consultation in the development of guidelines for REIA comprise three steps: (I) stakeholder identification; (II) drafting and reviewing the Regional EIA guidelines; and (III) pilot testing the guidelines in a specific aquatic ecosystem (fig. 22.2).

Step I, the identification of stakeholders (and the major focus of this chapter) defines the stakeholders and the institutional framework for

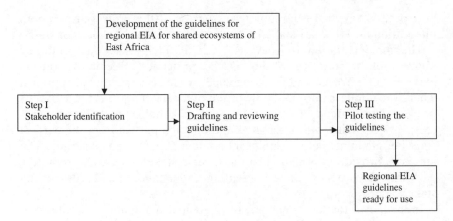

Figure 22.2 Steps in the development of guidelines for REIA

public participation. It also prescribes site visits to the aquatic ecosystem and national and regional workshops. In stakeholder identification, a wide range of institutions and individuals have been included, namely:

- sectoral government ministries and departments, including wildlife, water, tourism, forestry, and fisheries;
- local officials and members of the village development committees, elders, and resource users;
- academic and research institutions;
- NGOs and community-based organizations (CBOs);
- cultural institutions;
- programmes and projects;
- private sector (including businesses and professional societies);
- the EAC Legislative Assembly members;
- statutory agencies for natural resource management.

For each transboundary aquatic ecosystem, the ACTS has consulted various stakeholders to establish their specific mandates, roles, and responsibilities and to identify opportunities and constraints for promoting sustainable management of the transboundary aquatic ecosystems. Second, baseline information on the status of each aquatic ecosystem was obtained. Specific information collected included major resources and their status, causes and impacts of resource degradation, key management issues, priority areas for management, the major regional efforts to manage the resources, and forward-looking strategies for sustainable management. Major economic activities of the local people around each aquatic ecosystem (such as agriculture, fishing, livestock keeping and petty trading) were also captured.

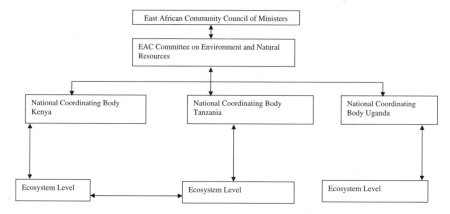

Figure 22.3 A four-tiered participatory framework for the development of guide-lines for transboundary EIA

The institutional arrangement for public participation in transboundary aquatic ecosystems was defined. This served several purposes: first, it defined the source of background information for informing the development of guidelines for REIA of shared aquatic ecosystems of East Africa; second, it enabled the identification of key clusters of stakeholder groupings in the partner states relating to transboundary aquatic ecosystems; third, it enabled a participatory framework to be developed for engaging all the stakeholders at an appropriate level in the development of the guidelines (see fig. 22.3).

A four-tier structure for the participatory framework was proposed. The first tier, at the top of the hierarchy, is the policy-making organ of the EAC, which would eventually recommend the guidelines to the Heads of State Summit of the EAC for endorsement. This is the Council of Ministers that makes decisions on the development of common EIA guidelines for shared ecosystems of East Africa.

The second tier is the Committee on Environmental and Natural Resources, which has four working groups for (1) terrestrial, (2) aquatic, (3) pollution, and (4) policy and legal and institutional frameworks. These working groups meet, discuss issues, and make recommendations that are forwarded to the Council of Ministers and eventually to the Heads of State.

The third tier constitutes individual national coordination bodies for Kenya, Tanzania, and Uganda under the aegis of the National Environmental Management Authorities (NEMAs) of Kenya and Uganda and the National Environment Management Council (NEMC)/Division of Environment (DoE) in Tanzania. These national coordination bodies are concerned with specific individual national EIA policies in their individual advisory and coordination capacities. Liaison among the three bodies is essential, especially if certain elements in their policies need consultation and harmonization as well as the domestic implementation of the guidelines, once they are promulgated by the EAC.

The fourth tier involves the border districts with transboundary aquatic ecosystems. These have stakeholders engaged in the management and exploitation of the aquatic resources. These include government departments (such as forestry, fisheries, water development, and tourism), as well as local and international NGOs implementing project activities in the catchments within individual countries. Also included are the resource users, including both the local communities (such as fishing communities and the fish-processing plants), as is the case in the Lake Victoria ecosystem. Other candidates considered in the fourth tier are the community-based organizations, such as the Health, Environment and Media Network (HEMnet); OSIENALA (Friends of Lake Victoria); and the East African Communities of Lake Victoria (ECOVIC); representation through business associations such as the fish-processors associations in Uganda, Kenya, and Tanzania; and researchers.

Donors play a significant role at all levels – including local, national, regional, and subregional levels – by facilitating programmes and projects being implemented in aquatic ecosystems.

ACTS also undertook site visits and consultations to each ecosystem to identify the most important natural resource issues in the aquatic ecosystems. These consultations enabled input into the process by the local people or resource users because they have useful knowledge (including information on natural resources in the area, resource management practices, cultural resources, and important religious sites). More specifically, local village groups and organizations that are involved in agricultural development programmes and fishing in the ecosystems were consulted. Further, local residents were also deemed to have insights and understanding regarding the nature and extent of natural resource degradation.

One case in point of the site visits and consultations is the visit to Lake Jipe, which highlighted the declining fisheries' potential and the degradation of the lake since the early 1960s. Until the 1950s, Lake Jipe was famous for its fishery business: it attracted many people from all corners of East Africa to invest in the flourishing fishing industry, because the lake was rich in various fish species, especially tilapia and catfish. Fishing was, however, artisanal, and has not changed since then. Some of those who were attracted to the fishing business performed the actual fishing, while others were middlemen who bought fish from the fishermen and sold it to consumers. One villager (Mr Iddi Kishavi, over 60 years of age), who was born in the area, recalls that, until the early 1950s, there were fishermen from as far away as Kisumu (Kenya), Mwanza, Bukoba, Musoma, Lake Nyasa, and Kigoma. Apart from fishing, they also practised subsistence farming (IUCN 1999). By the late 1960s the fishery began to decline as a result of the invasive typha and papyrus weeds, which made fishing too difficult. The local people call this weed *Magugu Maji* or *Makurubia* (on the Kenyan side) – a distortion of the water hyacinth. The weeds rapidly invaded the lake, to the extent that some fishermen fled to other places where fishing was profitable. As time passed, landing sites slowly became engulfed by weeds, making fishing more difficult. In due course, it became difficult to catch sizeable tilapia as the weeds increased. In response to the declining fish catches in the mid-1960s, many fishermen migrated to the newly built Nyumba ya Mungu dam, about 30 km south-east of Lake Jipe, where the fishery was more lucrative.

During the survey associated with this study the villagers on both sides of the lake (i.e. in Tanzania and Kenya) revealed that currently there is no commercial fishing because it is not easy to catch tilapia that exceed 2.5 inches in length. Fishermen suspect that the mature fishes hide in the thick growth of typha weeds, where they cannot be caught (Bisanda 2003).

Three national workshops (one in each country) and the regional workshop in Arusha, Tanzania have been conducted. The national workshops sought to share the findings of studies commissioned as the first step in developing guidelines for the EIA of shared ecosystems of East Africa (table 22.1). The participants of the national workshop in each country were drawn from sectoral ministries and departments (including forestry, water, fisheries, wildlife, local governments, finance and planning, and environment), local and international NGOs, academic and research institutions, and local CBOs.

The regional workshop sought to further the discussions on shared ecosystems of East Africa and to deliberate on forward-looking strategies for drafting the REIA guidelines. The participants in this workshop included the heads of sectoral departments and ministries, regional NGOs and CBOs, regional associations (such as the association of local authorities around Lake Victoria), East African communities around Lake Victoria, and key regional academic and research institutions.

One of the concerns expressed in the national and regional consultations was related to how the EAC partner states would incorporate the Regional EIA guidelines into domestic legislation and practice, as there was not yet any binding legal instrument that required them to do so. A protocol was suggested, but developing a protocol specifically for REIA guidelines is problematic. However, the existing regional instrument on environment matters – that is, the MOU among the three East African countries – is to be upgraded into a protocol; this protocol could, there-

Table 22.1 Commissioned studies to facilitate the development of REIA guidelines for shared ecosystems of East Africa

Study	Aims and specific focus
Constituency building and agenda setting (Stakeholder Analysis)	Define key players and stakeholders involved in the management of shared ecosystems in Kenya, Tanzania, and Uganda: – Establishing their mandates, constraints, and opportunities; – Identifying strategies to enhance their effectiveness; – Identifying entry/avenues for EAC participation in TBNRM
Review of the status of shared ecosystems (both aquatic and terrestrial) and socio-economic uses of shared ecosystems of East Africa	– Status information on the environment and resource; – Causes and effects of resource degradation; – Identification of issues central to sustainable management of resources; – Recommendations on forward-looking strategies
Review of national policies, laws, guidelines and procedures governing the conduct of EIA among the partner states	– Provide an overview of the current status of EIA in the EAC partner states; – Identify current common principles, approaches, and procedures in use in existing EIA guidelines, laws, and practices; – Identify gaps in national efforts towards implementation of EIA; – Propose forward-looking strategies for the conduct of REIA in the EAC partner states
Identification and review of best practices in regional EIA for shared ecosystems of East Africa	To document lessons learnt from the following: – Current EIA practices in the EAC; – Other regions and international experiences

EAC, East African Community; EIA, environmental impact assessment; REIA, regional environmental impact assessment; TBNRM, transboundary natural resource management.

fore, embrace the regional guidelines. This approach is fully supported by the EAC Secretariat and its Committee on Environment and Natural Resources. The role of the ACTS, just as in the development of the guidelines, would be to support the EAC in preparing this protocol.

The regional workshop also helped to identify issues to be considered when drafting the guidelines for transboundary EIA. Community participation was identified as one of the issues where definition of communities, methodologies for involving them, access to resources, creating awareness, and management practices are supposed to be inspected in detail. With regard to management practices, participants highlighted the need to account for co-management, sharing of benefits, indigenous knowledge and practices, self-policing, voluntary compliance, traditional and cultural practices, and formal and informal management systems.

Conclusions

This chapter has attempted to highlight the progress made to date towards the development of environmental impact guidelines for shared ecosystems of East Africa. The EAC partner states have put in place the policies and legislation with provisions for public participation in the management of natural resources, and of aquatic resources in particular. This is reflected in the national environmental and other related sectoral policies, as well as in the national constitutions and framework environment laws, particularly those for Kenya and Uganda. Sectoral laws on forestry, fisheries, water, and wetlands (for example) have also embraced public participation.

The review of national policies and laws and the regional framework on public participation has provided a number of implications for the development of guidelines for regional EIA of transboundary aquatic ecosystems. It has also informed the identification of the steps to be followed in developing the guidelines. These steps include stakeholder identification, drafting and reviewing the regional EIA guidelines, and pilot-testing the guidelines in specific aquatic ecosystems. The second two steps are to be undertaken in the next phase of the project. Owing to the geographical scope, it was impossible to involve all the stakeholders because of resource limitation (time and funds); the first step, therefore, centred on stakeholders who have a legitimate interest in transboundary aquatic resources and who should be involved in particular project-development processes and programmes within a given aquatic ecosystem.

The first phase of the project has yielded some specific lessons in developing regional EIA guidelines for shared resources in East Africa. First, that brokering regional consultations and consensus on regional issues takes time and consumes financial resources. Part of the reason is that ACTS did not have an adequate understanding of the EAC, its structures, and its functions, which are fairly complicated. Consequently, this affected the time necessary to implement the project. Before implementation, it affected ACTS's budgeting, as well as the extent of consultations in the partner states and with other stakeholders. Once the project had started, it was difficult to move expeditiously, because of the need to ensure that the scope of consultations had been broadened.

Second, the first phase of the project has shown that national positions on issues of shared ecosystems are complex, and that some technocrats from the partner states still do not see the benefits of pursuing common regional approaches – for example, through EIA procedures, policies, regulations, and laws. However, engaging the National Environment Management Authorities (NEMAs) of Kenya and Uganda, the National Environment Management Council (NEMC) of Tanzania, and the East African Community Secretariat, has helped to overcome the challenges encountered. These authorities are, by law, the custodians of their national environments and are responsible for EIA in their countries. Therefore, to improve regional (particularly EAC) processes, ACTS has learnt the following:

- The EAC should have the final decision regarding the selection of experts or consultants to undertake assignments with regional initiatives.
- There is a need to engage relevant authorities, such as the NEMAs and NEMC, from the onset. This engagement should include identification of the required teams, drafting of the terms of reference (ToRs) and definition of tasks, and keeping these authorities apprised of the project progress through appropriate channels.
- The EAC Secretariat should be involved in planning and running the national and regional workshops. This enables the regional activities to be truly an EAC Secretariat-led initiative, with organizations such as ACTS being seen as partners.

Third, there is room for increasing the "spin-off" transboundary natural resource management (TBNRM) opportunities. These opportunities include demonstrating in economic terms the costs and benefits of transboundary natural resources management and policy implications, valuing natural resources to aid decision-making, developing regional tourism, integrating local communities into the management of shared ecosystems, facilitating support from the EAC Secretariat and the governments of the partner states, and working toward the restoration of Lake Jipe.

The EAC partner institutions, such as ACTS, also have roles to play. Most notably, they can continue to engage in policy research and analysis and feed such information to the CBOs and NGOs working with local communities and local government authorities in shared aquatic ecosystems for effective conservation and management.

REFERENCES

African Centre for Technology Studies (ACTS). 2000. *"Plan of Action for the Development of Regional Guidelines for Environmental Impact Assessment of Shared Ecosystems of East Africa."* Proposal Submitted to the Regional Development and Economic Support Officer (REDSO), USAID. Nairobi, Kenya.

Bisanda, S. 2003. *"Socio-Economic Uses of Shared Ecosystems of East Africa."* Report submitted to the African Centre for Technology Studies (ACTS). Nairobi, Kenya.

Braack, L.E.O., and Greyling, T. 2003. "Stakeholder Consultation and Participation in Conservation Planning." In: L.E.O. Braack, T. Petermann, and F. Lerise (eds). *Proceedings of Workshop on Transboundary Protected Areas: Guiding TBPA Approaches and Processes in East Africa*, 24–28 February. Mweka, Tanzania: Inwent.

Bruch, Carl, Wole Coker, and Chris VanArsdale. 2001. "Constitutional Environmental Law: Giving Force to Fundamental Principles in Africa." *Columbia Journal of Environmental Law* 26:131–211.

Constitution of Kenya Review Commission. 2002. *"The People's Choice."* The Report of the Constitution of Kenya Review Commission, Short Version. September. Nairobi, Kenya: Government Printer.

DoE (Division of Environment). 2002. *Tanzania Environmental Impact Assessment Procedures and Guidelines.* Vol. I. Vice-President's Office. United Republic of Tanzania.

EAC (East African Community). 1998. *Memorandum of Understanding on Cooperation in Environment Management.* October 22. Arusha, Tanzania: EAC Secretariat.

EAC (East African Community). 2000. *The Treaty Establishing the East African Community.* Arusha, Tanzania: EAC Secretariat.

IUCN. 1999. *Lake Jipe Cross-border Planning Workshop Report*, 13–15 October. Moshi, Tanzania. East Africa Regional Representative Office. Nairobi, Kenya.

Kairu, J. 2003. *"The Status of Shared Terrestrial Ecosystems of East Africa."* Report Submitted to the African Centre for Technology Studies (ACTS). Nairobi, Kenya (unpublished).

Ministry of Natural Resources and Tourism. 1998a. *National Forest Policy.* Dar es Salaam, Tanzania: Government Printers.

Ministry of Natural Resources and Tourism. 1998b. *The Wildlife Policy of Tanzania.* Dar es Salaam, Tanzania: Government Printers.

NEMA (National Environment Management Authority). 2001. *State of Environment Report for Uganda.* Kampala, Uganda: NEMA.

NEMC (National Environment Management Council). 2002. "Tanzania Environmental Impact Assessment Procedures and Guidelines" (Draft). Republic of Tanzania. Vol. 1. March.

NES (National Environment Secretariat). 1994. *National Environment Action Plan*. Nairobi, Kenya: Government Printer.

Planning Commission. 1999. *The Tanzania Development Vision 2025*. Dar es Salaam, Tanzania: President's Office, Planning Commission.

Planning Commission. 2000. *Tanzania Assistance Strategy – A Medium Term Framework for Promoting Local Ownership and Development Partnership*. Dar es Salaam, Tanzania: President's Office, Planning Commission.

Republic of Kenya. 1977. *The Wildlife (Conservation and Management) Act*. Chapter 376 Laws of Kenya (1985). Nairobi, Kenya: Government Printer.

Republic of Kenya. 1994–1996. *National Development Plan*. Nairobi, Kenya: Ministry of Planning and Economic Development.

Republic of Kenya. 1997–2001. *National Development Plan*. Nairobi, Kenya: Ministry of Planning and Economic Development.

Republic of Kenya. 1999. *Sessional Paper No. 1 on National Policy on Water Resource Management and Development*. Nairobi, Kenya: Government Printer.

Republic of Kenya. 2000. "Draft Forest Bill." Nairobi.

Republic of Kenya. 2002. *The Water Act. Kenya Gazette Supplement No. 107* (Acts No. 9). Nairobi: The Government Printer.

Republic of Tanzania. 1983. *National Environment Management Council Act*. Act No. 19 of 1983. Dar es Salaam: Ministry of Tourism, Environment and Natural Resources. United Republic of Tanzania.

Republic of Tanzania. 1998a. *The National Beekeeping Policy*. Dar es Salaam: Ministry of Tourism, Environment and Natural Resources, United Republic of Tanzania.

Republic of Tanzania. 1998b. *The National Tourism Policy*. Dar es Salaam: Ministry of Tourism, Environment and Natural Resources, United Republic of Tanzania.

Republic of Uganda. 1994. *The National Environment Policy*. Entebbe, Uganda: Government Printers.

Republic of Uganda. 1995a. *The Constitution*. Entebbe, Uganda: Government Printers.

Republic of Uganda. 1995b. *The National Environment Statute*. Entebbe, Uganda: Government Printers.

Republic of Uganda. 1995c. *The National Policy for the Conservation and Management of Wetlands*. Entebbe, Uganda: Government Printers.

Republic of Uganda. 1995d. *The Water Policy*. Entebbe, Uganda: Government Printers.

Republic of Uganda. 1995e. *The Wildlife Policy*. Entebbe, Uganda: Government Printers.

Republic of Uganda. 1998. *EIA Regulations*. Entebbe, Uganda: Government Printers.

Republic of Uganda. 2000. *The Draft Forestry Policy*. Entebbe, Uganda: Government Printers.

Republic of Uganda. 2001. *The Fisheries Policy*. Entebbe, Uganda: Government Printers.

Twongo, T. 2003. "The Status of Shared Aquatic Ecosystems of East Africa."

Report Submitted to the African Centre for Technology Studies (ACTS). Nairobi, Kenya. Unpublished.

UNCED. 1992. *Rio Declaration on Environment and Development.* Report of the United Nations Conference on Environment and Development, Rio de Janeiro, 3–14 June. Annex 1.

UNEP. 1992. *Convention on Biological Diversity.* United Nations Environment Programme.

VPO (Vice-President's Office). 1997. *National Environmental Policy.* The United Republic of Tanzania. Dar es Salaam, Tanzania: Government Printers.

VPO (Vice-President's Office). 2000. *Poverty Reduction Strategy Paper.* Dar es Salaam: Ministry of Finance, United Republic of Tanzania.

23

Access to justice through the Central American Water Tribunal

Juan Miguel Picolotti and Kristin L. Crane

Introduction

Valuable watersheds and environments of Latin America are being contaminated and degraded. Rivers and underground water, vital for both
human populations and biodiversity, are used as dumps of black water,
rubbish, and other kinds of waste from agricultural and industrial processes. Water-quality degradation and reduced access have led to more
deaths in Central America than violent conflicts or natural disasters
(Sequeria 1999). Many communities face these threats to their environment, culture, and lifestyle without being able to find redress, and without
even knowing what options they have to pursue those responsible for the
environmental and social damage that they suffer.

In response, civil society in the region has embarked upon a parallel
judicial process to prosecute those responsible for degrading water resources in the region. The Central American Water Tribunal (CAWT)
is an example of civil society pursuing such an agenda. It has taken on
controversial cases, such as the contamination of Laguna del Tigre
National Park by Anadarko Petroleum Corporation, as detailed below,
and has made an impact. This chapter explains the origins of the Tribunal in the context of previous international tribunals that have influenced
the CAWT, as well as the Tribunal's objectives that are important in
understanding how the Tribunal functions in Latin America.

The CAWT represents an international, autonomous instance of social and environmental justice, created by a coalition of non-governmental organizations (NGOs) from Central America with the collaboration of an interdisciplinary team of lawyers and scientists from around the world. The CAWT was established in 1998 in San José, Costa Rica. The work of the Tribunal is important because of the pressing water-related challenges in Central America. Although water resources are plentiful, one-third of the total population does not have potable water (Sequeria 1999). The Pan-American Health Organization (PAHO) estimates that 15 million Central Americans do not have access to clean water, or have to pay 10–20 times more for potable water than the well-off sectors of society because of subsidies. Water-quality problems are a leading cause of infant mortality: the PAHO approximates that one Central American child dies every minute from acute diarrhoea (Sequeria 1999).

The CAWT is a non-jurisdictional tribunal characterized by its promotion of public participation that enables all interested actors to participate in a process to facilitate resolution of conflicts arising from water resource utilization in Central America. In addressing such disputes, the Tribunal seeks to enforce the relevant human rights and environmental law instruments at the local and international levels.

The functions of the Tribunal are to encourage the participation of civil society, deliver verdicts in cases, put in place policies that will prevent further abuse of water resources, and study previous cases on water pollution and analyse the positions of the defendants (CAWT 2000a). Recently, the Tribunal has placed emphasis on two areas – conflict resolution and prosecution of entities that are polluting water resources. The main strategy of the Tribunal is to offer a venue for action by groups that do not usually have access to traditional avenues of justice, and to those who are directly affected by the degradation of these resources. The groups that experience the effects of water resource degradation include groups composed of small farmers, households, and families headed by women. These groups are affected the most because water contamination usually occurs in rural areas, not urban centres. The money required to treat the water to ensure that it is potable cannot be provided by small, rural communities. At the same time, agricultural practices are disrupted because productivity is affected by the decreasing water quality. Women are especially affected because traditionally "women play a central part in the provision, management and safeguarding of water" (Dublin Statement 1992).

The CAWT is not a judicial tribunal; thus, it does not have the authority to issue binding sanctions. Its resolutions, although not legally binding, are founded upon ethical and legal considerations, and the success of its resolutions rests on established principles of coexistence; respect for individual and collective human rights; environmental rights; and recognition of the importance of all living forms consecrated in different international treaties, conventions, and declarations, as discussed below.

The establishment of CAWT was inspired by a few specific instruments. These include the Fresh Water Treaty, the Dublin Statement, the Convention on Biological Diversity, and the San José Declaration. The 1993 Fresh Water Treaty, concluded by NGOs, led to the establishment of the Central American Fresh Water Treaty in June 1998 (Fresh Water Treaty 1993). The Dublin Statement was adopted during the International Conference on Water and Environment in 1992. The numerous governments, international organizations, and members of civil society that attended the Conference decided that the situation of world water resources was critical and that a new focus was needed on the use, assessment, development, and management of fresh water resources. It was recognized that fresh water is a "finite and vulnerable resource" (Dublin Statement 1992). The 1992 Convention on Biological Diversity (CBD) is influential because it stressed that natural resources must be used sustainably. One of the most important aspects that inspired the establishment of the CAWT was the recognition that states both have sovereignty in their territories and must ensure that their resources are used responsibly without damaging the environment. The 1996 San José Declaration was adopted during the Conference on Water Resources Assessment and Management Strategies for Latin America and the Caribbean. The Declaration established strategies for reaching an equilibrium between the supply and demand of water.

The CAWT is also informed by similar experiences with the Rotterdam Tribunal (1983), the Second International Water Tribunal in Amsterdam (1992), and the National Water Tribunal in Brazil (1993). The 1983 Rotterdam Tribunal was the first water tribunal, and it adopted policies to control water pollution. The Rotterdam Tribunal was a civil-society endeavour that sought to respond to pollution in the Rhine River Basin. In 1992, the Second International Water Tribunal took place in Amsterdam; it addressed claims that governments, corporations, and industrial sectors were responsible for water contamination. The National Water Tribunal of Brazil was founded in Florianopolis in 1993 to investigate cases of water pollution from hydroelectric dams, mining, radioactivity, and fertilizers (CAWT 2000b). The previous tribunals, by means of cooperative initiatives, analysed complaints of contamination and other impacts on water systems in different parts of the world.

All of the above-mentioned declarations, tribunals, conventions, and treaties are considered hard or soft instruments of environmental law. In this context, the CAWT supports the implementation of international agreements and builds upon other, similar, efforts in other regions.

Tribunal organization

The Tribunal administration is composed of local organizations widely recognized for their dedication to sustainable development, biodiversity protection, and addressing the social effects of its degradation. The Tribunal's headquarters are in San José, Costa Rica, where the Tribunal administers the cases and plans its sessions.

The Scientific–Technical Commission of the Tribunal comprises an interdisciplinary group of professionals, academics, and students. Their role is to select the cases to be addressed in the Tribunal's subsequent session. This process occurs following the submission of cases.

It is worth noting that the Tribunal takes only two cases per country in Central America in each session, which accounts for a maximum of 12 cases. However, the Tribunal can hear more than two cases from a single country when other countries do not submit their allocated two cases, as long as the limit of 12 cases established by the Tribunal is not exceeded.

The jury of the CAWT consists of nine influential representatives – one juror from each member country and three others from South American countries and Europe. The jurors are professionals from civil society and are selected on the basis of their proven excellence in the areas of public service, education, science, human rights, and environment. For example, six Central American judges representing each country of the region (Costa Rica, Nicaragua, Panama, Guatemala, El Salvador, and Honduras) plus a Cuban judge from the Institute of Epidemiological Studies of Cuba, a Brazilian judge, and a Spanish judge (President of the Water Tribunal of Spain) comprised the jury of past hearings. Three Judges of Ordinary Justice from Costa Rica have also collaborated in this process.

Tribunal processes

Submission and prima facie evidentiary requirements

Cases must be submitted in writing and can be sent via fax, mail, electronic courier, or in person to the administrative office of the Tribunal in Costa Rica. Each case must be accompanied by evidence sufficient to determine that the case is based on violations of local and international environmental and/or human-rights norms. It is worth noting that the Tribunal considers the identification of the victims of water mismanagement in each particular case of utmost importance.

Potential complainants

The cases can be commenced either collectively or individually by activists, community leaders, professionals, or any other groups on behalf of people who suffer from improper water management. Any body of water that has experienced a negative change due to irrational or negligent use that affects a population in any of the member countries may be considered.

Potential accused parties

All natural or legal persons, including states and national or transnational corporations, that, owing to their operation, have contaminated, overextracted, improperly managed, or degraded aquatic habitats or water can be accused. In addition, anybody that supports or condones those activities that lead to the above problems can be charged.

Procedure

The Scientific–Technical Commission administers the selection process. An important consideration is ensuring that the 12 cases of the session will be representative. The Commission selects the cases that it will consider, on the basis of the actual or potential environmental damage, the geographical importance of the area, social and physical impacts on the affected community, and the nature and extent of population affected or threatened (CAWT 2000a). The complainants are notified as to whether their cases have been selected, and are obliged to maintain confidentiality until the accused are officially notified.

The cases that have been selected should have scientific evidence, studies, testimonies, and media reports to attest to the damage and provide the greatest impact. In the event that a complainant does not have sufficient resources or finances to supply the necessary evidence to present a solid case to the Tribunal, assistance can be provided.

Although it is not a requirement that the complainant should exhaust domestic judicial remedies before bringing a case before the Tribunal, the Commission asks the complainant to attach, if available, documentation regarding prior actions brought before their national judicial and administrative bodies. This step also seeks to ensure that the selected cases have a well-developed body of evidence and are able to make a significant impact.

Communicating with the accused parties

After confirming the scientific and technical merits of each case, the Tribunal notifies the named accused parties (individuals, groups, businesses, or governments). The Tribunal allows a period of two months for the accused parties to prepare their defence and evidence, and to present their response during a Tribunal hearing.

Functioning of the Tribunal

Once the Tribunal has acknowledged the lawsuit and the response of the accused party(ies), two possible procedures may ensue. First, a process of conflict resolution can take place, in which the Tribunal aims to reconcile differences between complainants and accused parties. The purpose of this measure is to seek a set of commitments to resolve the environmental and social problems identified. The Tribunal acts as mediator in this process to reach a private agreement among the parties. Second, a judgement hearing may be necessary in those cases in which reconciliation was not successful. The Tribunal establishes a hearing date at which the arguments of both parties are to be heard and the judgement pronounced.

The most common practice is to give parties the opportunity to reach a "friendly" and private agreement. The advantage of this practice is that it eliminates the need for the Tribunal to issue a public judgement, avoiding the consequent ethical and political implications. It is the most desirable outcome, as it also requires fewer resources and is less damaging to the accused party. A formal "guilty" verdict can entail significant negative international attention, and accused parties are well aware of the impacts of such negative publicity. The organizations that make up the Tribunal are also prepared to contribute to the cause by increasing international media attention to enhance the impact of the decisions of the Tribunal. Governments can also be charged as environmental offenders, and this has political implications. It is usually preferable for a government to avoid these negative effects to their political image by reaching a private solution. Further, governments can be charged for aiding the activities that have led to the degradation of water systems. Some financial liaisons have been kept private for reasons of negative publicity. Financially, it is good practice to try to solve the problem through friendly agreements. If the process does not have to continue to the next level, fewer resources have to be devoted to the same case and they can be used in other areas or to pursue environmental offenders.

First experiences of the Tribunal: Cases in 2000–2001

The CAWT held its first round of judgement hearings in August–September 2000. During this period, the Tribunal acknowledged the 11 cases selected by the Scientific–Technical Commission. In this first period, two cases were selected from each country of the region, except for Honduras which submitted only one. The cases and their complete judgements can be found on the Tribunal's new Web page (http:// www.tragua.com). Three cases are showcased here to demonstrate the diversity of issues that the Tribunal has addressed relating to environmental and human-rights matters.

Oil activities in Laguna Del Tigre, Peten (Guatemala)

The Mayan Biosphere is an ecologically important reserve of wetlands and tropical forests. It contains more than 2 million hectares of diverse flora and fauna and extremely important archaeological ruins. In the middle of the Reserve is the Laguna del Tigre National Park, which is Central America's largest protected wetland. It is listed under the 1971 Ramsar Convention on Wetlands of International Importance, ratified by the Government of Guatemala in 1990. Oil drilling is not allowed in the Mayan Reserve, but permits were granted to Basic Resources International, a subsidiary of Anadarko Petroleum Corporation, since the contract was introduced into the Reserve by a "grandfather" clause (which created an exemption based on circumstances previously existing). Basic started drilling in 1985. The International Finance Corporation (IFC) helped to expand Anadarko's activities in 1993, through a US$20 million loan to increase oil extraction by 30 per cent and to build a pipeline.

When Anadarko announced that it would begin drilling in the Laguna del Tigre Park, it did not indicate the magnitude of its plans or the severity of its impacts. The company insisted that the pipeline would decrease the environmental harm because transporation would be reduced and the likelihood of spills would be reduced. Instead of minimizing environmental damage, the pipeline construction has resulted in clear-cutting of primary rain forest, building of additional roads, and subsequent increased migration of landless peasants into the area. Because the pipeline is above ground, it has been exposed to numerous guerrilla attacks, and leaks have occurred.

There have been numerous ecological consequences of the petroleum development. These include the discharge of acid water in pools, destruction of vegetation, soil erosion, and disruption of water drainage and flow patterns. The complainants maintain that there is direct contamination from the production of oil. Hydrogen sulphide is generated and, when burned, becomes sulphur dioxide, affecting the biosphere and human health (CAWT 2000a).

In 1996, Basic applied for another loan from the IFC to construct a second pipeline. In order to receive the IFC loans, Basic was required to have secured approval from the public and encourage their participation in the planning and decision-making process. In addition, Basic was required to carry out a full environmental impact assessment (EIA). Basic agreed to perform a second EIA before receiving the second loan, but the NGO Conservation International alleged that the EIA was unsatisfactory because it minimized the adverse effects that increased park access would have on the area and it did not consider alternative routes for the pipeline (Mollman 2000).

The 1995 World Bank Natural Habitat Operational Policy described Laguna del Tigre Park as a "critical natural habitat" and recommended that the Bank should refrain from financing the project unless "comprehensive analysis demonstrates that overall benefits from the project substantially outweigh the environmental costs" (Mollman 2000). Basic, however, commenced construction of the second pipeline before receiving approval.

A case was launched by Madre Selva, a Guatemalan environmental group, charging Basic with ignoring the environmental consequences of its activities, and the CAWT insisted that the Guatemalan government halt Basic's activities immediately. Despite having ignored previous national lawsuits, Basic recognized the weight of the Tribunal's warnings and issued a statement arguing that there is no scientific basis for the claim that Laguna del Tigre National Park is a site governed by the Ramsar Convention, and that

the Ramsar Convention Declaration establishes the sovereign right of the signatory States' right to decide on the use and management of the Ramsar site lands, as well as the State's right to substitute and/or modify the registered wetlands area, for reason of public utility or need. (Mollman 2000)

The CAWT ruled that Anadarko should immediately halt its activities and pay compensation for the damage, as deemed necessary by an independent consulting group and the government. The Government of Guatemala was also cited for violating the international obligations in the Ramsar Convention for protecting wetlands and for not fulfilling its obligations to the people of Guatemala. The CAWT also recommended that the government implement measures to ensure that various legal regulations that were violated by oil exploration in the Laguna del Tigre area are followed in the future (CAWT 2000c).

Gold mining in the North Atlantic Autonomous Region (Nicaragua)

Hemconic S.A., also known as Greenstone Resources Ltd, operates four gold mines in the municipality of Bonanza in the North Atlantic Autonomous Region of Nicaragua. The area of the mine is surrounded by four rivers and rural, mestizo (mixed race), and indigenous populations use these water basins for all their needs. In 1994, the Nicaraguan Government granted Hemconic S.A. mining rights on 12,400 hectares for a period of 50 years (Global Mining Campaign 2003).

The mining techniques utilized are low-technology, and the installations are in poor condition. For example, deterioration of equipment has caused leaks of toxic chemicals. These chemicals found their way into the soil, groundwater, water systems, and the atmosphere. After many complaints had been received from the area, the Center of Water Resources Laboratory carried out testing in 1999 and found that cyanide levels were above 0.1 milligrams per litre and that copper levels were higher than the standards established in the Decree No. 33-95 (Regulations for the Control of Pollution Caused by Domestic, Industrial and Agricultural Discharges of Contaminated Waters) and the World Health Organization standard of 0.07 milligrams per litre (Global Mining Campaign 2003). Hemconic S.A. is also charged with violating 380 labourers' rights through "deplorable work conditions, absence of proper work conditions and exposure to toxic and hazardous substances" (CAWT 2000c).

The CAWT received the complaint from the Humboldt Center, alleging environmental contamination and dangerous and reckless mining practices. The CAWT ruled against Hemconic and the Nicaraguan Government and resolved:

1. To reprimand the Nicaraguan Government for not protecting the population and not ensuring compliance with environmental regulations. In addition, the Government was censured for not addressing the risk to the water supply that has resulted in health problems and environmental troubles in ecologically sensitive areas.
2. To compel appropriate authorities to put in place the necessary measures to stop the harmful mining practices of allowing contaminated sediments to be discharged without proper treatment into the environment.
3. To order Hemiconic to pay reparation costs to affected communities (CAWT 2000c).

Highway construction and landfilling in Panama Bay (Panama)

The Corredor del Sur is a highway built by Ingenieros Civiles Asociados (ICA). It is 19.5 km in length and stretches across the downtown and eastern districts of Panama City. A portion of highway passes over the Panama Bay and is built on a rockfill located 50 metres from the coastline. The project received funding approval from the IFC in 1998 but has had numerous problems with the consultation process, project design, and implementation (Solis and Saladin 2000).

With regard to the project design, landfills are used instead of columns. The landfill construction obstructs the marine currents that disperse sewage flowing from Panama Bay. Thus, faecal matter collects along the Corredor del Sur, increasing pollution and health risks for the people living along the coast. To reduce the faecal matter in the area, pipes were built to carry the sewage to the other side of the Panama Bay. The piping system uses gravity and not a pump, which leads to an additional problem: when the tides are high, the sewage could change direction and flow back into the pipes of people's homes. In the event that the system does not work, the ICA has an agreement with the government to be able to create a landfill between the coast and the causeway, in which case it would become owners of very valuable property (Solis and Saladin 2000). The significance of the agreement is that, in effect, by poorly designing the mitigation plan, the environment suffers and the contractors benefit financially.

The EIA was inadequate: the original assessment considered only the impact of the Corredor del Sur highway and not the effects of blocking the ocean current or the impact of the landfills. The IFC approved the project before insisting that an additional EIA be done. One was completed, but it did not fully represent the effects of the landfills or the effects on the health and environment due to construction, resettlement, and natural resource use.

The resettlement and compensation that was promised also has proved to be inadequate. Communities were not allowed to collectively bargain, and not every household was compensated for the value of their land and home. Of the families that did relocate, many lost their livelihoods because they lost access to the shore. The local fishing industry has also suffered from the faecal contamination (Solis and Saladin 2000).

The case was brought to the CAWT by a Panamanian NGO, and the Government of Panama was charged with negligence in permitting the construction of the Corredor del Sur, landfill real-estate projects, and the subsequent environmental damage. The CAWT blamed Ingenieros Civiles Asociados for the environmental damage and risks that were caused by the construction of the Corredor del Sur Highway without adequate plans having been made to "avoid potential backflow and flooding with serious damage and sanitary, environmental, material, and moral risks ... because of non-compliance with the established regulations in the Political Constitution and the Laws of Panama" (CAWT 2000c). The IFC was also blamed for approving the loan before having received an appropriate EIA and causing "serious damage to the health of Panamians, disrespecting the environment, and violating the Constitution and the Laws of the Republic."

The CAWT recommended:

- That ICA Panama was to compensate citizens who were affected by the Corredor del Sur and to consult with citizens on how to implement new mitigation measures to solve the problem of flooding and faecal sedimentation and implement a programme to ensure that affected citizens are consulted and are able to participate in the decision-making process.
- That IFC should give ICA the necessary resources to pay for an independent company to monitor and control the activities, so that it could make accurate reports available to the public to ensure greater transparency. In addition, that the IFC policies and guidelines should be changed to comply with its mission of fighting poverty.
- That the contract with ICA should be renegotiated to prevent further landfill construction.
- That a coordinating institution should be developed to evaluate urban development plans including sanitation (CAWT 2000c).

The highway was finished in February 2000, and the IFC has maintained that there are no fundamental problems in the project, while conceding that the consultation process could be improved. ICA received permission from the Panamanian Government to continue with the landfills and real-estate development. There is still strong opposition to the project and debate about the hydrological impacts of the landfill and its impacts on sanitation in Panama Bay.

Second round of cases (2002–2003)

At the time of writing, the Tribunal was receiving cases for its next round of hearing and judgement. The Tribunal has received more than 85 cases since its creation – a significantly larger number than the limit of 12 per year. The Scientific–Technical Commission has yet to select the cases to be presented to the Tribunal.

Four of the cases presented to the Scientific–Technical Commission are noted here, including two from Costa Rica. There is a complaint against the Boruca Hydroelectric Project, which threatens to disrupt the traditional way of life of indigenous communities in south-eastern Costa Rica. The second complaint is against the Meliá Resort Project in the Guanacaste Province, where there is a possibility that groundwater may be overexploited, which may affect the availability of fresh water for 13 populations in the area.

In El Salvador, a complaint has been submitted to the Scientific–Technical Commission involving the planning and eventual construction of the Río Torola Hydroelectric Project in the San Miguel Department.

This project could affect a considerable rural population and their livelihoods, with the potential displacement of thousands of people.

In Nicaragua, a complaint involving a tilapia-feeding project (as part of a project for cultivating freshwater fish for export) was presented to the Tribunal. This fish is not native to the area and, if introduced in great numbers, could disrupt the equilibrium of the natural ecosystem in Lake Nicaragua, which is Nicaragua's main freshwater resource.

Conclusions

The decisions of the Tribunal, despite not being legal, have great impact because ethical implications can carry a lot of weight. The trials of the CAWT have received significant media coverage in the region, proving that, despite its non-binding status, its decisions are important. In the case of the oil drilling in the Laguna del Tigre National Park, two cases were launched against Basic Petroleum, one by the Guatemalan Human Rights Ombudsman and another case by 50 citizens. Both cases were taken to the Guatemalan Supreme Court, which ruled on neither of the cases. Magal Rey Rosa, a representative of the Madre Selva Group that accused Basic in the Tribunal, commented thus: "I don't know if the courts are co-opted, corrupt or incompetent, but the fact of the matter is that BASIC's operations are illegal, and that the cases should have been ruled upon long ago" (Mollman 2000). Basic Petroleum recognized the authority of the CAWT by issuing a statement (which is more than it had done for the two previous denunciations). The lack of available information about Basic's actions following the decision of the Tribunal prevents the authors of this chapter from evaluating the changes brought about by the decisions. However, the fact that Basic did recognize the accusations of the complainants marks a moral victory.

Rosa's statement leads to another interesting point – the inability of many domestic court systems to try companies implicated in environmental exploitation. Often, it is the government itself that is implicated, making it more difficult for the courts to rule in favour of the plaintiffs. In these circumstances, it is necessary to have an impartial, external, mediating body that has the ability to rule on complaints against governments and multinational corporations alike. The CAWT, as an ethical tribunal, is such a body.

The CAWT also addresses the lack of connection between the law and the reality of law. As was seen in previous cases, laws exist in Central America to prevent the irrational use of water, but they often are not implemented. The CAWT brings these laws and legal instruments to the attention of the national and international communities. This process proves that the existence of laws asserting the right to clean water do not necessarily mean that justice is served; these rights need to be actively enforced and protected. The CAWT provides a forum in which people's involvement in their rights can be stimulated.

A great problem in Latin America is that civil society often does not trust its governments and judicial systems because of a history of corruption and patronage. People see the state as irrelevant and a distant force, one that frequently acts against their rights. This view challenges the concepts of justice. An institution such as the CAWT is important because it is not affiliated with any government; in fact, it charges governments with violations of domestic and international law. Moreover, the Tribunal is an instrument that anybody can use to ensure that justice is served, allowing even the powerful to be accused.

The Tribunal is unable to force sanctions, but its decisions are a type of moral sanction and have symbolic efficacy. It is able to serve justice when its verdicts are recognized, giving it legitimacy. The number and the quality of the groups and individuals that contact the Tribunal also indicate its importance. Many people come searching for justice and a solution, and the Tribunal is able to provide them with access to justice.

The Tribunal has been functioning for only a short period of time, and it is too early to be able to measure the effects of its decisions because change can be a lengthy process. The Tribunal has made an impact, but not necessarily because of the changes that the decisions have generated. The primary impact of the CAWT lies in its ability to generate and focus international attention on a particular issue in such a way that it could affect the reputation of companies and governments. In this respect, the CAWT is capable of preventing and remedying infringements of the law.

Although the results and effects of institutional arrangements such as CAWT are currently limited, the development of processes such as this is important in advancing a shift in favour of public participation and access to justice. There is great potential for advancing participation in water resource management through this means. If developed effectively, this form of independent tribunal can also contribute to better management of biodiversity and mitigate the effect of environmental degradation on the most vulnerable communities.

REFERENCES

CAWT (Central American Water Tribunal). 2000a. *Procedures*. Internet: ⟨http://tragua.com/procedures.html⟩ (visited 11 September 2003).
CAWT (Central American Water Tribunal). 2000b. *Previous Tribunals*. Internet: ⟨http://tragua.com/previoustribunal.html⟩ (visited 11 September 2003).

CAWT (Central American Water Tribunal). 2000c. *Verdicts.* Internet: ⟨http://tragua.com/veredics_text_eng.htm⟩ (visited 19 September 2003).

Dublin Statement on Water and Development (Dublin Statement). 1992. Internet: ⟨http://www.wmo.ch/web/homs/documents/english/icwedece.html#introduction⟩ (visited 17 November 2003).

Fresh Water Treaty. 1993. Internet: ⟨http://habitat.igc.org/treaties/at-21.htm⟩ (visited 17 November 2003).

Global Mining Campaign. 2003. "Nicaraguan Case Study: Mining in Bonanza, Autonomous Region of the North Atlantic." *The Global Mining News.* March 18. Internet: ⟨http://www.globalminingcampaign.org/theminingnews/assets/pdf/bonanza.pdf⟩ (visited 17 November 2003).

Mollman, Marianne. 2000. "Guatemalan Oil Debacle." *Multinational Monitor* 21(12).   Internet: ⟨http://multinationalmonitor.org/mm2000/00december/oil.html⟩ (visited 17 November 2003).

Sequeria, Maricel. 1999. "Defending Water Rights." *IPS.* May 17. Internet: ⟨http://domino.ips.org/sid/EnrissDb.nsf/0/4403ec632b24a1f6802567b00042fa3e?OpenDocument⟩ (visited 13 September 2003).

Solis, Felix Wing, and Claudia Saladin. 2000. "Panama's Corredor Sur: Turning the Bay of Panama into a 'Fecal Swamp.'" September. Internet: ⟨http://www.ciel.org/Ifi/ifccasepanama.html⟩ (visited 17 November 2003).

Conclusion

24

Conclusion: Strategies for advancing public involvement in international watershed management

Carl Bruch, Libor Jansky, Mikiyasu Nakayama, and Kazimierz A. Salewicz

Introduction

This volume has highlighted many instances – but by no means all – in which access to information, public participation, and access to justice are now included in international and regional agreements concerning transboundary watercourses, as well as policies and practices of international, regional, national, and sub-national institutions. Experiences with these norms, institutions, and practices are likely to affect how other watercourses involve the public in decision-making. However, both the success and the full implementation of such provisions depend on several factors.

Factors affecting the development of participatory frameworks

In developing and implementing norms and mechanisms for public involvement in the management of transboundary watercourses, it is important to look at the context of each particular watercourse. This can include an analysis of the bordering countries, local legal systems, and existing national or regional initiatives on public participation.

Experiences in transboundary watercourses vary greatly, depending on a range of geopolitical, historical, and social factors. When there are only a few riparian nations, agreements on transboundary watercourses are more likely to include the public. For example, the 1909 agreement between Canada and the United States on the management of their boundary waters and the North American Great Lakes included public-participation provisions that remain unmatched in many contemporary agreements. Conversely, rivers with numerous riparian nations (such as the Nile) are likely to raise more conflicts, and public participation often lags. Similarly, where communities straddle a watercourse, there may be more incentive to develop a management system that accounts for the interests of the counterparts on the other side of the watercourse (Milich and Varady 1998).

A related factor is the degree to which nations share a cultural, historical, and social background. With a common basis there can be greater trust, not only at the government level but also at the popular level. As a result, one notices that the United States–Canada and Kenya–Tanzania–Uganda agreements, for example, evolve more rapidly and include stronger provisions for public participation than those for many other watercourses.

A highly sensitive international context can make international agreements more difficult to reach, and governmental officials more reluctant to open the door to third parties whom they perceive as posing a danger of either compromising their own position or of confusing the relationship. A context can become sensitive through economic or political instability, including warfare (Eriksen 1998). The international context could also become sensitive owing to actual, imminent, or prospective overburden of the available water, particularly where there is a historically dominant water user. In contrast, areas such as Southern Africa generally present a comparatively stable economic and political environment in which the demand for available water is not yet as severe as, for example, with the Nile River or Aral Sea. As a result, there can be more room to negotiate and to involve the public.

Existing regional initiatives on public involvement can also be of assistance in furthering participation in transboundary watercourse management. Although some of these initiatives are non-binding, they may provide nations with a framework for addressing the governance of watercourses. These initiatives promote several specific tools that advance public participation, many of which are discussed below. The initiatives recommend practices such as environmental impact assessment (EIA), including transboundary EIA (TEIA), public meetings at an early stage in a project, free access to public records, regular reports by the government on the status of projects that may affect the public, and access to environmental information by citizens of neighbouring countries that may be affected by local decisions. These tools have been accepted widely for public participation domestically and may significantly increase public involvement – and, ultimately, the success of projects – in international watercourses.

Approaches for advancing public involvement

Eriksen suggests a general strategy when starting cooperative management of transboundary watercourses that also may apply to the context of public involvement:

focus[ing] on water quality issues avoids contention around water allocation. Water quality is also usually a concern shared by all riparians in some way. Co-operation on scientific assessments on a drainage basin and processes within it has been a starting point for basinwide co-operation. (Eriksen 1998)

It might also be prudent to start with transboundary watercourses that flow between two (or perhaps three) nations only and are not politically sensitive.

In many contexts, public involvement may be viewed by both governments and civil society as a means by which opponents of particular projects or activities may seek to stall or halt the proposed action. This view has some basis in experience: where the public does not have formal channels for providing input, or for having decision makers incorporate or respond to their input, protest and confrontation often are the primary avenues remaining for people to express themselves. In developing and implementing approaches to facilitate civil-society engagement, consideration should be paid to ways by which to facilitate more constructive forms of public involvement. This may take the form of a participatory priority-setting exercise or co-management. Such constructive participatory processes can foster a more congenial and collaborative relationship between governments and civil society.

This is not to say that confrontational approaches need to be eschewed; rather, there is a spectrum of participatory processes from collaborative to confrontational. To the extent that there are clear benefits of public involvement, as illustrated through collaborative processes, governments may be more willing to provide information and opportunities for public participation – even if confrontation sometimes results.

Seeking constructive and collaborative approaches for public involvement has implications for both governments and international institutions, on the one hand, and for civil society, on the other hand. For collaborative participation to work effectively, decision makers need to seek the input of civil society early in the process, when the decision can be changed or modified to reflect the various perspectives of civil society. It may be obvious but, in order for civil society to believe that their participation will make a difference (and therefore to become engaged in the process), the decision makers need to listen to civil society and also to be willing to modify the proposed action to reflect the priorities of civil-society. At the same time, civil-society institutions must show that they are willing to work constructively with the institutions, not just as critics but as collaborative stakeholders. This may mean, for example, a focus on finding alternatives and solutions rather than criticism.

Access to information can be promoted through a number of discrete mechanisms, many of which are relatively low cost. Making information available on request obviates the need for a sizeable staff and infrastructure. Imposition of a reasonable fee (to cover copying, for example) can further reduce the burden on the authority, although even a modest fee in many developing countries could mean that such information would be functionally unavailable except to institutions and relatively well-off individuals. Establishing a resource centre is a more expensive endeavour, but it could constitute a project that foreign donors would support and, in the long run, could reduce the overall burden on staff who might otherwise have to respond seriatim to requests that could otherwise be addressed through a resource centre. Another, more inexpensive, option is to develop a website. Producing a periodic "state of the river" report poses certain difficulties; however, these have been overcome in many developing countries to date. For example, the report could be restricted to a brief account, and there is also the possibility of publishing the report every two years rather than annually, again reducing the production and printing costs. Such a report could focus on water-quality issues, draw upon a modest number of sampling points, and grow from there.

As a first step to developing public participation in the management of international watercourses, EIA can be developed at the national level and harmonized through the region or along watercourses (Cassar and Bruch 2004; Sikoyo, chap. 22, this volume). As it is unlikely that the river management bodies will have the funds necessary to conduct detailed EIAs or lengthy public hearings on them, the riparian nations through the watercourse authority could require project proponents to conduct an EIA for projects likely to have a significant environmental impact, and then open the discussion to the public. This is the case for projects financed by most international financial institutions (Bernasconi-Osterwalder and Hunter 2002). One easy step is to open meetings of river management authorities to the public. This costs relatively little, and the public could participate as either silent observers or participating, but non-voting, observers.

Access to justice measures can be difficult because they often entail modification of procedures related to national judicial systems. Acknowledging this challenge, there are various incremental steps that can be undertaken to improve access to justice. For example, nations in a region can encourage broad interpretations of standing to facilitate access to their courts both by their nationals and by others who may be affected, particularly those living in other riparian nations. The East African nations did just this when they adopted their 1998 Memorandum of Understanding (MOU) on Environment Management, which also emphasized the imperative of cooperation in managing Lake Victoria.

In developing these norms – which give a voice to citizens, NGOs, and local governments – it will be necessary to balance the roles of international, national, and local actors in the management of transboundary watercourses (Milich and Varady 1998; Avramoski 2004). Moreover, it is important to develop culturally appropriate approaches (Kaosa-ard et al. 1998; Faruqui, Biswas, and Bino 2001; Avramoski 2004). National and international actors are essential to ensuring that local control does not lead to parochial dominance and unsustainable abuse of natural resources; and the participation of local actors is necessary for the norms and institutions to be relevant (and thus implemented) on the ground.

Promising mechanisms and practices

This volume has highlighted many ways in which nations and international institutions have developed and implemented mechanisms for promoting and ensuring public involvement in the management of international watercourses. In addition to some of the more established mechanisms, a variety of approaches are emerging that are likely to improve public involvement in the years to come. These may be refinements or extensions of established mechanisms, whereas in other cases they are new mechanisms (Bruch 2004).

As mentioned elsewhere in this volume (Bruch, chap. 18), Internet-based tools have become important for disseminating information relating to international watercourses. Additionally, tools such as e-mail, listservs, and chat rooms increasingly provide avenues to solicit public comment and otherwise to engage the public in the decision-making processes. As Internet connectivity continues to grow, particularly in developing nations, Internet-based tools are likely to gain more relevance and prominence.

Decision support systems (DSSs) present another tool for improving public access to information about potential impacts of decisions on international watercourses and for engaging the public in the decision-making process. A particularly innovative approach to making DSSs publicly available is to develop Internet-based DSSs, which has been facilitated by the development of faster computers and servers, and by broadband Internet access (Salewicz, chap. 19, this volume).

Although EIA is well established in national laws and international declarations, and the institutions to conduct EIAs continue to develop, there are a few specific ways in which EIA is likely to improve, particularly with regard to international watercourses. First, the expansion of EIA norms and methodologies to explicitly address transboundary impacts is an important step to improving basin-wide management of international watercourses (Knox 2002; Cassar and Bruch 2004). In many instances, international instruments and institutions have called for the development of TEIA, and this has been applied in a variety of circumstances. Considering the diverse experiences thus far, a comprehensive review of TEIA experiences could be instructive and could improve the ongoing development and operationalization of TEIA norms, institutions, and methodologies.

Another way in which EIA is being extended is to provide a participatory framework for analysing possible impacts of proposed plans, policies, programmes, and regulations. Many regions and countries are in the comparatively early stages of developing and implementing strategic environmental assessment (SEA), which could also provide a framework for improving public involvement in the development of norms governing international watercourses (Kravchenko 2002; Sikoyo, chap. 22, this volume).

A third way in which EIA can be improved is by examining the effectiveness of EIA methodologies. There is a growing body of literature examining the accuracy and effectiveness of EIAs, particularly in domestic contexts (Nakayama, Yoshida, and Gunawan 1999; Nakayama et al. 1999; Nakayama, Yoshida, and Gunawan 2000; Bruch 2004; Nakayama, chap. 17, this volume). By comparing predicted with actual impacts, EIA methodologies can be improved and made more effective. Applying the lessons learned from such comparative analyses could improve EIAs at both the national and transboundary levels, and this constitutes a continuing research need.

Developments in access to justice are likely to be more incremental. Initiatives such as the Aarhus Convention, which liberalize standing requirements and impose the obligation of non-discrimination in granting standing to citizens of other countries, provide a framework for opening up domestic courts. In many instances, though, such opportunities are only starting to be utilized. Granting public access to international tribunals is another development on the horizon. Although many significant developments have been made in the past decade (Bernasconi-Osterwalder and Hunter 2002; Gertler and Milhollin 2002; Jean-Pierre

2002; Di Leva, chap. 10, this volume; Garver, chap. 12, this volume; Picolotti and Crane, chap. 23, this volume), this progress has slowed. In some instances, it is simply a matter of the mechanisms maturing; in other instances, countries have been cautious about opening up dispute resolution processes too far to the public (Gertler and Milhollin 2002). Nevertheless, considering the substantial momentum and continuing pressure for transparent, participatory, and accountable governance, it is likely that international institutions will continue to develop approaches to ensure public access to tribunals and fact-finding bodies.

There are a number of other experiences, particularly at the national and sub-national levels, in promoting public involvement in water management that could be adapted and applied to different international watercourses (Bruch 2001; Avramoski 2004). Indeed, parts IV and V of this volume examine a variety of such experiences. Some of these promising approaches include coordinated local management authorities to supplement regional and national authorities (Gitonga, chap. 13, this volume), applying alternative dispute resolution techniques to contentious public consultations (Orton, chap. 21, this volume), and adaptive management frameworks (Volkman, chap. 20, this volume).

Exchange of experiences is a critical step towards identifying potential approaches to improve public involvement in the management of international watercourses. To this end, this volume has highlighted emerging norms, institutional approaches, and tools to enhance public access to information, participation in decision-making processes, and access to justice. However, consideration of whether and how to adapt these promising approaches is only one step; significant efforts are needed by way of capacity building. This includes technical training, development of resources to provide ongoing technical assistance, and financial and technology transfers. Mobilization of these resources is essential for the creation of structures and capacity for effectively engaging the public. With these resources, the public, governments, and international organizations together are poised to realize dramatic improvements in the management of international watersheds.

Acknowledgements

The authors gratefully acknowledge research assistance from Mark Beaudoin, Angela Cassar, Dorigen Fried, Samantha Klein, Molly McKenna, Turner Odell, Seth Schofield, Julie Teel, Elizabeth Walsh, and Jessica Warren. Support for this research was provided by the Carnegie Corporation of New York, the United States Agency for International Development, the John D. and Catherine T. MacArthur Foundation, and the

Richard and Rhoda Goldman Fund. This chapter builds upon ideas first developed in Carl Bruch. 2001. "Charting New Waters: Public Involvement in the Management of International Watercourses." *Environmental Law Reporter* 31:11,389–11,416.

REFERENCES

Avramoski, Oliver. 2004. "The Role of Public Participation and Citizen Involvement in Lake Basin Management." Internet: ⟨http://www.worldlakes.org/ uploads/Thematic_Paper_PP_16Feb04.pdf⟩ (visited 14 April 2004).

Bernasconi-Osterwalder, Nathalie, and David Hunter. 2002. "Democratizing Multilateral Development Banks." In Carl Bruch (ed.). *The New "Public": The Globalization of Public Participation*. Washington, DC: Environmental Law Institute.

Bruch, Carl. 2001. "Charting New Waters: Public Involvement in the Management of International Watercourses." *Environmental Law Reporter* 31:11,389–11,416.

Bruch, Carl. 2004. "New Tools for Governing International Watercourses." *Journal of Global Environmental Change – Human and Policy Dimensions* 14(1):15.

Cassar, Angela Z., and Carl E. Bruch. 2004. "Transboundary Environmental Impact Assessment in International Watercourses." *New York University Environmental Law Journal* 12:169–244.

Eriksen, Siri. 1998. *Shared River and Lake Basins in Africa: Challenges for Cooperation*. Nairobi: ACTS Press.

Faruqui, Naser, Asit K. Biswas, and Murad Bino (eds). 2001. *Water Management in Islam*. Tokyo: United Nations University Press.

Gertler, Nicholas, and Elliott Milhollin. 2002. "Public Participation and Access to Justice in the World Trade Organization." In Carl Bruch (ed.). *The New "Public": The Globalization of Public Participation*. Washington, DC: Environmental Law Institute.

Jean-Pierre, Danièle M. 2002. "Access to Information, Participation and Justice: Keys to the Continuous Evolution of the Inter-American System for the Protection and Promotion of Human Rights." In Carl Bruch (ed.). *The New "Public": The Globalization of Public Participation*. Washington, DC: Environmental Law Institute.

Kaosa-ard, Mingsarn, K. Rayanakorn, G. Cheong, S. White, C.A. Johnson, and P. Kongsiri. 1998. *Towards Public Participation in Mekong River Basin Development*. Bangkok: Thailand Development Research Institute Foundation.

Knox, John H. 2002. "The Myth and Reality of Transboundary Environmental Impact Assessment." *American Journal of International Law* 96:291.

Kravchenko, Svitlana. 2002. "Promoting Public Participation in Europe and Central Asia." In Carl Bruch (ed.). *The New "Public": The Globalization of Public Participation*. Washington, DC: Environmental Law Institute.

Milich, Lenard, and Robert G. Varady. 1998. "Managing Transboundary Resources: Lessons From River-Basin Accords." *Environment* 40:10.

Nakayama, M., T. Yoshida, and B. Gunawan. 1999. "Compensation Schemes for Resettlers in Indonesian Dam Construction Projects – Application of Japanese 'Soft Technology' for Asian Countries." *Water International* 24(4):348–355.

Nakayama, M., T. Yoshida, and B. Gunawan. 2000. "Improvement of Compensation System for Involuntary Resettlers of Dam Construction Projects." *Water Resources Journal* 80–93 (September).

Nakayama, M., B. Gunawan, T. Yoshida, and T. Asaeda. 1999. "Involuntary Resettlement Issues of Cirata Dam Project." *International Journal of Water Resource Development* 15(4):443–458.

Abbreviations and acronyms

3WWF	Third World Water Forum
a.k.a.	also known as
ACE	[US] Army Corps of Engineers
ACTS	African Centre for Technology Studies
ACWF	America's Clean Water Foundation
ADF	African Development Fund
ADR	alternative dispute resolution
AFDB	African Development Bank
ANGOP	Angola Press Agency
ATSDR	Agency for Toxic Substances and Disease Registry
AWIRU	African Water Issues Research Unit
AWRA	American Water Resources Association
BCM	billion cubic metres
BECC	Border Environment Cooperation Commission
BMPs	best management practices
BMUs	beach-management units
CAA	Clean Air Act
CAC	Citizen Advisory Committee (to the Executive of the Chesapeake Bay Program)
CADSWES	Center for Advanced Decision Support for Water and Environmental Systems (University of Colorado)
CAO	Compliance Advisor/Ombudsman
CAWT	Central American Water Tribunal
CBC	Chesapeake Bay Commission
CBD	Convention on Biological Diversity

CBF	Chesapeake Bay Foundation
CBOs	community-based organizations
CEC	Commission for Environmental Cooperation
CEO	Chief Executive Officer
CEQ	Council on Environmental Quality
CIA	Central Intelligence Agency
CILSS	Comité Inter-Etats pour la Lutte contre la Sécheresse dans le Sahel
CIMS	Chesapeake Information Management System
CIPS	Centre for International Political Studies
CLCs	Comités Locaux de Coordination
CLEAN	Children Linking with the Environment Across the Nation
CMAs	catchment management agencies
CNC	national coordination committee (for CLCs)
COMECON	Council for Mutual Economic Cooperation
CREP	Conservation Reserve Enhancement Program
CREST	Core Research for Evolutional Science and Technology
CSU	Colorado State University
DAD	Decide, Announce, and Defend
DANCED	Danish Cooperation for Environment and Development
DEOs	District Environmental Officers
DEPOs	District Environment Protection Officers
DFRD	District Focus for Rural Development
DoE	Division of Environment (Tanzania)
DoI	Department of the Interior
DRBC	Delaware River Basin Commission
DRC	Democratic Republic of the Congo
DRFN	Desert Research Foundation of Namibia
DPRC	Danube River Protection Convention
DSS	decision support system
DU	Ducks Unlimited
EAC	East African Community
ECOVIC	East African Communities of Lake Victoria
EIA	environmental impact assessment
EIS	environmental impact statement
E-LAW	Environmental Law Alliance Worldwide
ELI	Environmental Law Institute
EMCA	Environmental Management and Coordination Act [Kenya, 1999]
ENA	National Environment Strategy [Angola]
ENWC	Eastern National Water Carrier
EPA	Environmental Protection Agency
EPRI	Economic Policy Research Institute
ESA	Endangered Species Act
EU	European Union
FERC	Federal Energy Regulatory Commission

FRIEND	Flow Regimes from International Experimental Network Data
GATT	General Agreement on Tariffs and Trade
GCPBA	Grand Canyon Private Boaters Association
GEF	Global Environment Facility
GEMS	Global Environment Monitoring System
GIS	geographic information system
GLERL	Great Lakes Environmental Research Laboratory
GLIN	Great Lakes Information Network
GLINDA	Great Lakes Information Network Data Access
GLU	Great Lakes United
GLWQA	Great Lakes Water Quality Agreement
GUI	graphical user interface
GWP	Global Water Partnership
HEC	Hydrologic Engineering Center
HEMnet	Health, Environment and Media Network
HTTP	hypertext transfer protocol
IAC	Implementation Advisory Committee
IAHS	International Association of Hydrological Sciences
IBRD	International Bank for Reconstruction and Development
IBWC	International Boundary and Water Commission
IBWT	International Boundary Waters Treaty
ICJ	International Court of Justice
ICLARM	International Center for Living Aquatic Resources Management
ICLEI	International Council for Local Environmental Initiatives
ICPDR	International Commission for the Protection of the Danube River
ICPRs	international common-pool resources
ICRC	International Committee of the Red Cross
ICSID	International Centre for Settlement of Investment Disputes
ICTULA	Information and Communication Technology Use with Local Agenda 21
IDA	International Development Association
IDB	Inter-American Development Bank
IDPs	internally displaced persons
IFC	International Finance Corporation
IJC	International Joint Commission
ILA	International Law Association
ILC	International Law Commission
IMADES	Instituto des Medio Ambiente y el Desarrollo Sustenable [Institute of Environment and Sustainable Development]
IMF	International Monetary Fund
INBA	International Nile Basin Association
INE	National Institute of Ecology (Mexico)
INRC	Integrated Natural Resources Conservation
IOE	Institute of Ecology
IRAS	Interactive River–Aquifer Simulation
IRBM	integrated river-basin management

IRC	International Water and Sanitation Centre
IRDNC	Integrated Rural Development and Nature Conservation
IRIS	Interactive River Simulation
IRN	International Rivers Network
ISP	Inter-American Strategy for the Promotion of Public Participation in Decision Making for Sustainable Development
IUCN	International Union for the Conservation of Nature (now known as IUCN – The World Conservation Union)
IW:LEARN	International Waters Learning Exchange and Resource Network
IWRA	International Water Resources Association
IWRM	integrated water resources management
JPAC	Joint Public Advisory Committee
JSPS	Japan Society of Promotion of Science
JST	Japan Science and Technology Corporation
KCS	Kalahari Conservation Society
KMFRI	Kenya Marine and Fisheries Research Institute
LBPTC	Limpopo Basin Permanent Technical Committee
LHDA	Lesotho Highlands Development Authority
LVFO	Lake Victoria Fisheries Organization
LVFRP	Lake Victoria Fisheries Research Project
M&E	Monitoring and Evaluation
MAS	Mission d'Aménagement du Bassin du Fleuve Sénégal [Basin Development Mission]
MAWRD	Ministry of Agriculture, Water and Rural Development [Namibia]
MDB	Multilateral Development Bank
MDBC	Murray–Darling Basin Commission
MDBMC	Murray–Darling Basin Ministerial Council
MGDP	Maun Groundwater Development Project
MIGA	Multilateral Investment Guarantee Agency
MOU	Memorandum of Understanding
MRC	Mekong River Commission
n.d.	no date
NAAEC	North American Agreement on Environmental Cooperation
NACEC	North American Commission for Environmental Cooperation
NAFEC	North American Fund for Environmental Cooperation
NAFTA	North American Free Trade Agreement
NBI	Nile Basin Initiative
NEAP	National Environment Action Plan [Kenya]
NEMA	National Environmental Management Authority [Uganda]
NEMC	National Environment Management Council [Tanzania]
NEPA	National Environmental Policy Act [US]
NePAD	New Partnership for Africa's Development
NES	National Environment Secretariat [Kenya]
NES	National Environment Statute [Uganda]
NESDB	National Economic and Social Development Board [Thailand]
NGOs	non-governmental organizations

NIEHS	National Institute for Environmental Health Sciences
NNF	Namibia Nature Foundation
NOAA	National Oceanic and Atmospheric Administration
NPDES	National Pollution Discharge Elimination System
NPF	National Patriotic Front [Hungary]
NRT	nutrient removal technology
OAS	Organization of American States
OAV	Organisation Autonome de la Vallée (Valley's Autonomous Organization)
OBSC	Okavango Basin Steering Committee
OECF	Overseas Economic Cooperation Funds
OERS	Organization of Riparian States
OKACOM	Okavango River Basin Commission
OMB	Office of Management and Budget
OMVS	Organisation Pour La Mise en Valeur du Fleuve Sénégal (often referred to as the Senegal River Development Organization)
OSIENALA	Friends of Lake Victoria
PAHO	Pan-American Health Organization
PASIE	Programme d'Atténuation et de Suivi des Impacts Environnementaux (Programme for the Mitigation and Monitoring of Environmental Impacts)
PATH	Process for Analyzing and Testing Hypotheses
PCBs	polychlorinated biphenyls
PGNA	National Environmental Management Programme [Angola]
PIR	Project Implementation Review
PLN	Perusahaan Umun Listrik Negara [Indonesian National State Electric Company]
PMTF	Programme Management Task Force
PRSP	Poverty Reduction Strategy Paper [Tanzania]
PRTRs	Pollutant Release and Transfer Registers
RAM	random access memory
RBO	river basin organization
REC	Regional Environment Center
RECONCILE	Resources Conflict Institute
REDSO	Regional Economic Development Support Office [of USAID]
REIA	regional environmental impact assessment
REPSI	Resources Policy Support Initiative
RFF	Resources for the Future
RISDP	[SADC] Regional Indicative Strategic Development Plan
RSAP	Regional Strategic Action Plan
RTKNet	Right-to-Know Network
SADC	Southern African Development Community
SAIEA	Southern African Institute for Environmental Assessment
SAP	Strategic Action Plan
SARDC	Southern African Research and Documentation Centre
SAV	submerged aquatic vegetation
SCN	Seattle Community Network

SEA	strategic environmental assessment
SEM	Manantali Development Company
SIA	social impact assessment
SIDA	Swedish International Development Authority
SIL	Summer Institute of Linguistics
SIWI	Stockholm International Water Institute
SMEC	Snowy Mountains Engineering Corporation [Australia]
SOGED	Agency for the Management and Development of the Diama Dam
SOGEM	Société de Gestion de l'Energie de Manantali
SOLEC	State of the Lakes Environment Conference
TAC	total allowable catch
TAC	Toxics Advisory Committee
TAS	Tanzania Assistance Strategy
TBNRM	transboundary natural resource management
TCM	total catchment management
TCP/IP	transmission control protocol/internet protocol
TCTA	Trans-Caledon Tunnel Authority
TDRI	Thailand Development Research Institute
TEIA	transboundary environmental impact assessment
TMDL	total maximum daily load
ToR	terms of reference
TPTC	Tripartite Permanent Technical Committee
TRI	Toxic Release Inventory
UI	user interface
UNCBD	United Nations Convention on Biological Diversity
UNCCD	United Nations Convention to Combat Desertification
UNCED	UN Conference on Environment and Development
UNCLNUIW	United Nations Convention on the Law of the Non-Navigational Uses of International Watercourses
UNCLOS	United Nations Conference on the Law of the Sea
UNDP	United Nations Development Programme
UNECA	United Nations Economic Commission for Africa
UNECE	United Nations Economic Commission for Europe
UNEP	United Nations Environment Programme
UNESCO	United Nations Educational, Scientific, and Cultural Organization
UNFCCC	United Nations Framework Convention on Climate Change
UNITA	União Nacional para a Independência Total de Angola [National Union for the Total Independence of Angola]
USAID	United States Agency for International Development
USEPA	United States Environmental Protection Agency
USGAO	United States General Accounting Office
USGS	United States Geological Survey
VPO	Vice-President's Office [Tanzania]
WCD	World Commission on Dams
WHYCOS	World Hydrological Cycle Observing System

WSIS	World Summit on the Information Society
WSSD	World Summit on Sustainable Development (2002 Johannesburg)
WTO	World Trade Organization
WUAs	water user associations
WWF	Worldwide Fund for Nature (formerly World Wildlife Fund)
ZACPLAN	Zambezi [River] Action Plan
ZAMCOM	Zambezi Basin Commission
ZRA	Zambezi River Authority

List of contributors

Peter Ashton, Council for Scientific and Industrial Research (CSIR), Division of Water, Environment and Forestry Technology, P.O. Box 395, Pretoria 0001, South Africa. Tel: +27-12-8412237; Fax: +27-12-8413789; E-mail: pashton@csir.co.za

Ruth Greenspan Bell, Director, International Institutional Development and Environmental Assistance, Resources for the Future, 1616 P Street, NW, Washington, DC 20036, USA. Tel: +1-202-328-5032; Fax: +1-202-939-3460; E-mail: bell@rff.org Website: www.rff.org/iidea

Carl Bruch, Senior Attorney, Environmental Law Institute, 1616 P Street NW, Suite 200, Washington, DC 20036, USA. Tel: +1-202-939-3240; Fax: +1-202-939-3868; E-mail: bruch@eli.org; Website: http://www.eli.org

Angela Cassar, Visiting Scholar, Environmental Law Institute, 1616 P Street NW, Suite 200, Washington, DC 20036, USA. Tel: +1-202-939-3800, +1-202-729-7683; E-mail: angela@wri.org, a.cassar@pgrad.unimelb.edu.au

Prachoom Chomchai, Emeritus Professor of Economics, Chulalongkorn University (Bangkok), Coordinator, Mekong Development Research Network (MDRN), 1131/298 Bangkok Housing Cooperative Building, 19th Floor, Apartment 1916, Dusit, Bangkok, 10300, Thailand. Tel: +66-2-243-1234; Fax: +66-2-243-7423; E-mail: pchomchai@hotmail.com

Kristin L. Crane, Intern, Center for Human Rights and the Environment

493

(CEDHA), General Paz, 186, 10A,
Córdoba 5000, Argentina. Tel:
+54-351-425-6278; E-mail:
kristin@cedha.org.ar

Charles E. Di Leva, Lead Counsel,
ESSD and International Law, The
World Bank Legal Department,
Mail Stop MC 6-601, (Room
MC6-449), 1818 H Street, NW,
Washington, DC 20433, USA. Tel:
+1-202-458-1745; Fax:
+1-202-522-1573; E-mail:
cdileva@worldbank.org

Aboubacar Fall, Interim Executive
Secretary, African Development
Bank, 1936 HLM5, P.O. Box 6740,
Dakar Etoile, Senegal. Tel:
+221-835-51-30; E-mail:
a.fall@afdb.org,
asdam@cooperation.net

Tomlinson Fort III, Principal,
Integrated Science & Technology,
Inc., 102 Pickering Way, Suite 200,
Exton, PA 19341, USA. Tel:
+1-484-875-3037; E-mail:
tfort@integratedscience.com

Geoffrey Garver, Director,
Submissions on Enforcement
Matters Unit, North American
Commission for Environmental
Cooperation, 393, rue St-Jacques
Ouest, Bureau 200, Montréal
(Québec) H2Y 1N9, Canada. Tel:
+1-514-350-4300; E-mail:
Ggarver@ccemtl.org

Nancy Gitonga, Director of Fisheries,
Kenya Fisheries Department,
Museum Hill, P.O. Box 8187,
Nairobi, Kenya. Tel:
+1-254-2-3744530, 37442320/49;
Fax: +1-254-2-3743530; E-mail:
kgitonga@wananchi.com

Rebecca Hanmer, Director,
Chesapeake Bay Program, US
Environmental Protection Agency,
410 Severn Avenue, Suite 109,
Annapolis, MD 21403, USA. Tel:
+1-410-267-5709; Fax:
+1-410-267-5777; E-mail:
hanmer.rebecca@epa.gov

Roy A. Hoagland, Virginia Executive
Director, Chesapeake Bay
Foundation, Capitol Place, 1108 E.
Main Street, Suite 1600, Richmond,
VA 23219, USA. Tel:
+1-804-780-1392; Fax:
+1-804-648-4011; E-mail:
rhoagland@cbf.org; Website:
http://www.cbf.org

John Jackson, Director Emeritus,
Great Lakes United, 17 Major
Street, Kitchener, Ontario, N2H
4R1, Canada. Tel: +1-519-744-7503;
Fax: +1-519-744-1546; E-mail:
jjackson@web.ca; Website:
http://www.glu.org

Libor Jansky, Senior Academic
Programme Officer, United Nations
University, Environment and
Sustainable Development, 53-70,
Jingumae 5-chome, Shibuya-ku,
Tokyo, 150-8925, Japan. Tel:
+81-3-3499-2811; Fax:
+81-3-3499-2828, or
+81-3-3406-7347; E-mail:
jansky@hq.unu.edu; Website:
http://www.unu.edu

Bradley C. Karkkainen, University of
Minnesota Law School, 229 19th
Avenue South, Minneapolis, MN
55455, USA. Tel: +1-612-624-5294;
E-mail: bradk@umn.edu

Michael Kidd, Professor of Law,
School of Law, University of Natal,
Pietermaritzburg, Private Bag X01,
Scottsville 3209, South Africa.

Tel: +27-33-260-5382; Fax: +27-33-260-5015; E-mail: kidd@nu.ac.za

Mikiyasu Nakayama, Associate Dean and Professor, United Graduate School of Agricultural Science, Tokyo University of Agriculture and Technology, 3-5-8 Saiwai-cho, Fuchuu-city, Tokyo 183-8509, Japan. Tel: +81-42-367-5667; Fax: +81-42-360-7167; E-mail: mikiyasu@cc.tuat.ac.jp

Marian Neal, Council for Scientific and Industrial Research (CSIR), Division of Water, Environment and Forestry Technology, P.O. Box 17001, Congella 4013, South Africa. Tel: +27-31-242-2330; Fax: +27-31-261-2509; E-mail: mneal@csir.co.za

Mary Orton, The Mary Orton Company, LLC, 2254 Morning Mesa Avenue, Henderson, NV 89052-2627, USA. Tel: +1-702-914-8066; Fax: +1-702-914-8466; E-mail: mary@maryorton.com; Website: http://www.maryorton.com

Juan Miguel Picolotti, Legal Officer, Center for Human Rights and the Environment (CEDHA), General Paz, 186, 10A, Córdoba 5000, Argentina. Tel: +54-351-425-6278; E-mail: juan@cedha.org.ar

Nevil W. Quinn, Senior Lecturer, Centre for Environment and Development, University of Natal, Pietermaritzburg, Private Bag X01, Scottsville, South Africa 3209. Tel: +27-(0)33-260-5664; Fax: +27-(0)33-260-6118; E-mail: quinnNW@nu.ac.za

Kazimierz A. Salewicz, Tamariskengasse 102/121, 1220 Wien, Austria. Tel/Fax: +43-1-947-3894; E-mail: kaz_salewicz@yahoo.com

George Michael Sikoyo, Research Fellow, African Centre for Technology Studies (ACTS), P.O. Box 45917, Nairobi, Kenya. Tel: +254-2-524714; Fax: +254-2-524701; E-mail: G.Sikoyo@cgiar.org; Website: http://www.acts.or.ke

Hans van Ginkel, Rector, United Nations University, 53-70, Jingumae 5-chome, Shibuya-ku, Tokyo 150-8925, Japan. Tel: +81-3-3499-2811, +81-3-5467-1224; Fax: +81-3-3499-2828, +81-3-3499-2810; E-mail: rector@hq.unu.edu; Website: http://www.unu.edu

John Volkman, Stoel Rives LLP, 900 SW 5th Avenue, Suite 2600, Portland, OR 97204, USA. Tel: +1-503-294-9809; Fax: +1-503-220-2480; E-mail: jmvolkman@stoel.com

Index